AF560402

RURAL DEVELOPMENT AND NON-FARM SECTOR

RURAL DEVELOPMENT AND NON-FARM SECTOR

Editors

ANIL KUMAR THAKUR
PARMANAND SINGH

Published on behalf of
THE INDIAN ECONOMIC ASSOCIATION

REGAL PUBLICATIONS
New Delhi - 110 027

RURAL DEVELOPMENT AND NON-FARM SECTOR

ISBN 978-81-8484-247-0

Typeset by
RAHUL COMPOSERS
New Highway Apartments, Lakshmi Niwas
760, Pocket-D, Lok Nayak Puram, New Delhi - 110 041

Printed in India at
Printed in India at MAYUR ENTERPRISES
WZ Plot No. 3, Gujjar Market, Tihar Village, New Delhi - 110 018

Published by
REGAL PUBLICATIONS
F-159, Rajouri Garden, New Delhi - 110 027
Phone : 45546396, 25435369
E-mail : regalbookspub@yahoo.com, regaldeepbooks@yahoo.com

Contents

PART II

NON-FARM SECTOR AND EMPLOYMENT OPPORTUNITIES

Part IV
NON-FARM SECTOR : INFLUENCES ON INDUSTRIES, SPECIFIC GROWTH

Non
endeavou
been a
activities
Rural Inc
rampant
poverty a
labour fo
activities
force in e
rights for
making th
has attach
The non-
women an
at their d
far baptiz
employme
villages and
become
modernis
process of
conditions
rural red
transform
economy

Preface

Non-farm sector employment has been envisaged in this endeavour as an engine of rural development in India. It had been a missing sub-sector consisting of both secondary activities and tertiary activities in rural areas of the country. Rural India is facing heavy burden of population pressure and rampant disguised unemployment, thus adding the menace of poverty and miseries in villages. Besides, there are bulk of labour force participating in the labour market, but their activities has not been reckoned so far as contributing labour force in economic well being. But they are working days and nights for household chores, caring and rearing of children and making them ready for work in the labour market. Karchedi has attached much value of women's work for labour force. The non-farm sector growth will help these weaker sections, women and minorities and provide them gainful employment at their door step. These labour force participation has been so far baptized as invisible labour or missing labour force. Their employment in service activities, manufacturing households in villages and handicraft industries in cottages will help them become economically strong and provide the rural populace a modermised living conditions. It will also help value addition process of the rural product and help in improving the wage conditions and economic well-being of the vast majority of rural redundant labour force. It will thus, help rural transformation process thus leading the traditional rural economy from subsistence-based economy to industrial economy attempting rural development in sustainable direction where ecological and environmental hazards are less pronounced.

There has been a lot of studies on this issue in the past, the most important among these are Okafor (183), Hazel and Brown (1987), Hearth (1996) at international level in India. T.S. Papola (1993), A.N. Sharma, Rohim Nayar (2005) has attempted on this score. The present edited book "Rural Development and Non-Farm Sector" is an attempt to make a step forward in this rural transformation process.

ANIL KUMAR THAKUR
PARMANAND SINGH

List of Contributors

A. Kuruswamy, P.G. Department of Economics, Vivekananda College, Agasteeswaram, Kanyakumari (T.N.).

A. Shanmugasundaram, Asst. Professor, Kandaswami Kandar's College, Velur-Namakkal (T.N.).

Ajay Kumar ao, Asstt. Teacher, Middle School, Barauriya, Bettiah.

Angad Singh, Head, Department of Economics, Satish Chandra College, Ballia (U.P.).

Anupama C. Mareguddikar, Post-Graduate in Commerce, Karnatak University, Dharwad (Karnataka).

Apurba Saikia, Selection Grade Lecturer, Department of Economics, Furkating Colllege (Assam).

Arifa Saleem, Research Scholar, AMU, Aligarh (U.P.).

Ashish Kumar, Department of Training and Placement, Anand Engg. College, SGI, Agra (U.P.).

Ashok R. Rathod, Research Scholar, Department of Economics, Karnatak University, Dharwad (Karnataka)

Awadhesh Kumar Sinha, Department of P.A., Sir Ganeshdatt Memorial College, Patna (Bihar).

B. Eswar Rao Patnaik, Retd. Sr. Reader in Èconomics, S.B.R.G. Women's College, Berhampur (Orissa).

Bh. Anjaneya Reddy, P.A. to Registrar, Acharya Nagarjuna University, Nagarjuna Nagar, Guntur (A.P.).

Bhawana Jha, Senior Lecturer, MAM College, Naugachia, T.M. Bhagalpur University, Bhagalpur (Bihar).

Birendra Kumar Jha, HOD Economics, D.B.K.N. College, Narhan, Samastipur (Bihar).

C. Mythili, Research Scholar, Department of Economics, Bharathiar University, Coimbatore (T.N.).

Ch. Masenamma, Research Scholar, Department of Economics, Andhra University, Visakhapatnam.

Chandra Prakash Azad, Centre for Regional Studies, T.M. Bhagalpur University, Bhagalpur (Bihar).

D.R. Bachhav, Reader and Head, Business Economics, K.T.H.M. College, Nashik (Maharashtra).

Deepshikha Singhal, Reader in Economics, Agra College, Agra (U.P.).

Dhana Laxmi Patnaik, Professor Bhagabata Patro, Department of Economics, Berhampur University, Bhanja Bihar (Orissa).

Dilip Arjune, H.O.D. of Economics, Commerce, Science College, Jalana (Maharashtra).

E. Neela Perumal, P.G. Department of Economics, Vivekananda College, Agasteeswaram, Kanyakumari (T.N.).

G. Rajalakshmy, Head of the Department of Business Economics, L.S. Raheja College of Arts & Commerce, Santacruz (West), Mumbai (Maharashtra).

G. Sridevi, Assistant Professor, Department of Economics, University of Hyderabad, Emailgummadi (A.P.).

Harindra Kishor Mishra, Department of Economics, Nagendra Jha Mahila College, Laheriasarai, Darbhanga (Bihar).

J. Narayan, Professor in Economics Department, University of Allahabad, Allahabad (U.P.).

J. Sreenivasulu, Research Scholar, Department of Economics, Sri Venkateswara University, Tirupati (A.P.).

Jeegyasa, Reader, Department of Economics, Aligrah Muslim University, Aligarh (U.P.).

K. Govindarajalu, Professor, Department of Economics, Bharathiar University, Coimbatore (T.N.).

K. Nageswara Rao, Research Fellow, Department of Economics, Acharya Nagarjuna University, Nagarjuna Nagar, Guntur (A.P.).

K.G. Suresh Kumar, Associate Professor, A.V.K. College for Women, Hassan (Karnataka).

M. Koteswara Rao, Chairman, Board of Studies in Economics (PG), Department of Economics, Acharya Nagarjuna University, Guntur (A.P.).

M. Reddi Ramu, Assistant Professor, Department of BS&H, Siddhartha Institute of Engineering & Technology, Puttur, Chittoor (Dist.) (A.P.).

M.V. Narasimha Sarma, Professor and Registrar, Acharya Nagarjuna University, Nagarjuna Nagar, Guntur (A.P.).

Mahendra Ranawat, Principal, BNPG Girls College, MLS University, Udaipur (Rajasthan).

Mallikarjun G. Naik, Lecturer in Economics, B.N. Jr. College, Dandeli, Haliyal, Karwar (UK) (Karnataka).

Mini Goyal, Department of Business Management, Punjab Agricultural University, Ludhiana (Punjab).

Mohan C. Patel, Principal, N.S. Patel Arts College, Anand (Gujarat).

Mohana Bandkar, Head of the Department of Business Economics, Lala Lajpatrai College, Mahalakshmi, Mumbai (Maharashtra).

N. Saravanakumar, Teaching Assistant in Economics, PG Department of Economics, Arulmigu Palaniandavar College of Arts & Culture, Palani, Dindigul District (T.N.).

N. Visalakshi, Research Scholar, Department of Economics, Andhra University, Visakhapatnam.

Nagaraja, G., Assistant Professor, Department of Economics, Sri Venkateswra University Post-Graduate Centre, Kavali, Nellore District (A.P.).

Nahid Akhtar Siddiqi, Research Scholar, Department of Economics, Aligarh Muslim University, Aligarh (U.P.).

Niharika Srivastava, Asst. Professor in Economics Department, P.B.P.G. College, Pratapgarh City (U.P.).

P. Mahendra Varman, Assistant Professor, Department of Econometrics, University of Madras, Chennai (T.N.).

P. Purna Chandra Rao, Asst. Professor, Department of Tourism & Hospitality Management, Acharya Nagarjuna University, Nagarjuna Nagar (A.P.).

P. Subramanyachary, Assistant Professor: Department of MBA, Siddhartha Institute of Engineering & Technology, Puttur, Chittoor (Dist.) (A.P.).

P. Tara Kumari, Professor, Department of Economics, Andhra University, Visakhapatnam.

Pappu Yati, Research Scholar, Magadh University, Bodh Gaya (Bihar).

Parmanand Singh, Principal, B.N.M. College, Barhiya, T.M.B.U., Bhagalpur (Bihar).

Prankrishna Pal, Reader and former Head, Department of Economics, Rabindra Bharati University, Kolkata (W.B.).

Prashantrao M. Yadwad, Assistant Professor, Department of Economics, GES's JSS Degree Arts, Science & Commerce College, Gokak, Belgaum Dist. (Karnataka).

Pratibha Goyal, Department of Business Management, Punjab Agricultural University, Ludhiana (Punjab).

R. Kumar, Guest Lecturer, Department of Economics, Presidency College (Autonomous), Chennai (T.N.).

R. Venkatachalam, Asst. Professor, Erode Arts College, Erode (T.N.).

R.B. Bhandwalkar, Reader, P.G. Department of Economics, Amolokchand Mahavidyalaya, Yavatmal (Maharashtra).

R.K. Chaudhary, Sr. Lecturer, Department of Economics, M.J.K. College, Bettiah (Bihar).

R.R. Gawhale, Head, P.G. Department of Economics, G.S. College, Khamgaon Dist. Buldana (Maharashtra).

Ram Naresh Thakur, HOD P.G. Centre, Samastipur College, Samastipur (Bihar).

Rana Naseer, Research Scholar, Department of Economics, Aligrah Muslim University, Aligarh (U.P.).

S. Rathnakumari, Department of Economics, Sri Venkateswar University, Tirupathi, Chittoor (Dist.) (A.P.).

S. Rengarajan, Associate Professor, P.G. & Research Department of Economics, Sir Theagaraya College, Chennai (T.N.).

S.M. Jawed Akhtar, Associate Professor, Department of Economics, Aligarh Muslim University, Aligarh (U.P.).

Sachita Nanda, Research Scholar, Department of Economics, University of Hyderabad (A.P.).

Sajena M. Pillai, Associate Professor, C.P. Patel & F.H. Shah Commerce College, Anand (Gujarat).

Sehar Shabzad, Research Scholar, AMU, Aligarh (U.P.).

Shehroz A. Rizvi, Reader, Department of Economics, Aligrah Muslim University, Aligarh (U.P.).

Sukhjeet K. Saran, Department of Economics and Sociology.

Suvash Chandra Pradhan, Department of Economics, Biju Pattnaik College of Science & Education, Jayadev Vihar, Bhubaneswar (Orissa).

T.V. Raman, Research Scholar, Department of Economics, MLS University, Udaipur (Rajasthan).

Introduction

I. THE CONCEPT AND BACKGROUND

The Rural Development is ingulfed with heavy population burden on agriculture and a vast ocean of redundant labour force either going lost even if they seems performing, they are contributing zero marginal productivity and hence popularly know in economic literature as disguised unemployment.

To bring agriculture of South East Asian Agriculture from this overburdened pressure of population a non farm activity mechanism was evolved to give them side by gainful employment opportunities at their door step and thus helping agricultural sector both the base of activities via additional opportunities and through non-farm activities—such as handicrafts, food processing, carpentry, pottery and other market linked activities with value addition in core agricultural product thus giving the surplus labour force in agriculture gainful job opportunities *vis-a-vis* let loosing a process of modernization and standardization of agricultural produce and its efficient market linkages *via* creation of SHGs through a net work of micro-credit facilities.

This strategy of rural development got much popularity as Non-farm sector growth *via* creating off farming activities in rural areas. It thus has helped much to ease agricultural of its population burden and helped agriculture modernize and standardize its product and farming community new opportunities for additional income and employment.

Besides, in achieving its economic goal Non-farm sector growth has also been helpful in empowering the down trodden groups and weaker sections *via* modernizing their household non-farming activities and providing them modern know how and equipment support. Thus, this helped rural industries with a fertile hinterland with adequate infrastructural and raw material support to grow and prosper enough to engulf the whole twittering masses with adequate work options converging thus the social milieu churning for development demand with adequate pressure on the political class. This demand of Economic Development is converting the rural economy in terms of agrarian structure, occupational structure and living and working condition of labour. This can be possible only when dependence on farm activities decreases and non-farm activities like household, industrial operations, servicing, construction and rural artisan class work culture and work ethics and the very organization of work take a moving trend towards modernization and production for the market. So, the growth of Non farm sector demands two things commercialization *vis-a-vis* modernization of agriculture at the same rate.

The increased performance in these non-farm activities will lead to greater participation of labour in rural non-farm activities. This is an indication of development in rural economy.

The developmental were withal, that new economic regime imposed on India *via* (L.P.G.) liberalization, privatization and globalization has resulted in declined G.D.P. in primary sector i.e. rural/agricultural sector though overall growth rate has taken a sweep forward trend. On the other hand the over all economy has experienced lower employment growth. Agriculture being almost saturated, off-farm jobs are to be created in rural areas and semi-urban areas or kasbai areas through the creation of new production centers (food processing, ancillary industry, receiving based on transport and construction works etc.) New agro-industry linkages are to be growth; otherwise, the goal of more jobs in the long-run will be a cry.

Besides, we also have heard the talk of where has missing work force has gone. We have a lot of workers, whole working hours are not reckoned or paid due attention or even if, they are not paid adequately, a vast majority of women's household work come in this category. Informal sector growth will help them find remunerative job in alternative occupations just out of their door step. So non-farm activities are helpful in providing additional job and employment opportunities to the rural working folk.

1.1. Rural India : Road to Progress

India does not live in Mumbai, Delhi and Kolkata. India lives in villages and without self-reliance or Gram Swaraj in villages, India's political independence will remain incomplete. Gandhi perceived the tremendous natural and human potentialities of villages contributing to the country's development. Gandhi was of the opinion that the only way to bringing hope and good living to the people in rural areas was by making the villages central focus of economic development programmes.

Since poverty concentrates in rural India 27.1% poor lives in villages, i.e. 280 million rural poor make larger size than India had in 1947 constituting 13 Australia. Income poverty is 26% and nutritional poverty is 56%. According to Dr. Swaminathan's food security Atlas—"India has largest absolute number of undernourished people in the world. Rural India has an employment problem rather than an underemployment problem village workers earn very little per day or month despite working very long hours and so on.

According to National Social watch 48% people in 13 major states of India viz A.P. Tamil Nadu, Maharashtra, Gujarat, Arunachal Pradesh, Assam, Jharkhand, Bihar Chhattisgarh, M.P., Orissa, Rajasthan and Uttar Pradesh do not get two meals a day properly. No doubt a survey conducted by National Council of Applied Economic Research shows that our efforts for rural development have began to bear fruits, and income levels and quality of life in rural India have registered in upward trend but much needs to be in this direction on Gandhian lines. Once Gandhi in his talk with an European philosopher, (Roma Rona) turned Gandhian reputed, told her when see told Gandhi that rural India seems sleeping in deep slumber. Gandhi refilled instantly rural India has vast potential.

Have you deep Sargaso Sea water which apparently seems dormant with tonsils and sea grass on upper water layers, but in its within it is the meeting points of vast ocean currents. In the like fashion rural proletariat are humming with activities, but they are so absolute and low paying, that even in activity, they seems earning little for their lives and fishes. That is, why they seem shaken, and despaired and dismayed. Village Swaraj with use of their Global meet resources will give them life and blood. The non-farm activity through emboldened growth of non-farm sector is, thus the need of the hour.

APJ Abdul Kalam has proposed the concept of PURA (Providing Urban Amenities in Rural Areas) in the vision 2020 project initiated by him and recited that in the very inauguration of International Seminar on "Resurgent Bihar" its identification dimension and eradication on 17-19 Feb. 2006 at Patna, where he restated the PURA vision of making rural areas laden with Urban Amenities. By making rural areas so attractive to investors as cities, then rural areas too will generate urban style employment to halt rural-urban migration. In line with experiment Kalam's vision of India's path to progress through the concept of PURA where in to progress through the concept of PURA where in envisaged independent cluster of villages in rural areas by building self-sustainable business enterprises through four types of connectivity (Physical, electronic, knowledge, economic and proper infrastructural facilities.)

The theme of PURA apart from concentrating on reinforcing agriculture, will emphasize on agro processing, development of rural craftman ship, dairy, fishing, silk production, so that non-farm revenue for the rural sector is enhanced. This should be based on the core competence of the region. The rural economy will also be driven by renewable energy such as solar, wind, bio-fuel and conversion of municipal waste into power. In this approach the aim is to provide *sustainable development* with the core competence of rural sector.

PURA is, thus a vehicle for sustainable development. It is to improve the quality of life of over 220 million people living below poverty level in rural areas and makes special suggestions to remove urban congestion also. This sustainable rural development programme is for a nation of billion plus people (121 crores). In India, development of rural sector is very important. The Government, private and public sectors have been taking up rural development. During the last few decades, it is our experience that these initiatives start well just like heavy rains resulting into multiple streams of water flows. As soon as the rain stops, few days later all the streams become dry because there is no water bodies to collect surplus water and store it at the right place. For the first time PURA envisages an integrated development plan with employment generation as the focus, driven by provision of habitat, health care, education, skill development, physical and electronic connectivity and marketing. PURA is thus a perfect design of rural development strategy full of sustainability potential giving adequate weight non-farm activity at an ample scale. Thus, proper rural development through PURA is desirable and is a must for sustainable economic development of the country.

1.2. Rural Non-farm Sector Employment and Enterprises in Rural India—The Forgotten Sector

Fisher Thomas and Vijay Mahajan has called rural non-farm sector as the forgotten sector in India, though it had been the wheels of growth expanding on which ancient India thronged as the mightiest global power of the world. To him RNFS comprises all non-agricultural activities, mining and quarrying, household and non-household manufacturing, processing, repairs, electricity, gas and water supply, construction, wholesale and retail trade, restaurants and hotels, transport, storage and warehousing, communication, financial and business services, community, social and personal services and other services in villages and rural towns of up to 50,000 population, undertaken by enterprises varying in size from household to own account enterprises; and all the way to factories. (Fishar Thomas and Mahajan Viyay (1996))

As per Government of India (1998) there were a total of 17.71 million rural enterprises of which 3.20 million or 18.1% were engaged in agricultural activities such as live stock rearing (dairy, poultry, piggery) forestry, fishing and agricultural services such as custom of hiring of tractors and managing of irrigation systems. The remaining 14.51 million or 81.9% of the rural enterprises were non-agricultural in nature. These comprised of a wide range of activities including mining and quarrying, manufacturing both within and outside households, electricity, gas and water supply construction, wholesale trade, retail trade, restaurants and hotels, transport, storage warehousing, communications, financial and real estate and business services and community social and personal services.

The last category itself comprises a broad range all the way from school teaching to hair cutting and laundry services. In terms of employment, we learn from GOI (1998) that nearly 40 million persons worked in RNFS enterprises of these 73.9% were adult males, 22.5% were adult females, while 3.6% were children. This amounted to 1.4 million children below the age of 15 working in RNFS enterprises. Over three quarters or 76.8% of the enterprises were own account enterprises *numbering 13.6 million* while 23.2% of the enterprises were establishments. Here by it is worth-noting that enterprises operating with the help of household labour only are termed as own account enterprises, which are enterprises without any hired worker, in contrast to establishments which employ at least one hired worker.

A vast majority of the RNFS Enterprises are small and dispersed and employment mainly self and family labour. There is, thus a strong case for waiving labour legislation applicable to them. In terms of number of persons employed per enterprise establishments RNFS averaged 4.65 as compared to 1.5 persons in other agricultural establishment. Thus, RNFS establishments are a major source of wage employment and it needed to be encouraged.

In terms of nature of activities, the larger number of RNFS enterprises was in retail trade totalling 5.23 million. The next largest categories were community. Social and personal services (3.82 million) and manufacturing (3.52 million), Transport and storage (0.48 million), were also significant (Vijay Mahajan and B.Y. Raju, 2005).

The rural Non-farm sectors are widely dispersed and labour-intensive. It requires small amount of initial capital and are ideally suited to provide jobs for the rural poor. However, the rural non-farm sector has not received as much policy and programme support as has been the case in other countries such as China (D. Rajasekhar, 2005) and moreover there has been a definite decline in the quality of employment in rural India. The two prominent and inter related trends are the *decline in self-employment* and increasing casualisation.

Due to unavailability of estimates regarding growth in output or income of the rural non-farm sector, one can get an indication of the performance of this sector from the changes in its share in employment. There has been a diversification of employment in rural India with a share of rural non-farm employment having increased by 7.2 percentage points or by 0.33 percentage points per annum) over twenty-two years between 1977-78 to 1999-2000. The annual rate of this shift is nearly the same between different periods during the pre-reform as well as post-reform periods, whereas one would expect the rate of shift towards rural non-farm activities to have increased in the past-reform period. This could well be attributed to the slowing down of agricultural growth and insufficient infrastructural support to rural non-farm activities during the post-reform period. This is reflected in the slow pace of reduction in rural poverty during the post-reform period of 1990s when compared to 1980's.

This is not surprising, in view of the continued decline in the public investment in agriculture and the squeeze in public expenditure on rural development, in general following fiscal compression during the past-reform period. However, since productivity and wage rates in rural non-farm sector are much higher than in agriculture, the above facts do indicate that the shares of rural non-farm sector in output and incomes over the last two decades have been significantly higher than its share in employment (Hanumantha Rao, 2005).

India's most significant feature of economic development since independence and particularly after reform had been the decreasing proportion of work force dependence on agriculture but the pace of decrease is much less than expected. Despite significant decline in the share of income of the primary sector and its substantial increase in that of the tertiary sector activities, their employment shares have remained stable, while the share of national income originating in agriculture has dropped from over 50% in 1950 to less 25% in current and again lesser in 2008-09, 2009-11 to 19.7%. But the share of labour force engaged in agriculture, which was about 70% in 1951, Still remains over 60%. The consequence is the ratio of per worker domestic product in non-agriculture to that in agriculture which was about 2 in 1950's to now well over 4. This shows the widening gap between incomes in agriculture and non-agriculture work. This widening gap in incomes in agriculture and non-agriculture is the major source of poverty in rural India and major cause of persistent rural poverty (Abhijeet Sen and Praveen Jha, 2005).

Thus, we can't deny the RNFS in generating employment and reducing poverty in rural area because of the limited absorptive capacity of the urban sector and near saturation in the agricultural sector in further absorption of workers. The importance of the rural non-farm sector in reducing rural-urban differences in income levels, bringing greater integration between rural and urban economies and generally contributing rural and over all development is also now recognized.

I.3. Rural Non-farm Sector as an Engine of Increasing Labour Force Participation and Women Empowerment

The LPG regime has led rural economy of India a multivaried nature of transformation. This transformation is structural and has converted rural economy into modernized one with ICT boom. This has led to decline in the size of labour force dependent on agriculture on the one hand and reduction of poverty on the other. Labour force of both male and female has declined during 1993-94 to 2004-05. But the male labour force has declined more than female labour force during the period under consideration.

(a) LFPR vis-a-is Non-farm Economy

Male-Female. According to NSSO report labour force participation is the proportion of persons in the age group of 15 years and above who are either working i.e. employed on the usual principal status and subsidiary sates (PS or SS) or seeking or available for work, i.e. presently unemployed.

Estimates on this basis by NSSO, reveal that in rural India LFPR has declined marginally from 87.6% in 1993-94 to 85.4% in 1999-2000 and then increased marginally to 85.9% in 2004-05 for male. It has also declined from 48.8% in 1993-94 to 45.6% in 1999-2000 and increased to 49.4% in 2004-05 for rural female.

Thus, sex-wise analysis by NSSO estimates shows that male LFPR is more than female one in India and its all the constituents states during the period 1993-94 & 1994-2000 and 2004-05.

LFPR variation is very wide in Indian states. In rural India male LFPR has declined in India and its almost all constituents states except Assam, Bihar, Gujarat, Orissa and Punjab during 1993-2005.

Female LFPR also has declined in almost all constituents states excepting—Assam, Gujarat, Haryana, Kerala, Orissa, Punjab and U.P. during this period under study.

Also, not worthy is the fact, that irrespective of sex LFPR has increased in Assam, Gujarat, Orissa and Punjab during period under study.

Male participation rate is lowest in Haryana in 1993-94, 1999-2000 and 2004-05. A.P. was ranked highest in 1993-94 and then lost its prime place to Karnataka in 1999-2000 and to Gujarat in 2004-05.

On the other hand, female LFPR was lowest in Assam in 1993-94, West Bengal in 1999-2000 and Bihar in 2004-05. It was highest in Himachal Pradesh in 1993-94, 1999-2000 and 2004-05.

Thus, Gujarat and Himachal Pradesh have utilized more labour force of male and female respectively for the development of the rural economy during the reform era in the period 1993-94, 1999-2000 understudy.

It makes the analysis clear on this point that LFPR for male and female is not uniform in rural India and in its various constituents states. It had been low in states where infrastructure is poor and high where it is rich. Besides the pace of modernization of rural economy is also one of the contributory factor in designing the dimension of LFPR. So, we find that a clear-cut disparity in LFPR clearly exists among various states which tells the varying path of rural transformation *via* non-farm sector growth in rural India is taking place.

(b) Employment Status vis-a-vis Non-farm Sector Growth

Labour is employed in different types of activities. These

activities are agricultural and non-agricultural, i.e. Non-farm. Labour is also classified as self-employed and wage labour—both regular and casual. Mode of employment of labour (male and female) in rural India has been expectedly changing during the reform period. Estimates of NSSO data, different rounds reveal that *in rural India the incidence of self employment has been consistently declining over time.* It has declined from 65.9% in 1971-72 to 56.7%, in 1993-94 and to 56.3% in 2002 in the case of male labour, it has increased to 57.6% in 2004-05, the respective 65.5%, 51.3%, 49% and 56.4% in the case of female labour.

Thus, male self-employment is more than the female one during the period under analysis. But regular salaried employment has been declining for male (12.15% in 1972-73 to 9.1% in 2004-05) and rising for female (4.1% in 1972-73 to 4.8% in 2004-05). On the other hand, the incidence of *casual labour give a noticeable picture during the period under study. It has increased gradually from 20.5% in 1972-73 to 33.3% in 2004-05 for male.* For female it has increased from 31.4% in 1972-73 to 46.4% in 2002 then declined to 38.9% in 2004-05.

Thus, females are casually employed more than male during the period 1993-94, 2003-2004-05) this is presumably due to the fact that during the reform period various types of activities mostly non-farm have been generated in favour of rural female be so that they can utilized more, relative to male at the low wage rate (Pal, P.K., 2002).

Thus, we can see growth of non-farm sector as an additional area regular employment creation for women workforce large and *via* that way a income and employment security to them and by empowering women folk more participation more evolvement in the development activities.

But crux of the issue is casualisation of female work force. At first incidence we see feminization of agriculture and at the other end we witness casualistaion which is engagement of women of lower wage rate in a censual manner.

Casualisation of female labour is more than that of male one in rural India and in all its constituent states. Indian states have exhibited wide variations in the employment of labour during the reform periods. Estimates reveal that the incidence of self-employment for male has increased overtime in almost all the sates excepting A.P., Haryana, H.P., Karnataka, M.P.,

Punjab, Rajasthan, T.N. and West Bengal while that for female has also increased in all states excepting H.P., Karnataka and Punjab during 1993-94 to 2004-05. It has varied from 39.9% in Kerala to 71.7% in U.P. in 1993-94 for rural male and from 39.8% in T.N. to 71.67% in U.P. in 2004-05. Kerala (34.6%) and H.P. (95.6%) have ranked respectively lowest and highest in 1993-94 in respect of the rural self-employment.

In 2004-05, the respective states are Punjab (35.4%) and H.P. (94.47).

Similarly in the case of regular salaried employment male employment has also increased in almost all the states in India excepting Assam, Bihar, Karnataka, Maharashtra and West Bengal during the period 1993-94, 2000 to 2004-05.

This is also true of female employment excepting Assam, Maharashtra and West Bengal.

In most of the states the incidence of casual labour for both male and female has increased at the cost of self-employment. In case of female casualisation is more intense.

The index of casualization which is defined as the number of casual wage labour for every one hundred of regular salaried wage labour is used to measure incidence of casualisation, let loosen by the new economic regime.

Estimates of casualisation show that in rural India the index has gradually increased from 182 in 1972-73 to 455 in 1997 and then declined to 366 in 2004-05 for male and from 766 in 1972-73 to 1948 in 1997 and then remarkably declined to 810 in 2004-05 for female.

Also irrespective of sex, the index has varied across the states in India during 1993-94 to 2004-05. The index is lowest in Haryana (104 in 1993-94 and 110 in 2004-05) and highest in Bihar (840 in 1993-94 and 1173 in 2004-05) for male, for female lowest in H.P. (76 in 1993-94 and 30 in 2004-05 highest in Bihar (3973 in 1993-94 and 2065 in 2004-05). Thus rural labour is casually employed more in Bihar and less in H.P. during the period of reform.

However, sex wise analysis reveal that the casualisation of female labour is more than that of male ones in rural India and in all it) the constituent sates except H.P., Kerala, Punjab, Haryana, West Bengal during the period 1993-94, 2004-05) under study.

Thus, female labour finds jobs of casual and/or contractual nature or on a daily basis more relative to male ones at a low wage rate (Pal 2002).

(c) Rural Employment : Farm and Non-farm by Sex

Rural labour is employed in various alternative sectoral activities like—(a) Primary sector consisting of cultivation, agricultural activities, forestry fishing, mining and quarrying; (b) Secondary sector including manufacturing, servicing, repairing in both household and non-household industries and construction; (c) Tertiary sector consisting of transport, trade and commerce, storage and communication and services.

The last two sectors jointly call non-farm sector, while the former is the farm sector the latter SS and TS is together Non-farm sector. This sector has been missing since the advent of Britishers as in rural India and resultant delay of rural industries and commercial activities in villages.

The total labour force utilized in these two sectors; the FS and NFS are to a great extent inter-dependent. Modernization of farm sector in post-green revolution period has brought about a series of changes in rural India villages are linked with cities and towns through a wide net of transports, markets are developed and expanded, farm outputs are transported, agricultural machinery and repairing shops are opened, small electrical units are set-up, etc. All these necessitate increasing absorption of rural labour in NFS.

NSSO data reveals that employment in NFS has increased all through in India at the cost of farm employment both for male and female. For male it has increased from 16.4% in 1972-73 to 21.8% in 1983 to 25.2% in 1993-94 and 33.2% in 2004-05.

While for female it has increased from 10.1% in 1972-73 to 12.1% in 1983, to 13.6% in 1993-94 and to 18.2% in 2004-05, Thus, male non-farm employment is more than female ones in rural India due to expansion of non-farm activities more for male relative to female.

The supply of labour force to NFS demand economic and educational development. So irrespective of sex Indian states have exhibited wide variation in NFE during the reform period.

Estimates reveal that rural NFE for male has increased in all the constituent states in India during 1993-94 to 2004-05 at cost of farm employment while for female it has increased in almost all the states excepting Karnataka and West Bengal during the period 1993-94 to 2004-05. The NFE is lowest in M.P. (20.3% in 2004-05) and highest in Kerala (45.7% in 1993-94 and 61.8% in 2004-05). The respective figures for female are lowest in H.P. 5.3% in 1993-94 and Maharashtra (9.0% in 2004-05) and highest in West Bengal (54.7 in 1993-94) and Kerala (56.2% in 2004-05).

Thus, among the states only Kerala has significantly enjoyed more the absorption of rural NFE for both male and female due to the generation of the Non-farm activities, and rural industrialization during the reform period of 1993-94 and 2004-05.

Sex-wise analysis reveals the fact that female employment in NFS is more than male ones only in Punjab and West Bengal during 1993-94 and 2004-05. This is presumably due to the fact that during the reform period various types of activities mostly non-farm activities have been generated more in these two states compared to other states in favour of rural female so that they can be utilized more relative to male at a low wage rate.

RNFE and Women

If rural development is an attempt aiming at improving living standards of the low income population living in rural areas and making the process of development sustaining, the women have key role to play in farming systems throughout the world. Women are involved not only in household chores and child rearing in rural areas but are a major source of labour for food production, and account for large proportion of economic activity. In majority of the developing countries women play two major roles in the rural area. At first they own household responsibilities for child rearing, food preparation and other chores and on the other role they are paid or unpaid workers in agriculture and off-farm activities. In many areas women manage the affairs of the household and the farm both.

When surveys are taken about the true extent of involvement of women in agriculture-related activities the estimates are under statement as women describe their principal occupation as house-wife. They are, then counted as economically inactive and this "invisibility" of female employment ignores women and sometimes affects then, when policies and programmes are taken.

In African countries women play larger role in farming, nearly all tasks connected with food production are left to women, while men may tend livestock, and produce cash crops and when men migrate to elsewhere the entire administration of the household is often left to women. In Malawi over two-thirds of those working full time in farming are women.

In most of the areas in the world, women are important to agriculture. Households headed by women make up to 20 to 25 precent of rural households in developing countries, excluding China and Islamic countries. Women typically care for animals particularly chickens and pigs and tending garden vegetables and other food crops in Latin America. In the Caribbean countries women work as cash labourers on plantations, sugar, and fruit producing areas, earn a substantial portions of household income. In Asia for example, Nepal female farming systems are known in which women have 50% household income in the subsistence farm.

Social, cultural and religious factors, population pressures, farming techniques; off-farm opportunities; colonial history; and many other factors determine the role of women in farming systems. As off farm job opportunities, population pressures and farming technique change, women assume very different role even when physical conditions of the area are same.

Shifting cultivation with hand labour tend to lend itself more to female labour then does settled cultivation with a plough, but in settled agriculture and cash crops men typically do more work than previous, but often the role of women in farm work still dominates.

Krishnaraj Maitheyi (*EPW*, 2005), "Food security : How and for, Whom ?" has highlighted the contribution of women in agricultural work and it seems to be rising judging from the number of women workers in crop production and if we include the work involved in livestock, poultry, fisheries, water conservation, forestry and work related to common property resources the contribution of women to agriculture would far exceed than of men. Male migration from rural to urban, translate into marked increase in agricultural work for women, including wide range of farm tasks, a heavier workload, and less time for domestic tasks and childcare.

This phenomenon in Indian condition is called feminization of agriculture, i.e. that is increasing burden of work on women and lower compensation.

Indian agriculture has been emphasized by : (i) increased involvement of women in agriculture at the beginning of this century compared to earlier period; (ii) underestimation of women's contribution to agriculture due to faculty definition. Women in this situation continue to be immersed in agriculture and bear heavy burden of farming tasks with less access to non-farm work than men.

Feminization of agriculture is not only specific to India but also all over Asia, South Asia, South-East Asia and in parts of west Asia, work participation of women is mostly related to family enterprises and local work available, their upward mobility for employment is highly restricted.

Women participate in work in four ways, according to their caste and class position in Paddy-rice labour markets at the end of twentieth century. Harris and White (2005) has pointed out :

(i) Women from pauperized, female and female headed and/or low caste petty producing and trading household are confirmed to seasonal operations, to subsistence orientation.
(ii) Female casual wage workers from asset less class from the largest substratum of labour in grain milling and pre-milling process.
(iii) In small family firms, unwaged family members have provided that part of the wage to labour in trading firms which takes the form of prepared food.
(iv) A typical milling and trading firm is owned by the makes of a joint family and practices a diversity of trading and transformation activities, women are used for caste-based reproduction and expansion of firms first by means of their dowries on marriage and second through the practice of fictitious registration of a trading company in a woman's name.

The relation of gender inequality with poverty is obvious. A disaggregated analysis of sex ratios (used as a proxy for gender quality) across districts in India in fact, suggest that women's survival chances decline with prosperity. Given this disadvantage, faced by women in terms of lower literacy levels, lower wage jobs in the labour market, restriction on mobility and responsibility for children and household maintenances. Men have generally moved into non-farm activities leaving women behind in rural areas to tend the land. In Zimbawe 68%, Tangania 52%, India 85%, of women workers are dependent on agriculture (Rao, 2005). For forty years (1961-2001) the rice production grow worldwide at the rate greater than that of population and doubled. The division of tasks in rice production shows a structural diversity. The control over commercial assets is dominated by men and technological change displaces female labour disproportionately to that of men.

In the light of above facts regarding discriminated position of women labour in agricultural activities, non-farm employment will be a boon for them, if provided with adequate technological know-how *vis-a-vis* strong infrastructural base. This will help enhancing the income position of women and help in giving them a food security net to the household. Thus, it will help the rural development process in two ways *via* empowering the women through additional gainful employment and a food security net to the household. Thus, it will-help the rural development process two ways *via* empowering the women and through a food secured family. When there is no place for malnutrition and malnourished childhood.

Thus, rural non-farm sector employment has very big role to play in the rural development of India. As a result of it many studies in India and outside has come up to underline the role of RNFS in generating additional income and employment. Some of notable among them are Okafor (1983), Hazel and Brown (1987), Hearth (1996), Krishnamurthy (1984), Vaidyanathan (1986), Besant Rakes (1989), Kshyap and Desai (1990, 1991), Basant and Parthasarthy (1991), Unni (1991), Basu and Kashyap (1992), Papola (1993) are note-worthy.

The present edited book "Rural Development and Non-farm Sector" has altogether 41 chapters covering various dimensions of non-farm sector role in the development of rural economy of India.

Chapter by Dr. Parmanand Singh, Awadhesh Kumar Sinha, Pappu Yati, Prankrishna Pal, S. Rengarajan, Birendra Kumar Jha and Ram Naresh Thakur explains NFS as an engine of growth for rural development *via* giving additional income generating opportunities to rural toiling masses specially women.

Prof. Subramanyachary and M. Reddi Ramu considers NFS as a mechanism for sustainable development in India and Dr. D.R. Bachhav and Dilip Arjune underline its importance in modernizing agriculture and improving the living conditions of rural folk. In Deepshika Singhal and Asish Kumar have tried to look into the issue of RNFSE *via* processed food sector growth in rural India.

Besides, we have a number as the pivot chapters in Part I dealing RNFS in of Rural economic health and it will provide adequate strength to the rural working mass adequate strength to face the challenges of globalization *via* producing low cost utility services and artistic handicraft goods. There are altogether twelve chapters in this part concentrating on various aspects of rural development and the role of RNFS in revolutionizing the so far stagnant rural economy and leading it towards modernization and better living conditions.

Part II of this book concentrates primarily with RNFS role in creating new employment opportunities, they are altogether eight chapters in this part. Article by Suvash Chandra Pradhan analyses the various opportunities and job alternative provided by RNFS to the Rural proletariat especially the downtrodden and weak, minorities and females.

K.G. Suresh Kumar analyses the relationship between labour force growth and requirements/need of rural non-farm sector growth in India *via* providing additional employment in non-farm activities of small scale, home-based industrialization and handicrafts article production. It thus, has helped both goals of employment and income security to so far redundant labour force and much helpful in increasing the labour force participation rate.

Besides we have also chapters in this part by Chandra Prakash Azad and Bhawana Jha dealing with role of RNFS in Jharkhand and Bihar—Which is a classic piece of comparative trend analysis of this sub sector on employment issue. The article by P. Tara Kumari, Ch. Masenamma and N. Visalakshi dealing with NREGA impact on rural labour employment and non-farm sector growth.

Part III in this book is based on micro-level experiences of RNFS growth in constituent states of India.

Prof. R. Kumar's chapter deals with RNFs role in Tamil Nadu. Prof. S. Rathnakumari and J. Sreenivasulu explains inter-Mandal variation in RNFS employment in Kurnool district of A.P. using explicit effort to incorporate RNFS into a growth paradigm on Mellor's (1976) Agricultural Growth Linkages (AGL) : forward and backward and consumption linkages from agriculture to the Non-farm sector. Forward linkages include those to processors and distributors in the non-farm sector (such as processing, transport and marketing of agricultural activities). These linkages are quite strong and the strength of such forward linkages depends crucially on the choice and location of the processing technology involved. The backward production linkages of agriculture to RNFSE (as rural input suppliers) are also crucial to agricultural performance and there may be backward linkages to those who supply and maintain produced goods. Finally consumer demand linages are generated as a result of increasing farm incomes. The authors have very neatly explained the (AGL) linkages and growth of RNFSE in various Mandals of Kurnool district in A.P.

There are also chapter in this part explaining NGOs role in extending RNFSE *via* empowering women workers. Mallikarjun G. Naik and Ashok R. Rathod's article is mentionworthy. Besides, this part has also chapters concerning Orissa, Punjab, Tamil Nadu.

We have also chapters regarding region specific non-farm growth in the poultry farming in Namakkal district of Tamil Nadu by A. Shanmugasundaram and R. Venkatachalam which worth mention.

In this part, there are altogether thirteen chapters analysed line with Apurba Saikia on North-Eastern Region.

In Part IV chapters converging rural development domain *via* industry specific explanation in RNFSE has been analysed. Chapters by R.K. Chaudhary and Ajay Kumar Rao explains prospect of Tourism in Bihar in expending non-farm employment to rural toiling masses *via* restaurant activities, transportations, resorts stays and Hotels, etc.

Mahendra Ranawat and T.V. Raman seeks to highlight the case of handicraft industries R.B. Bhandwalkar and R.R. Gawhale discuss it in small scale industrial growth perspective. Nagaraja G. discusses the special case of coir industry in this concern. G. Rajalakshmy and Mohana Bandkar discuss RNFSE in the context of food processing industry. The micro-credit programme as a service mechanism to boost RNFSE has been discussed by P. Mahendra Varman. There are altogether eight chapters discussed in this part.

The book thus, with a deep discussion on the issue of rural development and *via* RNFSE growth is lively and trys to spell the truth about transforming rural economy in the wake of globalization and Indian economy is trying its best to capture this opportunity and it has helped India converting into a vibrant emerging new economy of the world even in midst of global economic crisis.

References

Datta, R. (2001), Economic Liberalization and its Implications for Employment in India, "The IEA Conf. Volume, Vellor.

Pal, P. K. (2002), "Globalization and Labour Utilisation in Rural India : An Inter-state Analysis, IEA Conference Volume, Trivandrum.

NSSO (2003), Household Consumer Expenditure and Employment-Unemployment Situation in India, Government of India.

NSSO (2004), Employment and Unemployment Situation in India, Government of India.

NSSO (1997), Employment and Unemployment in India, 1993-94, Government of India.

NSSO (2000), Employment and Unemployment in India, 1999-2000, Government of India.

Dev. Mahendra (2006), "Ensuring Equitable Development Through Reforms", *The Hindu*, Jan. 21.

Harris-Barbara and White (2005), Commercialization, Commodityfication and Gender Relations in Past Harvest Systems for Rice, *EPW*, XL (25) June 18-24., 2530-47.

Paris Thelma, Abha Singh, Toyceluris and Mahabub Hassain (2005), "Labour Out Migration, Livelihood of Rice Farming Households and Women Left Behind", *EPW*, XL (25) June 18-24, 2522-29.

Kodoth, Praveen (2005), "Fostering Insecure Livelihood : Dowry and Female Seclusion in Left Development Context in West Bengal and Kerala, *EPW*, XL (25), June 18-24, 2555-62.

Dhingra, Ishwar (2006), The Indian Economy : Environment and Policy, Sultan Chandra and Sons, New Delhi pp. 367-78.

Krishnaraj, Mathereyi (2005), "Food Security : How and for Whom ?", *EPW*, XL (25), June 1825-2508-12.

Mishra, Jagnanath (1997), My Vision for India's Rural Development, Vikas Publishing House, New Delhi.

Hashim, S.R. (2009), Employment and Unemployment in a Society in Transition in "Growth, Employment and Labour Market", (Ed.) by Krishnamurthy and R.P. Mamgain, Danish Book, Gazipur.

Manjundappa (1981), The Presidential Address in IEA from Diamond to Platinum (ed.) by R.K. Sen.

Sen, Abhijeet and Jha, Praveen (2005), Rural Employment—Patterns and Trends from National Sample Survey.

Rajshekhar, D. (2005), Micro Finance and Rural Non-farm Sector Some NGO Experiences, *Ibid*.

Papola, T.S. and Sharma, A.N. (2005), "Towards a Policy Agenda for Rural Non-farm Sector in Rural Transformation in India" by R. Nayyar and A.N. Sharma, Institute for Human and Development, New Delhi.

PART I

NON-FARM SECTOR AS AN ENGINE OF GROWTH DEVELOPMENT

CHAPTER

1

Stagnation in Farm Economy, Non-Farm Sector as an Strategy of Additional Employment Generation and Position of Dalit

An Explorative Analysis

PARMANAND SINGH, AWADHESH KUMAR SINHA
AND PAPPU YATI

THE ISSUE

"Indian agriculture is at the cross-roads economically, ecologically and technologically, socially and nutritionally. The business as usual approach in the farm sector has led to unprecedented human calamity, the beginning of which are being witnessed in the form of farmers' suicides in several parts of the country including Punjab, the heart land of intensive agriculture" (Report of National Commission on Farmers, 2004).

Indian agriculture's share in GDP growth have seen substantial decline from 31.3 percent in 1991 to 19.6 percent in 2005-06, whereas dependence of workforce on agriculture decreased only marginally from 64.8 percent in 1991 to 58.40 percent in 2001.

The reform measures of early 1990 did not target agriculture directly, they have an indirect impact on the sector via reduction in developmental role of the government and its capacity to generate resources for investment.

We see therefore, a sharp reduction in rates of custom and excise duties and other taxes and we see a decline in the share of states in central taxes as a result of it in recent years.

With dwindling state resources, the expenditure of state governments on agriculture and allied activities as a percentage of net state domestic product (NSDP) has declined.

As a result, the worst victim is that section of population who dependents on agriculture for their livelihood base, wage rate and levels of living.

Since the inception of new economic policy, we see a number of policy changes in agriculture in terms of reduction of import duty on agricultural products, removal of minimum export price, lifting of quantitative restrictions on agricultural products; removal of minimum export price and entry of foreign direct investment in food processing and marketing up to 100 percent equity; direct investment of corporate sector in highly capital-intensive farming like floriculture, private sector participation in agricultural Research and Development technology transfer, extension and marketing; changes in laws of tenancy and leasing of land in order to achieve economies of scale, rationalization of stamp duties on agricultural land transaction; and promotion of contract farming. All these policy changes have adversely impacted our farming community and farming process.

ADVERSES OF THE NEW ECONOMIC REGIME ON AGRICULTURE

Farmers today are experiencing slowdown in agricultural productivity and a rising trend in input cost, with adjacent votality of the market because of trade liberalization. Failure of technology, markets and crops side by side heavy debt burden has added to farmers' distress and they are facing shortage of income and employment. To ameliorate farming fraternity out of this impasse non-farm averness have been shorted out as the 10th plan strategy to give an income option to those working population dependent on agriculture for their levels of living, wage rate and employment.

The cost of cultivation increased manifold due to rises in price of diesel, fertilizer, power, seeds, labour and other farm inputs (Singh, 2005). Due to inadequate agriculture, extension services from state government, farmers largely depend on the input dealer who not only supply inputs (seeds, pesticides, fertilizers, etc.) to them but also provides credit and advice for usages of inputs. This supply induced extension services extended by him induces farming community to apply more quantities of the purchased inputs that enhance the credit requirement and indebtedness among them.

Diversification of rural economy is thus regarded in this context an essential component of rural transformation. Growth of non-farm sector is seen by many as an important element of the strategy for alleviation of rural poverty. An expanding non-farm sector contributes to higher rural incomes by providing additional opportunities for employment and income opportunities in rural areas. It also helps, in raising the income levels of the remaining workers in the agricultural sector by reducing population pressure on land as well as through its impact on the agricultural wages which is used to at its lowest in its absence in rural areas.

The pace of structural transformation in rural economy, that followed in the past green resolution period, was however not uniform across various states and regions. Nor the factors associated with this process were identical : the same thing can be said about the impact of the process of diversification of rural income, employment and poverty. While in some areas RNFS was associated with dynamic forces operating in rural economy, leading to higher rural incomes and declining poverty levels, in other regions it reflected what has been called distress diversification. Again, different components of non-agricultural sector of rural economy were affected differently, some even experiencing a decline. Besides rural communities, with case of landlessness are affected aversely even if there is RNFS to ameliorate their miseries is concerned because of low level of skill, entrepreneurial ability. Their participation rate in rural non-farm sector activities used to be very low as result they have suffered from low wage rate option and poor and manual type of work even in this sector.

In this present paper an effort has been made to re-examine the participation rate of schedules caste workers and their wage rate levels and levels of living taking data base support from population, census 1961, 1971, 1981, 1991 NSS. Employment/unemployment survery-1987-88, 1993-94 C.S.O., Delhi. Rural Labour inquiry report on large earning 1974-75, 1983-84, 1987-88, 1993-94. Labour Bureau, Ministry of Labour and Consumer Expenditure Survey, NSSS 5th Round 1993-94 report.

TABLE I

Sector-wise Annual Compound Growth Rates in GDP at Factor Cast of 1993-94 Prices

Period	*Agri.*	*Industry*	*Services*	*Total*
1981-82 to 1990-91	3.09%	7.65%	6.71%	5.58%
1992-93 to 2003-04	2.53%	6.27%	7.90%	6.12%

Sources : Based on data collected from Handbook of statistics on Indian Economy.

Not only the GDP share of agriculture has declined but the value of output of foodgrains grew only 1.13 percent per annum. The value of output of fruits, vegetables and livestock grew 5.27 percent and 3.79 percent respectively. As a result of it the share of food growers in total agricultural income declined from 33.06 percent in 1991 to 28.16 percent in 2003-04. As a result, the foodgrain growers, situation worsened, compared to fruit and vegetable and livestock growers; which till now covers a small area and require greater skill in production methods. Besides, it can also be seen as an evil design to snatch family land from corn to non-corn production in the name of diversification in agricultural production.

TABLE 2

Trends in Value of Output of Agriculture and Allied Activities

Year	VOP of Agr. Allied Sectors	Total food-gains	Total cereals	Total pulses	Total oil seeds	Total fruits	Total Live stock Product
(1)	(2)	(3)	(4)	(5)	(6)	(7)	(8)
1990-91	251885	33.06	27.9	5.17	8.44	13.55	23.38
1992-93	261903	31.78	27.29	4.49	8.74	14.03	24.4
1993-94	271839	31.2	26.68	4.52	8.86	14.13	24.63
1998-99	323017	29.98	25.59	4.4	8.65	16.58	24.73
1999-2000	324298	30.27	26.43	3.83	7.23	16.69	25.39
2003-04	346538	28.16	24.52	3.65	8.14	17.58	27.56
AC GR (%)	2.50%	1.13	1.34	-0.14	0.35	5.27	3.79

RURAL NON-FARM SECTOR AS A STRATEGY TO GIVE ADDITIONAL EMPLOYMENT AND INCOME AVENUES

Diversification of rural economy is regarded as an essential component of rural transformation, not only for the purpose of easing out the population burden on agricultural sector but also from the angle of additional employment and income generation avenues. Besides, it will also help diversification of employment structure in rural areas which as basically a corn production centric and of mono-cropping nature. The strategy of increasing role of non-farm sector is seen by many as an element of strategy for alleviation of poverty of the masses.

An expanding non-farm sector contributes to higher rural incomes by providing additional opportunities for employment and income opportunities in the rural area. It also helps in raising income levels of the remaining workers in the agricultural sector by reducing population pressure on land as well as through its impact on agricultural wages.

The face of rural transformation picked up in rural economy in the post-green revolution period, but the rate of diversification was space centric limited only to green revolution belt of the country. The factors that contributed much to this process was not identical. The same can be said about the diversification of rural incomes and declining poverty levels. In green revolution zones it was seen as booster to growth areas and increased quantum of income and additional number of gainful employment.

In non-green revolution region it reflected what has been called distress diversification. It is because of these reasons poorly addressed sick states of Indian Union, witness this phenomenon in a more pronounced form. Again the different components of non-agricultural sectors of rural economy were affected differently, some even experiencing a decline. This above divergencies in the process of growth of RNFS call for regionally, sectorally and a disintegrative analysis, of its effects even on different social groups. The quinquenniel surveys on employment and unemployment by NSS, the compound growth rate in rural non-farm employment during 1977-83 was

4.6 percent per annum while it was 3.28 percent per annum during 1983-94 and 2.14 percent during 1993-2001.

Thus, in the reform era his sector has shown a declining trend, which goes to prove that fact of the country, particularly its vast rural sector till now has not been infrastructurally prepared to make an easy growth stride of the rural non-farm sector.

THE TENTH PLAN STRATEGIES AND THE RURAL NON-FARM SECTOR

Assuming 1.6 percent population growth annually, the planning commission expects a quantum jump in per capita income growth, that is doubling the per capita income in the next ten years. From the above angle, and from the point of view of experience of Rural Development Programmes in the past, the Approach paper of the Tenth Plan suggests a specific set of objectives as follows :

1. Reorganisation of SGSY in to micro-finance to be run by banks.
2. Unemployment programmes to be replaced by a food for work programme.
3. Rural development fund should be used for enhancing the budgetary allocation of successful Rural Development Schemes that are being run by the state governments or for meeting the state contribution for donor assisted programmes for proverty alleviation.
4. Strengthening the economy of the marginal and small farmers so that they should also contribute to growth.
5. Special effects to encourage development of small industry and non-farm activities.

THE CENTRAL ISSUE OF THE POLICY OF NON-FARM SECTOR STRATEGY

The central issue policy involved in the above strategy of Tenth Plan is not that there is co-existence of high rate of participation of rural population in non-farm activities along with high incidence of poverty. This is the reality of the most developing countries and is also true of India as well. In the context of economic reform it has triggered debates on whether rural non-farm sector has grown or declined in-reform era. The data base shows an alarming situation with compound growth rate in Rural non-farm employment declining in 1977-83 from 4.06 to 3.8 in 1983-94 and 2.14 percent in 1993-2000.

That is why it bought the attention of the policy planners with higher doses of inclusiveness in 11th Five Year Plan.

But the Creation of a viable rural non-farm sectors poses various challenges. The big question related to the precautious that have been taken to assure that such strategies work efficiently.

The approach paper of the Tenth Five Year Plan can be summarised.

(a) As challenge of creating a large number of enterprises of a sustainable nature, which absorb a significant portion of people who enter the labour market every year. Achievement of such a goal requires *a major qualitative changes in the existing order of things relating to programme* design and implementation.

It has rightly been envisaged by the planning commission that the poors need not only be the beneficiaries of growth but they should be partners of growth process itself. But what is it that prevented the poor to be partners of growth so far. The non-farm sector growth and its effective expansion for the benefit of the poor requires this question addressed meaningfully.

Instead of confining to macro-level strategies alone we need to look at the problem from a more realistic and practical micro point of view. This can but be done in terms of a critical analysis of the performance of various sub-sectoral programmes. They are as follows.

(b) Rural Development Programmes

During the Tenth Plan an allocation of Rs. 7000 crores for poverty alleviation was made. The approach paper of the 10th plan, however, admits substantial failure regarding the way, these programmes have been implemented. The erstwhile programmes were merged with the new programmes called SGSY (Sawaran Jayanti Swarojgar Yojna) with effect from April (1994). As an innovative model of non-farm development, SGSY has been potential of creating substantial entrepreneurial skills and employment opportunities. However, evaluations have indicated that the programme itself does not have much trickle down effects. The two major constraints institutional and lack of participation was underlined as the major cause of concern while the institutional constraints are significant in most parts of northern India, where the feudal bondages are still prevalent, the SGSY does not visualise any change in the existing rural institutional structure. The rural activities and clusters identified under the programme are largely to the convenience of Zamindars and the local traders and to the limited understanding and vested interests of bureaucracy. This situation can be improved through enhanced participatory processes in the villages. While approach paper highlights the need for "unleashing the protective forces and entrepreneurial energies (page 10) and this can be achieved only through enhanced participatory process. While in 1970s and 1980s the only form of official response was formation of cooperatives, we need to look at more innovative form which are flexible and efficiency [M.M. Mathus, Director, Institute of Small Enterprises and Development, Member, Gupta Committee on Small Enterprise Development, Planning Commission.]

(c) Financial Sector Intervention

The rural non-farm sector growth strategy envisages a new role of financial intervention. The nationalization of 14 commercial banks in 1971 introduced a new commitment on the part of financial sector not only to see business in the activities of the poor people but also to play a proactive role to groom and nurture this sector. However, with recent reforms in the banking sector, the above priority was lost. The development banks, which were expected to cater to the needs of the small scale enterprises, themselves became passive agents of development, through a process of the blurring of activities of commercial banks and DFIS (Direct Financing Institutions services).

The burden has increasingly been passed on to self-help groups and NGOs. But beyond mobilization of the meagre savings of the rural areas, what entrepreneurial activities have been initiated by SHGS ? or Do they promote the market for consumer goods ? Two vital areas requiring urgent an attention in the growth of non-farm sector are : (1) Directed Credit, and (2) Transformation lending. It is high time that the performance and prospects of various actors such as DFI, NGOs and Commercial banks are closely analysed from this angle.

The NSS data base segregated on gender basis with regard to the effectiveness of existing policies regarding RNFs shows that employment growth in RNFE sector is more of a male phenomena also lesser to weaker scheduled castes. During 1993-94 and 1999-2000, there was a 2.8 percent increase in male employment in RNFE Rural Non-farm Enterprises Sector while the increase in the female employment was only 0.70%.

The scheduled caste workers (male) distribution in to different industrial category can be grouped in to farm and non-farm sector. The scheduled case share in employment reduced in non-farm sector from 24.27% in 1961 to 17.61% in 1991, while in total rural non-farm sector from 1987-88 it was 22.4%. While other caste employment in non-farm sector was 23.5% in reform era 1993-94 it was 20.54% for SCs and 20.56% for others sectors castes.

From the point of view of commodity approach to Development, RNFE growth must be looked from the angle of several special commodity boards/Agencies like KVIC, Coir Board, Central silk Board, the DC (Hadi crafts). These organizations have largely implemented their programmes through organizational instruments like cooperatives, which to a large extent have failed. While boards have looked at the developments of these sub sectors from the point of view of the respective commodities, the entreperennial capacity of the people involved in them have been largely neglected.

The cultural patricides by the Government of India through a few all India boards in 1950s had its ripples at the state level also. State Government have initiated a large number of welfare-oriented public sector corporations which obviously have played some positive role. However, more than being fruitful interventions in the non-farm sector, these have increasingly become parasatic. The major problem faced by the government today is to maintain such corporations through budgetary support.

Thus, what flows from the above discussion from the point of view of policy formulation is the fact that RNF's potential must be examined from two main angles.

(1) The tri-dimensional nature of the sector, i.e. the sub-sectoral, spatial and scale dimensions, which, make policy formulation easier.

(2) An 'understanding of the structure and internal consistency of the sector. Based on available data, it is possible to delineate the RNFs into : (1) High share, (2) High growth, and (3) Emergent sub-sectors.

Each of these sub-sectors requires an appropriate organizational form and institutional support structure. It is important that existing promotional policies of the government need to be thoroughly overhauled as per these characteristics. Such a meaningful flexible approach is likely to facilitate a more dynamic growth of the Rural Non-farm sector in the changing economic environment.

It helps to constitute a proto-industrialization to contribute to the process of industrialization. The more recent phenomenon of micro-privatization in terms of its contribution to capacity building and entreprenership. Micro-privatization is largely supply induced rather than demand driven. The larger number of telephone booths and pan shops that are opened up in the rural areas of the country does not assure that these activities remain sustainable. On the other hand, public policies themselves should be tailored to the rural demand as perceived by the poor people themselves. This is not simply an economic question, it is a political question as well.

RNFS AND POSITION OF DALITS IN INDIA

Dalits in pussuant of the India is constituted by SC, ST population and women workers mostly. Though the exploitative and discriminatory system prevalent in the country with regard to production process compels certain other social groups be included in this name like backwards, extremely backwards and minorities.

But peruse definition of Dalit and their subdued social position enforces the rule of underclass. It is the underclass people where in SC and ST comes mostly. Since ST are dependent mostly on forest economy. The SC are the prime concern of Dalit analysis in a farm economy context in order to make the analysis precise and pin-pointedly clear. Here also SCs concerns of employment, income and wages in non-farm sector in the rural economy of India is the prime concern of the analysis.

About three-fourth of all scheduled caste live in rural areas where the main source of income are either cultivation of agricultural land, wage labour or some kind of non agricultural employment. Access to agricultural land for cultivation and capital for undertaking non-farm self-employment is critical. Due to lack of access to land and capital, a large proportion of scheduled caste is dependent on wage employment. However between farm and non-farm sector, they depend for employment, mainly on agriculture. Their participation in rural non-farm activities is much less as compared to the other groups in rural area. In other words the diversification of the employment of the scheduled caste in favour of non-farm employment in rural area is much less.

The data base on employment pattern of scheduled caste are population census and national sample survey. However, employment data for scheduled caste are not available for all years. Table 3 present the distribution of scheduled caste workers (Male) in to nine industrial categories. The nine industrial categories can be grouped into farm and non-farm sector.

The Tables 3, 4 and 5 concerning workers by Industrial categories, occupational pattern and disparities in wage rates in non-Agriculture Employment explains the fact in 1991 of the total workers about 17.61 percent were employed in rural non-farm sector and about 83.52 percent in farm sector. Their participation in rural non-farm sector to the extent of 17.61 percent was less compared to 25.1 percent for the non-SC/ST

TABLE 3
Percentage of Workers by Industrial Categories—Rural (Male) : 1991

Category	*Scheduled Caste*				*Others*			
Cultivator	*1961*	*1971*	*1981*	*1991*	*1961*	*1971*	*1981*	*1991*
(1)	*(2)*	*(3)*	*(4)*	*(5)*	*(6)*	*(7)*	*(8)*	*(9)*
Agr. Lab.	32.83	50.24	46.82	49.27	12.12	19.19	18.3	20.14
Agriculture	75.73	85.18	83.52	82.52	75.86	79.26	76.74	74.88
Non-Agriculture	24.27	14.82	16.48	17.61	24.14	20.74	23.26	25.12

Source : Population Census : 1961, 1971, 1981, 1991.

TABLE 4
Occupational Pattern : Scheduled Caste and Others

(*in percentage*)

Occupational Category	*1987-88*		*1993-94*	
	SC	*Others*	*SC*	*Others*
Self-employed in Agriculture	18.9	43.3	19.12	42.42
Average Wage Labour	51.7	23.2	50.6	22.37
Self-employed in Non-agriculture	11.0	13.8	10.32	13.89
Non-Agr. Wage Labour	11.4	9.7	10.22	6.67
Total in Rural Non-farm Sector	22.4	23.5	20.54	20.56

Source : NSS employment/unemployment survey 1987-88 and 1993-94, CSO, Delhi. SC = Scheduled Caste, others = Non-SC/ST, Excluding, Scheduled Caste and Scheduled Tribe.

TABLE 5
Wage Rate of Rural Labour in Agriculture Current Price (1993-94)

(*in Rupees*)

States	*Non-Agriculture*			
	1974-75		*1993-94*	
	All	*SC*	*All*	*SC*
Andhra Pradesh	9.62	8.85	25.22	22.87
Assam	12.82	12.00	34.53	30.92
Bihar	11.97	10.48	24.04	25.30
Gujarat	10.20	10.66	35.34	28.68
Haryana	19.44	17.23	41.05	40.38
Himachal Pradesh	18.43	17.7	36.52	36.55
J & K	1928	18.38	34.11	31.01
Karnataka	12.75	12.71	28.47	30.45
Kerala	19.33	16.5	52.64	47.67
M.P.	7.93	8.3	22.43	22.57
Maharashtra	11.84	8.93	33.31	34.61
Orissa	7.83	7.98	20.75	16.42

Punjab	23.70	22.43	43.18	40.78
Rajasthan	13.09	12.67	34.88	34.41
Tamil Nadu	11.68	10.60	30.71	28.45
U.P.	13.7	13.01	30.26	28.14
West Bengal	15.95	15.75	29.97	29.11
India	14.21	13.53	32.48	30.62

Source : R. Labour Enquiry Report on Wage Economics, 1974-75, 1983.

population. Thus, the percentage of scheduled caste workers employed in non-farm sector in rural area was less by about 7.40 percentage point. Similar gap also existed in the earlier years, i.e. 1961, 1971, 1981. The another important point is that the percentage of workers employed in rural non-farm sector has remained constant round 2.5 percent although there was marginal decline in 1971 and 1981.

In the case of scheduled caste however there has been a decline in the percentage of workers employed in rural non-farm sector from 24.2 percent to 17.61 percent. In fact, there had been a significant decline in the percentage of workers employed in non-farm sector belonging to scheduled caste between 1961 and 1971 from 24.27 to 14.82 respectively. However, it shared a marginal increases from 16.48 to 17.61 percent during 1980's that is 1.03 percent increase. It is this trend during 1981's that encouraged for increased impress to this sector. However, during the reform period SCs to the total rural non-farm sector employment, the SC showed a declining trend from 22.4 percent to 20.54 percent between 1987-88 and 1993-94 respectively. (*Source* : NSS, Employment/ Unemployment Survey, 1987-88, and 1993-94, CSO, Delhi). The decline was thus 2.5 percent in SC non-farm sector employment.

Table 3 shows that in 1993-94 about one-fifth of the workers were employed in rural non-farm sector both among scheduled caste and others. So, there is no much difference in the participation in rural non-farm sector both among scheduled caste and others. So, we do not see much difference in the participation in rural non-farm activity between scheduled caste and others. However, there is a significant difference in the participation in non-farm sector of the scheduled caste as a wage labourer and self-employed. The percentage of the scheduled caste workers as a wage labourer in non-farm sector is around 10.2 percent as compared to 6.67 percent among the others. But their participation as self-employed workers was less compared to other groups. (18.90 percent SCs others 43.3 percent (1987-88) (19.12 percent SC, others 42.42 percent (1993-94)

Not only SC workers are lesser in self-employed in agriculture, their self-employment in non-agriculture category occupational pattern is 11.0 percent and for others 13.8 percent in 1987-88 and in 1993-94 for SC, it is 10.32 and 13.89 respectively for SC's and others. In the non-agricultural wage labour category the SCs share is 11.4 percent in 1987-88 and 10.22 percent in 1993-94. For others it is 09.7 percent in 1987-88 and 6.67 in 1993-94.

The scheduled caste workers suffers wage disparities in the case of wage rate in non-agriculture sector. In 1993-94 the wage rate received by scheduled caste male labour are however, lower than the national average. As against the national average of Rs. 32.98, the daily wage rate received by the scheduled caste labour is Rs. 30.62. In nine states, out of seventeen states wage rate received by scheduled castes male labour are also lower than the national average.

This includes A.P., Assam, Gujarat, J & K, Kerala, Orissa, Punjab, T.N., U.P. The real wage rates received by the scheduled caste labour varied from Rs. 16.00 to about Rs. 41 in Punjab and Rs. 47.67 in Kerala. So, backward states and regions have less potential to pay wages in real terms. It is because of this region levels of poverty is very high among scheduled caste population in backward regions. The monthly per capita expenditure for SC and other agricultural and other labourers in 1993-94 on food and non-food items. The monthly per capita expenditure (MPE incurred by scheduled caste labourer in 1993-94 is Rs. 139.52 which is low as compared to Rs. 150.14 for the other labourer (non-agriculture labour of scheduled caste.) For instance, the MPE on the food items in their case is Rs. 157.91 as compared to Rs. 128.82 for non-SC/ST labour. The per capita expenditure incurred by scheduled caste labour is low compared with other labour on all the items. The gap is particularly wide in the case of expenditure on meat, egg and fish and fruits.

With regard to MPE of scheduled caste agricultural labour is Rs. 70.09 as compared to Rs. 77.50 for other labour, the differences are particularly large in the case of miscellaneous goods, services and durable goods. In the case of items such as pan, tobacco, intoxicants, fuel and light, clothing and footwear the differences are minimal. In the case of non-agriculture SC labour the MPE is Rs. 90.23, which is lower than Rs. 106.62 for non-SC/ST labour. Like agricultural labour, the differences are pronounced in the case of expenditure on clothing, miscellaneous goods, services and durable goods.

The poverty ratio for scheduled castes and others labour for 1987-88 and 1993-94 at all India level for agricultural and other labour. In 1993-94 about 48.00 percent of scheduled caste households are below the poverty line in rural area as compared to 31.29 percent for the general population. What is striking is the variation in poverty ratio across household types. The incidence of poverty is about 60 percent among agricultural labour followed by 41.44 percent among non-

TABLE 6

MPE (Consumption on Food Total) and Non-food Items All India Rural

Social Group	*Consumption (Food Total)*	*(Non-Food Total)*
Agri. Lab		
SC	139.52	70.9
Others	150.14	77.5
All	144.17	73.24
Other Labour		
SC	157.91	90.23
Others	178.82	106.62
All	168.29	98.44
All Households		
SC	154.83	84.08
Others	189.01	113.07
All	177.78	103.65

Source : 5th Quinquennial Survey, NSS, Delhi, 1993-94, Report No. 422.

TABLE 7

Percentage of Persons Below Poverty Line by Household Type for Each Social Group

Category	*Self-emp. in Agr.*		*Self-emp. in Non-Agr.*		*Agr. Lab.*		*Non-Agr. Lab.*		*Others*		*All*	
	1987	*1983*	*1987*	*1993*	*1987*	*1993*	*1987*	*1993*	*1987*	*1993*	*1987*	*1993*
(1)	*(2)*	*(3)*	*(4)*	*(5)*	*(6)*	*(7)*	*(8)*	*(9)*	*(10)*	*(11)*	*(12)*	*(13)*
Rural	88	94	88	94	88	94	88	94	88	94	88	94
SC	41%	27%	41%	38.1	59	60	46	41.44	29	29	50	48.1
Others	27%	25%	31%	29.4	53	52.3	34	35.5	19	20.5	34.3	31.2

Source : Based on Consumption Expenditure Survey NSSO, 1987-88 and 1993-94.

TABLE 8
Occupational Structure of Main Workers by District and Caste (Bihar Case)

Occupational Structure	*SC/ST*		*Total*
	Male	*Female*	
Agri. and Allied	17.12	17.43	17.18
Agr. Lab.	50.97	61.90	53.10
Other Lab.	8.23	1.62	6.95
Service	9.95	1.98	8.40
Caste Occup.	1.46	4.94	2.13
Business	8.36	1.62	7.05
Others	3.91	10.51	5.20

Source : Action Aid DHD Servey, New Delhi (1999-2000).

agricultural labour. The level of poverty is low among the labour engaged in rural non-farm activities. The level is relatively low for persons engaged in self-employed activities in agriculture (37.7%) and in the non-agricultural sector 38.19 percent for each of these households type. However, the proportion of poor among SC households is much higher compared to their counterparts among the non-SC/ST group. The incidence of poverty is very high among the SC wage labourer, engaged in non-farm sector (41-44%) campared to the other population (37.71%). Similarly, the poverty level among the self-employed engaged in non-farm sector is also high among the scheduled caste (38.19) as compared to the others (29.49) percent. The high level of poverty among the scheduled caste in rural non-farm sector is mainly due to high unemployment rate and low wage earning.

The two tables depicting Bihar situation makes it clear that majority of SC workers are in agricultural labour category as the occupational pattern table concerning Bihar explains. They have lesser opportunity of education and enhancement of abilities. Majority of SC population are illiterate in Bihar (19.18%) and only 2.595 Graduate holders. In this situation due to capability building opportunity at their disposal they are engaged in Sundry/non-farm activities with low income

TABLE 9

Percentage Distribution of Workers by Educational Standards/Major Educational Status (Caste) : Bihar

Caste Groups	*Illi.*	*Primary School*	*Middle School*	*High School*	*Graduate College Level*	*Post-Graduate (Unv. Level)*
(1)	(2)	(3)	(4)	(5)	(6)	(7)
Forward						
Caste	15.6	16.55	2047	23.4	21.17	2.57
OBC-I	71.69	16.32	6.38	2.80	2.77	0.04
OBC-II	48.48	23.90	13.25	10.83	3.07	0.47
SC/ST	79.18	9.75	5.40	2.93	2.59	0.11
Muslims	73.88	15.10	4.89	3.17	2.39	0.58

Source : Action Aid DHD Servey, New Delhi (1999-2000).

generating potential. Therefore, the non-farm strategy and its success demand more fund for the social infrastructure investment which is more a political question rather than a contentious issue of economy analysis.

REFERENCES

Mathew, P.M., Workshop on Rural Transformation in India—Role of Non-farm Sector.

RNFS (2001): Policies and Strategies : Past Experiences and Emerging Perspectives, pp. 19-21, September-December.

Thorat Sukhadeo (2001): The Participation in SC in Rural Non-farm Sector: Analysis with respect to level of living, wage rate and poverty. A paper presented in international seminar on non-farm sector role and rural transformation in India, 19-21 September, 2001.

Thakur, C.P. (2001): Non-farm and Rural Development need for revised policy perspectives and effective governance : International workshop on Rural transformation in India, the role of Non-farm Sector, September 19-21.

NSS/Employment/Unemployment Survey : 1987-88 and 1993-94, CSO, New Delhi.

Population Census, 1961, 1971, 1981, 1991.

Rural Labour Enquiring Wage Earning : 1974-75, 1983-84, 1987-88, 1993-94, Labour Bureau, Simila, Ministry of Labour, Govt. of India.

CHAPTER

2

Non-Farm Employment and Rural Economic Development in India : Composition, Diversification and Growth

PRANKRISHNA PAL[1]

I. INTRODUCTION

The rural economy is primarily a farm economy. People here earn their livelihood mainly from farm activities. Over time the economy changes. So also the rural economy in terms of agrarian structure, occupational structure as well as the living and working conditions of the rural labour. Development of the rural economy occurs when dependence on farm activities decreases and non-farm activities like household industrial operations, servicing construction, etc. are increasingly performed. Increasing participation of labour in rural non-farm activities is indicative of such development of the rural economy.

In rural India, economic reforms were not designed. The GDP growth rate of the primary sector has declined, though the overall growth rate has risen. On the other hand, the overall economy has experienced lower employment growth. Agriculture being almost saturated, off-farm jobs are to be created in the rural and semi-urban region through the creation of new production centers (food processing, ancillary industry, repairing based on transport and construction works, etc.). New agro-industry linkages are to be grown; otherwise the goal of more jobs in the long-run will be a cry.

In this chapter an attempt has been made to examine the structural change in rural non-farm employment in terms of composition, diversification and growth across the states in India during 1972/73-2004/05. The changing scenario of inter-state rural labour force participation rate (male-female differentials) during 1993/94-2004/05 is discussed in section I. Section II contains the activity status of rural employment (i.e., the mode of employment) during 1993/94-2004/05. Section III discusses the rural farm and non-farm employment in India during 1972/73-2004/05 while the rural non-farm employment by occupations is discussed in section IV. Section V deals with the growth of rural non-farm employment in India. Section V contains the concluding remarks. The NSSO data, Govt. of India, different rounds are used in the analysis.

I. RURAL LABOUR FORCE PARTICIPATION RATE BY SEX

In the post-liberalization period the rural economy in India has been changing in many aspects. With the structural transformation and modernization of the rural economy the size of the labour force is expected to decline over time. Labour force of male and female has declined in rural India during 1993/94 to 2004/05. But male labour force has declined more than the female one during the period under period.

According to the NSSO Report, the Labour Force Participation Rate (LFPR) is defined as the proportion of persons in the age group 15 years and above who were either working (i.e. employed) on the usual Principal Status and Subsidiary Status (PS & SS) or seeking or available for work (i.e. presently unemployed). Estimates (Table 1) reveal that in rural India the LFPR has declined marginally from 87.6% in 1993-94 to 85.4% in 1999-2000 and then increased marginally to 85.9% in 2004-05 for male. It has also declined from 48.8% in 1993-94 to 45.6% in 1999-2000 and increased to 49.4% in 2004-05 for rural female. Thus sex-wise analysis shows that male LFPR is more than female one in India and its all the constituent states during the period under study.

TABLE 1

Sex-wise Rural Labour Force Participation Rate (PS & SS) in Indian States During 1993/94-2004/05

	1993-94		*1999-2000*		*2004-05*	
	Male	*Female*	*Male*	*Female*	*Male*	*Female*
A.P	90.3	72.5	88.3	66.4	86.7	65.5
Assam	84.7	26.0	84.1	24.9	87.2	32.8
Bihar	86.7	27.1	86.9	28.7	87.8	23.1
Gujarat	89.1	58.1	88.4	60.0	89.4	62.1
Haryana	80.2	44.2	78.0	31.9	80.2	48.4
H.P.	88.7	73.0	82.5	67.5	81.8	71.3
Karnataka	89.4	63.5	88.5	55.5	87.7	62.0
Kerala	82.6	35.8	81.4	36.3	80.9	41.8
M.P.	89.8	63.1	86.9	61.4	87.6	57.4
Maharashtra	85.9	70.7	83.9	64.1	83.5	65.7
Orissa	86.8	46.1	85.8	44.0	88.2	49.3
Punjab	84.3	32.8	82.7	41.5	84.4	47.5
Rajasthan	87.8	67.3	84.9	59.7	84.8	62.6
T.N.	86.9	65.1	85.7	57.4	85.3	63.1
U.P.	88.4	34.7	84.3	33.2	85.4	39.1
W.B.	89.7	29.1	86.3	24.4	86.6	26.8
India	87.6	48.8	85.4	45.6	85.9	49.4
DI(%)	3.27	35.0	3.33	33.26	3.19	29.20

Note : DI : Disparity Index.
Source : NSSO, Govt. of India, Different Rounds.

Indian states have exhibited wide variations in LFPR. In rural areas, male LFPR has declined in India and its almost all the constituent states excepting Assam, Bihar, Gujarat, Orissa and Punjab during 1993-2005. Also, female LFPR has declined in its almost all the constituent states in India, excepting Assam, Gujarat, Haryana, Kerela, Orissa, Punjab and UP during this period under study. Also, we note that irrespective of sex, the LFPR has increased in Assam, Gujarat, Orissa and Punjab during the period under study. Also, male LFPR was lowest in Haryana in 1993-94, 1999-2000 and 2004-05. Andhra Pradesh was ranked highest in 1993-94 and then lost its prime place to Karnataka in 1999-2000 and to Gujarat in 2004-05. On the other hand, female LFPR was lowest in Assam in 1993-94, West Bengal in 1999-2000 and Bihar in 2004-05. It was highest in Himachal Pradesh in 1993-94, 1999-2000 and 2004-05. Thus Gujarat and Himachal Pradesh have utilized more labour force of male and female respectively for the development of the rural economy during the period under study.

It thus follows from the above discussion that the LFPR for male and female had not been uniform in rural Indian states. It had been relatively low in some states while it had been relatively high in some other states. All this clearly indicates the existence of inter-state disparity in LFPR for male and female. Such inter-state disparity in LFPR is clearly revealed by the Disparity Index (DI) of LFPR as:

$$DI = 100\ [\Sigma\ (e_i - \bar{e})^2/(n-1)]^{1/2}/\bar{e},$$

where, e_i : LFPR (male/female) in rural Indian states

$i = 1, 2, ..., n$;

$\bar{e}$: LFPR (male/female) in India as a whole; and

n : Number of states in India.

Estimates (Table 1) reveal that in the rural economy of India the DI of LFPR for male has risen during 1993-2000 and then fallen during 1999-2005. But the DI of LFPR for female has fallen during 1993-2005. Interestingly, we note that the DI of female LFPR has been relatively more than that of male one during the period under study. This indicates that inter-state disparity has widened more for female LFPR than for male one in the rural India during 1993-2005.

III. MODE OF EMPLOYMENT BY SEX: ACTIVITY STATUS

In section I we have discussed the changing scenario of labour force in India and its constituent states by sex during 1993/94-2004/05. We shall now examine the activity status of employment in rural India and its constituent states by sex during 1972-2005.

Labour is employed in different types of activities. Activities are agricultural and non-agricultural (i.e., non-farm). Labour is classified as self-employed and wage (regular and casual) labour. Mode of employment of labour (male and female) in rural India has been expectedly changing during the reform period. Estimates (Table 2) reveal that in rural India the incidence of self-employment has been consistently declining over time. It has declined from 65.9% in 1972-72 to 56.7% in 1993-94 and to 56.3% in 2002 in case of male labour. It has increased to 57.6% in 2004-05. The respective figures are: 65.5%, 51.3%, 49% and 56.4% in case of female labour. Thus the male self-employment is more than the female one during the period under study. But regular salaried employment has been declining for male (12.1% in 1972-73 to 9.1% in 2004-05) and rising for female (4.1% in 1972-73 to 4.8% in 2004-05). On the other hand, the incidence of causal labour gives a noticeable picture during the period under study. It has increased gradually from 20.0% in 1172-73 to 33.3% in 2004-05 for male. For female it has increased from 31.4% in 1972-73 to 46.4% in 2002 and then declined to 38.9% in 2004-05. Thus female are casually employed more than male during the period under study. This is presumably due to the fact that during the reform periods various types of activities mostly non-farm activities have been generated in favour of rural female so that they can utilize more relative to male at a low wage rate (Pal, 2002).

Indian states have exhibited wide variations in the employment of labour during the reform periods. Estimates (Table 3) reveal that the incidence of self-employment for male has increased over time in almost all the states excepting

TABLE 2

Percentage Distribution of Employment (Principal Status) in Rural India by Activity Status and Sex during 1972/73-2004/05

Years	Sex	Activity Status				
		S.E.	R.E.	C.L.	Tota.	II.C.
1972-73	Male	65.9	12.1	20.0	100.0	182
1977-78	"	62.8	10.6	26.6	100.0	251
1983	"	59.5	10.6	29.9	100.0	282
1987-88	"	57.5	10.4	32.1	100.0	309
1993-94	"	56.7	8.7	34.6	100.0	398
1997	"	59.5	7.3	33.2	100.0	455
1999-2000	"	54.4	9.0	36.6	100.0	406
2002	"	56.3	8.9	34.8	100.0	391
2004-05	"	57.6	9.1	33.3	100.0	366
1972-73	Female	65.5	4.1	31.4	100.0	766
1977-78	"	62.1	2.8	35.1	100.0	1253
1983	"	54.1	3.7	42.2	100.0	1140
1987-88	"	54.9	4.9	40.2	100.0	820
1993-94	"	51.3	3.4	45.3	100.0	1332
1997	"	57.0	2.1	40.9	100.0	1948
1999-2000	"	50.0	3.9	46.1	100.0	1182
2002	"	49.0	4.6	46.4	100.0	1008
2004-05	"	56.4	4.8	38.9	100.0	810

Notes : SE : Self-employed, RE: Regular-salaried employed.
CL : Casual Labour
IC : Index of Casualization (= CL/RE × 100)

Source : NSSO, Different Rounds, Government of India.

Andhra Pradesh, Haryana, H.P., Karnataka, M.P., Punjab, Rajasthan, T.N., U.P. and West Bengal while that for female has also increased in almost all the states excepting H.P., Karnataka and Punjab during 1993/94-2004/05. It has varied from 39.2%

in Kerala to 71.7% in U.P. in 1993-94 for rural male and from 39.8% in T.N. to 71.67% in U.P. in 2004-05. Kerela (34.6%) and H.P. (95.6%) have ranked respectively lowest and highest in 1993-94 in respect of the rural self-employment. In 2004-05 the respective states are Punjab (35.4%) and H.P. (91.47%), Similarly, in case of regular salaried employment male employment has also increased in almost all the states in India excepting Assam, Bihar, Karnataka, Maharashtra and West Bengal during the period under study. This is also true in case of female employment excepting Assam, Maharashtra and West Bengal. In most of the states the incidence of casual labour for both male and female has increased at the cost of self-employment.

Interestingly, we note that the casualisation of female labour is more than that of male one in rural India and its all the constituent states. But the cases of self-employed and regular-salaried male employment are more than that female one in rural India and its all the constituent states during the period under study.

The incidence of casualization is measured by index of casualization, which is defined as the number of casual wage labour for every one hundred of regular-salaried wage labour. Estimates (Tables 2 and 3) reveal that in rural India the index has gradually increased from 182 in 1972-73 to 455 in 1997 and then declined to 366 in 2004-05 for male; and from 766 in 1972-73 to 1948 in 1997 and then remarkably declined to 810 in 2004-05 for female. Also, irrespective of sex, the index has varied across the states in India during 1993/94-2004/05. The index is lowest in Haryana (104 in 1993-94 and 110 in 2004-05) and highest in Bihar (840 in 1993-94 and 1173 in 2004-05) for male; lowest in H.P. (76 in 1993-94 and 30 in 2004-04), and highest in Bihar (3973 in 1993-94 and 2065 in 2004-05). Thus rural labour is casually employed more in Bihar and less in H.P. during the period under study. However, sex-wise analysis reveals that the casualisation of female labour is more than that of male ones in rural India and its almost all the constituent states excepting H.P., Kerala, Punjab and West Bengal during the period under study. Thus female labour finds jobs of casual and/or contractual nature or on a daily basis more relative to male ones at a low wage rate (Pal, 2002).

TABLE 3

Percentage Distribution of Rural Employment (Principal Status) by Activity Status and Sex in Indian States during 1993/94-2004/05

States	1993-94								2004-05							
	Male				Female				Male				Female			
	SE	RE	CL	IC	SE	RE	CL	IC	SE	RE	CL	IC	SE	RE	CL	IC
(1)	(2)	(3)	(4)	(5)	(6)	(7)	(8)	(9)	(10)	(11)	(12)	(13)	(14)	(15)	(16)	(17)
A.P.	48.6	8.2	43.2	527	43.5	1.9	54.6	2874	48.1	9.7	42.2	435	45.4	4.2	50.4	1200
Assam	58.3	14.0	27.7	198	17.4	33.7	48.9	145	70.7	9.6	19.8	206	50.2	15.2	34.5	227
Bihar	54.9	4.8	40.3	840	38.9	1.5	59.6	3973	61.8	3.0	35.2	1173	50.2	2.3	47.5	2065
Gujarat	46.9	9.7	43.4	447	44.5	2.0	53.5	2675	49.2	10.3	40.5	393	51.8	3.7	44.5	1203
Haryana	61.6	14.0	24.4	174	62.5	5.4	32.1	594	57.4	20.3	22.3	110	80.9	6.0	131.1	218
H.P.	71.2	14.1	14.7	104	95.6	2.5	1.9	76	59.1	19.5	21.4	110	91.4	6.6	2.0	30
Karnataka	56.8	6.7	36.5	545	48.0	3.1	48.9	1577	49.4	6.4	44.3	692	44.5	3.2	52.3	1634
Kerala	39.2	12.8	48.0	375	34.6	15.0	50.4	336	40.8	15.3	44.0	288	40.8	25.7	33.5	130
M.P.	61.7	6.0	32.2	528	55.8	1.7	42.5	2500	61.2	6.5	32.3	497	58.6	3.1	38.4	1239
Maharashtra	48.3	12.6	39.5	324	44.3	2.8	52.9	1889	50.3	12.4	37.3	301	49.2	2.7	48.1	1781

(Contd.)

TABLE 3 (*Contd.*)

(1)	(2)	(3)	(4)	(5)	(6)	(7)	(8)	(9)	(10)	(11)	(12)	(13)	(14)	(15)	(16)	(17)
Orissa	55.7	6.4	37.9	592	46.6	1.9	51.5	2710	55.8	7.6	36.5	480	52.4	3.2	44.4	1387
Punjab	54.3	13.3	32.4	244	42.1	18.4	39.5	215	48.7	17.6	33.7	191	35.4	32.3	32.3	100
Rajasthan	71.2	7.6	21.2	279	86.3	1.3	12.4	954	69.9	8.6	21.5	250	87.3	2.1	10.6	505
Tamil Nadu	40.7	12.8	46.5	363	38.0	6.2	55.8	900	39.8	13.3	46.9	353	45.0	6.8	48.2	709
Uttar Pradesh	71.7	5.9	22.4	380	74.2	1.5	24.3	1620	71.6	7.3	21.1	289	77.9	3.1	18.9	610
West Bengal	53.3	10.7	36.0	336	38.2	15.8	46.0	291	53.1	7.4	39.5	534	46.9	13.7	39.4	288
India	56.7	8.7	34.6	398	51.3	3.4	45.3	1332	57.6	9.1	33.3	366	56.4	4.8	38.9	810

Notes : SE: Self-employed, RE: Regular-salaried Employed, CL: Casual Labour, IC : Index of Casualization (= CL/RE × 100).

Source : NSSO, Govt. of India, Different Rounds.

III. RURAL EMPLOYMENT BY SEX : FARM AND NON-FARM

Apart from variations in mode of employment, variations in the sectoral activity (occupation) of rural labour are also observed. Rural labour is employed in alternative sectoral activities like: (a) primary sector (PS) consisting of cultivation, agricultural activities, forestry, fishing, mining and quarrying; (b) secondary sector (SS) including manufacturing, servicing and repairing in both household and non-household industries and construction; (c) Tertiary sector (TS) consisting of transport, trade and commerce, storage and communications and services. The last two sectors jointly we call the Non-Farm Sector (NFS) while the former is the Farm sector (FS). Therefore, total rural labour force is utilized in these two sectors. The FS and NFS are to a great extent interdependent. Modernization of the FS (in the Post-Green Revolution period) has brought about a series of changes in rural India. Villages are linked with the cities and towns through a wide net of transports, markets are developed and expanded, farm outputs are transported, agricultural machinery and repairing shops are opened, small electrical units are set-up, etc. All these necessitate increasing absorption of rural labour in NFS.

Estimates (Table 4) reveal that employment in NFS has increased all through in India at the cost of farm employment both for male and female. For male it has increased from 16.4% in 1972-73 to 21.8% in 1983, to 25.2% in 1993-94 and to 33.2% in 2004-05. While for female it has increased from 10.1% in 1972-73 to 12.1% in 1983, to 13.6% in 1993-94 and to 18.2% in 2004-05. Thus male non-farm employment is more than female ones in rural India due to expansion of the non-farm activities more for male relative to female.

Rural employment has varied across the states in India over time. It is expected that more and more a state is economically and educationally developed, more and more is the supply of labour in the NFS. So, irrespective of sex, Indian states have exhibited-wide variations in NFE during the reform period. Estimates (Table 5) reveal that the rural NFE for male has increased in all the constituent states in India during 1993/

TABLE 4
Rural Non-Farm Employment in India during 1972/73-2004/05

Years	*Male*		*Female*	
	FE	*NFE*	*FE*	*NFE*
1972-73	83.6	16.4	89.9	10.1
1977-78	81.1	18.9	88.3	11.7
1983	78.2	21.8	87.9	12.1
1987-88	75.2	24.8	85.1	14.9
1993-94	74.8	25.2	86.4	13.6
1997	76.3	23.7	87.8	12.2
1999-2000	71.9	28.1	94.4	15.6
2004-05	66.8	33.2	81.8	18.2

Notes : FE : Farm Employment, NFE : Non-Farm Employment.
Source : NSSO data.

94-2004/05 at the cost of farm employment while for female it has also increased in almost all the states excepting Karnataka and West Bengal during the period under study. Thus among the states, only in Karnataka and West Bengal NFE for female has decreased during the reform period. The NFE for male is lowest in Madhya Pradesh (13.4% in 1993-94), M.P. (20.3% in 2004/05) and highest in Kerala (45.7% in 1993-94 and 61.8% in 2004-05). The respective figures for female are: lowest in H.P. (5.3% in 1993-94) and Maharastra (9.0% in 2004-05); and highest in West Bengal (54.7% in 1993-94) and Kerala (56.2% in 2004-05). Thus among the states only Kerala has significantly enjoyed more the absorption of the rural NFE for both male and female due to the generation of the non-farm activities and the rural industrialization during the reform period of 1993/94-2004/05.

Sex-wise analysis also reveals that female employment in the NES is more than male ones only in Punjab and West Bengal during 1993/94-2004/05. This is presumably due to the fact that during the reforms period various types of activities mostly non-farm activities have been generated more in these

TABLE 5

Rural Farm and Non-Farm Employment in Indian States during 1993/94 to 2004/05

States	*1993-94*				*2004-05*			
	Male		*Female*		*Male*		*Female*	
	NFE	*FE*	*NFE*	*FE*	*NFE*	*FE*	*NFE*	*FE*
A.P.	76.6	23.4	83.9	16.1	68.0	32.0	79.5	20.5
Assam	78.0	22.0	84.6	15.4	69.3	30.5	82.7	17.3
Bihar	82.3	17.7	91.2	8.8	75.6	24.4	82.8	17.2
Gujarat	71.9	28.1	88.4	11.6	69.7	30.3	87.5	12.5
Haryana	60.8	39.2	84.7	15.3	50.9	49.1	80.9	19.1
H.P.	61.4	38.6	94.7	5.3	46.6	53.4	89.5	10.5
Karnataka	79.8	20.2	83.5	16.5	78.3	21.7	87.3	12.7
Kerala	54.3	45.7	51.4	48.6	38.2	61.8	43.8	56.2
M.P.	86.6	13.4	94.2	5.8	79.7	20.3	88.5	11.5
Maharashtra	75.5	24.5	91.6	8.4	71.7	28.3	91.0	9.0
Orissa	79.6	20.4	85.2	14.8	66.4	33.6	72.5	27.5
Punjab	67.9	32.1	65.5	34.5	54.4	45.6	48.4	51.6
Rajasthan	71.6	28.4	92.6	7.4	61.8	38.2	87.6	12.4
Tamil Nadu	64.2	35.8	77.7	22.3	59.3	40.7	73.2	26.8
U.P.	75.9	24.1	89.0	11.0	65.6	34.4	81.4	18.6
W.B.	64.4	35.6	45.3	54.7	64.3	35.7	54.1	45.9
India	74.4	25.6	85.2	16.8	66.8	33.2	81.8	18.2

Note : FE : Farm Employment, NFE : Non-Farm Employment.
Source : NSSO Data.

two states compared to other states in favour of rural female so that they can utilize more relative to male at a low wage rate.

IV. RURAL NON-FARM EMPLOYMENT BY SEX AND OCCUPATIONS

We have already said the NFS consists of Secondary Sector (SS) and Tertiary Sector (TS). The SS and TS are the sub-sectors of the NFS. These two sub-sectors are interdependent. So inter-sector shifting of labour occurs with the development of the rural economy in India. The activities or occupations of these sub-sectors are: (i) manufacturing, (ii) electricity, gas and water supply, (iii) construction, (iv) trade and commerce, (v) transport, storage and communications, and (vi) other services. These services in the rural economy are known as the rural non-farm activities and labour employed in these activities are called the rural non-farm employment (RNFE).

Let us now examine the importance of these non-farm activities and hence the sub-sectors for the generation of non-farm employment in reducing the disguised unemployment in the farm sector. Between the two sub-sectors, the share of the TS has all through been the highest for male in India and its almost all the constituent states (Table 6). At the national level, its share has increased from 54.3% in 1972-73 to 58.3% in 1993-94 and then declined to 54.2% in 2004-05. This is presumably due to the fact that employment rise in the TS is caused by employment rise in its components namely trade & commerce and transports. The share of trade and commerce has increased from 19% in 1972-73 to 21% in 1993-94 and to 25.1% in 2004-05. The respective figures for transports are : 6%, 8.7% and 11.4%. But employment in services has declined from 29.3% in 1972-73 to 27.8% in 1993-94 and 17.3% in 2004-05. It has lost its prime place now. On the other hand, the share of the SS has decreased from 45.7% in 1972-73 to 41.7% in 1993-94 and then increased to 45.8% in 2004-05. The decline in employment share in the SS is due to the declining share in manufacturing and electricity. Though maintained its prime place in the SS, the share of manufacturing has declined from 34.7% in 1972-73 to 27.8% in 1993-94 and to 24.2% in 2004-05. The share has also fallen for electricity: 1.3%, 1.2% and 0.6% in 2004-05. But the share of construction has risen from 9.7% in 1972-73 to 12.7% in 1993-94 and to 21% in 2004-05. Thus we observe that the TS is the growing sub-sector while the SS is the decling sub-sector in the non-farm sector for male RNFE. If we consider the activities of these sectors, the construction in the SS, the trade & commerce and transports in the TS are growing activities for the absorption of male RNFE. The declining activities are the manufacturing and electricity in the SS and services in the TS.

TABLE 6
Percentage Distribution of Rural Non-Farm Employment in India by Occupations During 1972/73-2004/05

Years	*Manuf.*	*Elect.*	*Const.*	*SS*	*Trade*	*Transport*	*Services*	*TS*	*Total*	*DI*
(1)	(2)	(3)	(4)	(5)	(6)	(7)	(8)	(9)	(10)	(11)
Male NFE										
1972-73	34.7	1.3	9.7	45.7	19.0	6.0	29.3	54.3	100.0	0.834
1977-78	33.9	1.0	9.0	43.9	21.2	6.9	28.0	56.1	100.0	0.837
1983	32.1	1.4	10.1	43.6	20.2	7.8	28.4	56.4	100.0	0.857
1987-88	29.8	1.2	14.9	45.9	21.0	8.1	25.0	54.1	100.0	0.879
1993-94	27.8	1.2	12.7	41.7	21.8	8.7	27.8	58.3	100.0	0.877
1997	27.4	1.3	13.5	42.2	20.7	8.4	28.7	57.8	100.0	0.878
1999-2000	26.0	0.7	16.0	42.7	24.2	11.4	21.7	57.3	100.0	0.893
2004-05	24.2	0.6	21.0	45.8	25.1	11.8	17.3	54.2	100.0	0.895
Female NFE										
1972-73	46.5	—	10.9	57.4	14.8	—	27.8	42.6	100.0	0.892
1977-78	51.3	—	5.1	56.4	17.2	0.8	25.6	43.6	100.0	0.736
1983	53.7	—	5.8	59.5	16.5	0.8	23.2	40.5	100.0	0.730
1987-88	47.0	—	18.1	65.1	14.1	0.7	20.1	34.9	100.0	0.807
1993-94	51.5	0.7	6.6	58.8	15.5	0.7	25.0	41.2	100.0	0.684
1997	50.8	0.8	9.0	60.6	9.0	0.8	29.6	39.4	100.0	0.678
1999-2000	49.3	—	7.7	57.0	14.8	0.6	27.6	43.0	100.0	0.755
2004-05	48.3	—	9.4	57.7	15.5	1.2	25.5	42.3	100.0	0.785

Notes : SS : Secondary Sector, TS : Tertiary Sector, DI : Diversification Index.

A reverse situation has happened in case of female RNFE. Here the SS is the prime sector for female employment. Its share has increased from 57.4% in 1972-73 to 58.8% in 1993-94 and then declined to 57.7% in 2004-05. So the share has remained more or less stagnant over 38 years. In the SS, manufacturing industry has employed 50% of total female RNFE. It has played a vital role for rural female employment due to expansion of household small-scale industry for female. On the other hand, the share of the TS has fallen from 42.6% in 1972-73 to 41.2% in 1993-94 and then increased to 42.3% in 2004-05. Female employment has not increased in trade, transport and servicing during the period under study.

Sex-wise analysis reveals that the SS has absorbed more female labour than male ones during the period the period of 1972/73-2004/05. The TS has absorbed more male labour than female ones during the period under study. Also, compared to male, female employment is more in manufacturing and services. This is presumably due to the fact that during this period various types of activities mostly non-farm activities have been generated in favour of female so that they can absorb more relative to male at a low wage rate.

Inter-state analysis reveals that Indian states have exhibited wide variations in RNFE. Estimates (Table 7) reveal that the share of the TS has risen only in three states namely, Gujarat, Rajasthan and West Bengal for male during 1993/94-2004/05. It has also risen in the states of A.P., Gujarat, Karnataka, Kerala, Maharashtra, Punjab, T.N., and West Bengal during this period. Thus the TS is the growing sector in these states for absorbing more RNFE. On the other hand, the SS is the growing sector in absorbing the RNFE (a) for male in all the constituent states excepting Gujarat, Rajasthan and West Bengal, and (b) for female in the states of Assam, Bihar, Haryana, H.P., M.P., Orissa, Rajasthan and U.P. during the period under study. Also, activity (occupation)-wise analysis shows that the share of manufacturing has risen in the states of Bihar, Haryana, H.P., Punjab and Rajasthan during 1993-2005 for male; Assam, Bihar, Haryana, Rajasthan and U.P. for female during this period. Manufacturing is thus the growing sub-sector in these states for employment generation and reducing poverty. Similarly, construction, trade & commerce, transport

TABLE 7

Percentage Distribution of Rural Male Non-Farm Employment by Occupations in Indian States During 1993/94-2004/05

States	*1993-94*									
	1	*2*	*3*	*SS*	*4*	*5*	*6*	*TS*	*Total*	*DI*
A.P.	27.3	0.4	11.1	38.8	23.2	9.4	28.6	61.2	100.0	0.681
Assam	10.0	1.4	3.6	15.0	36.8	9.1	39.1	85.0	100.0	0.720
Bihar	19.2	1.8	9.0	30.0	30.5	9.0	30.5	70.0	100.0	0.817
Gujarat	44.5	0.7	10.7	35.9	15.0	9.9	19.2	44.1	100.0	0.779
Haryana	14.0	1.8	16.8	32.6	18.4	14.3	34.7	67.4	100.0	0.840
H.P.	14.2	3.9	35.3	53.4	14.7	5.9	26.0	46.6	100.0	0.801
Karnataka	26.2	1.4	10.0	37.6	23.8	8.5	30.1	62.4	100.0	0.819
Kerela	21.7	0.8	16.4	38.9	26.0	16.1	19.0	61.1	100.0	0.76
M.P.	28.9	1.7	10.5	41.1	19.3	8.0	31.6	58.9	100.0	0.811
Maha.	27.3	1.2	13.8	42.3	18.0	11.1	28.6	57.7	100.0	0.850
Orissa	29.0	1.0	11.3	41.3	25.5	5.7	27.5	58.7	100.0	0.789
Punjab	19.3	4.7	15.0	39.0	19.6	13.3	28.1	61.0	100.0	0.897
Rajasthan	19.4	1.8	37.7	58.9	13.4	7.9	20.8	41.1	100.0	0.816
T.N.	35.8	1.1	10.0	46.9	17.6	11.1	21.4	53.1	100.0	0.843
U.P.	29.9	0:8	11.2	42.0	21.6	9.9	26.5	58.0	100.0	0.819
W.B.	33.4	0.3	7.9	41.6	25.8	11.5	21.1	58.4	100.0	0.798
India	27.7	1.2	13.0	41.9	21.7	10.3	26.1	58.1	100.0	0.840

(*Contd.*)

TABLE 7 (*Contd.*)

States	2004-05									
	1	*2*	*3*	*SS*	*4*	*5*	*6*	*TS*	*Total*	*DI*
A.P.	24.0	0.6	16.2	40.8	24.6	14.3	20.3	59.2	100.0	0.901
Assam	9.8	0.6	10.2	20.6	38.4	10.8	30.2	79.4	100.0	0.815
Bihar	21.0	0.4	13.1	34.5	36.1	11.9	17.5	65.5	100.0	0.861
Gujarat	33.7	0.6	12.5	46.8	22.5	14.9	15.8	53.2	100.0	0.875
Haryana	23.3	1.5	25.2	50.0	21.8	12.3	15.9	50.0	100.0	0.911
H.P.	14.6	5.2	35.9	55.7	15.3	11.4	17.6	44.3	100.0	0.917
Karnataka	23.5	0.4	14.3	38.2	27.2	15.7	18.9	61.8	100.0	0.893
Kerela	16.6	0.5	24.5	41.6	25.6	16.4	16.4	58.4	100.0	0.899
M.P.	22.0	1.0	22.5	45.5	25.5	5.8	23.2	54.5	100.0	0.875
Maharashtra	27.3	0.7	14.9	42.9	23.0	14.5	19.6	57.1	100.0	0.899
Orissa	25.4	0.6	20.3	46.3	27.2	9.2	17.3	53.7	100.0	0.882
Punjab	21.0	2.4	30.0	53.4	20.5	11.6	14.5	46.6	100.0	0.911
Rajasthan	19.7	0.8	37.3	57.8	17.3	9.7	15.2	42.2	100.0	0.861
T.N.	33.0	0.5	21.0	54.6	19.5	11.5	14.5	45.5	100.0	0.875
U.P.	28.5	0.3	22.4	51.2	24.7	9.0	15.1	48.8	100.0	0.869
W.B.	24.9	0.3	14.0	39.2	32.1	13.1	15.6	60.8	100.0	0.871
India	24.2	0.6	21.0	45.8	25.1	11.8	17.3	54.2	100.0	0.895

Notes : Column Heads: 1. Manufacturing, 2. Electricity, Gas & Water Supply, 3. Construction, 4 Trade & Commerce, 5. Transport, etc., 6. Services, SS: Secondary Sector, TS: Tertiary Sector, DI: Diversification Index.

Source : NSSO, Govt. of India, Different Rounds.

and services are the growing sub-sectors for rural non-farm employment generation in most of the states in India.

Across the states, the rural non-farm employment has varied. The RNFE for male has varied from 15% in Assam to 58.9% in Rajasthan in 1993-94 and from 20.6% in Assam to 57.8% in Rajasthan in 2004-05 in the SS. The respective figures in the TS are: 41.1% in Rajasthan and 85% in Assam in 1993-94; and 42.2% in Rajasthan and 79.4% in Assam in 2004-05. Thus Rajasthan and Assam have ranked highest in respect of the utilization of rural male in the SS and TS respectively and hence these are the growing states for these sectors. Under the SS, the manufacturing is the major industry which absorbs more male labour in the states of Gujarat, Karnataka, M.P., Maharashtra, Orissa, Tamil Nadu, U.P. and West Bengal. But in Rajasthan construction plays a major role for absorbing more male labour during the period under study. So manufacturing and construction under the SS are the growing sub-sectors while electricity, gas and water supply are the declining sub-sectors. On the other hand, under the TS, trade and commerce and services play an important role in almost all the states in India for the non-farm employment generation and hence reducing rural poverty. Thus these two sub-sectors are the growing sub-sectors while transport and storage and communication are less growing sub-sectors under the TS.

Similarly, the RNFE for female has varied across the states in India (Table 8). It is lowest in Punjab (21.7%) and highest in West Bengal (70.5%) in 1993-94. The respective figures are: Punjab (16.5%) and Orissa (77.3%) under the SS. But in the TS it is lowest in M.P. (31%) in 1993-94 and Orissa (22.7%) in 2004-05; and highest in Punjab (78.5% & 83.5%) in both the years. The RNFE for female has increased: (a) in the SS in the states of Assam, Bihar, Haryana, Orissa, Punjab, Rajasthan, T.N., U.P. and West Bengal, and (b) in the TS in the states of A.P., Gujarat, H.P., Karnataka, Kerala, M.P. and Maharashtra. Thus both the SS and TS are the growing sectors in these states for the generation of female employment in non-farm activities. Manufacturing under the SS, and trade and commerce and services under the TS are the growing sub-sectors in almost all the states in India during the period under study.

TABLE 8

Percentage Distribution of Rural Male Non-Farm Employment by Occupations in Indian States During 1993/94-2004/05

States	*1993-94*									
	1	*2*	*3*	*SS*	*4*	*5*	*6*	*TS*	*Total*	*DI*
A.P.	47.2	—	4.3	51.5	20.5	0.6	27.4	48.5	100.0	0.696
Assam	20.1	—	1.9	22.0	14.4	2.6	61.0	78.0	100.0	0.609
Bihar	48.9	—	2.3	51.2	27.3	—	21.5	48.8	100.0	0.810
Gujarat	46.6	0.9	17.2	64.7	6.0	—	29.3	35.3	100.0	0.764
Haryana	16.3	—	8.5	24.8	16.3	—	59.0	75.2	100.0	0.802
H.P.	34.0	5.7	11.3	51.0	13.2	—	35.8	49.0	100.0	0.878
Karnataka	60.0	—	4.2	64.2	12.2	1.2	22.4	35.8	100.0	0.674
Kerela	49.6	0.4	6.6	56.6	9.9	2.8	30.7	43.4	100.0	0.647
M.P.	62.0	—	7.0	69.0	15.5	—	15.5	31.0	100.0	0.765
Maharashtra	36.5	1.2	15.5	53.2	17.4	2.4	27.0	46.8	100.0	0.757
Orissa	45.3	—	9.5	54.8	23.0	—	22.2	45.2	100.0	0.905
Punjab	17.7	3.4	—	21.1	7.6	—	71.3	78.9	100.0	0.620
Rajasthan	19.0	—	44.5	63.5	10.8	—	25.7	36.5	100.0	0.913
T.N.	61.4	—	3.6	65.0	11.2	1.8	22.0	35.0	100.0	0.603
U.P.	42.7	—	2.7	45.4	21.0	2.6	31.0	54.6	100.0	0.775
W.B.	65.4	—	5.1	70.5	7.3	9.2	21.0	29.5	100.0	0.659
India	50.7	—	7.4	58.1	14.9	13.5	13.5	41.9	100.0	0.775

(Contd.)

TABLE 8 (*Contd.*)

States	1994-95									
	1	2	3	*SS*	4	5	6	*TS*	*Total*	*DI*
A.P.	45.8	—	5.4	51.2	23.9	—	24.9	48.8	100.0	0.868
Assam	22.4	—	8.6	31.0	9.2	2.3	57.5	69.0	100.0	0.727
Bihar	57.0	—	—	57.0	19.7	—	23.3	43.0	100.0	0.892
Gujarat	38.7	—	18.5	57.2	13.7	—	29.1	42.8	100.0	0.946
Haryana	35.3	1.1	7.9	44.3	10.0	1.6	44.1	55.7	100.0	0.712
H.P.	21.1	2.0	8.6	31.7	8.7	5.8	53.8	68.3	100.0	0.741
Karnataka	46.4	—	5.6	52.0	18.9	2.4	26.7	48.0	100.0	0.792
Kerela	42.0	0.5	3.2	45.7	10.1	3.2	41.0	54.3	100.0	0.674
M.P.	49.5	—	20.8	70.3	15.6	—	14.1	29.7	100.0	0.895
Maharashtra	33.0	—	13.6	46.6	25.0	1.1	27.3	53.4	100.0	0.862
Orissa	60.2	—	17.1	77.3	9.2	0.7	12.8	22.7	100.0	0.699
Punjab	15.2	—	1.3	16.5	8.3	4.7	70.5	83.5	100.0	0.584
Rajasthan	35.5	—	33.8	69.3	7.3	0.8	22.6	30.7	100.0	0.808
T.N.	56.3	—	7.8	64.1	16.4	1.5	18.0	35.9	100.0	0.739
U.P.	52.0	—	5.5	57.5	13.6	0.5	28.4	42.5	100.0	0.717
W.B.	63.2	—	1.3	64.5	9.4	0.5	25.6	35.5	100.0	0.586
India	48.3	—	9.4	57.7	15.5	1.2	25.5	42.3	100.0	0.785

Notes : Column Heads: 1. Manufacturing, 2. Electricity, Gas & Water Supply, 3. Construction, 4. Trade & Commerce, 5. Transport, etc., 6. Services, SS: Secondary Sector, TS: Tertiary Sector, DI: Diversification Index.

Source : NSSO, Govt. of India, Different Rounds.

Measures of Occupational Diversification

We have so far discussed the occupational distribution of rural non-farm employment for sexing India and its constituent states during the period under study. The occupational distribution of employment has changed during this period. Employment has risen in some occupations while it has declined in some others. We shall now measure the extent of variations in the occupational distribution of non-farm employment. For this we have estimated the *Occupational Diversification Index (DI)* based on Theil (1967) entropy measure which is defined as :

$$DI = E/\log n,$$

Where: $E = \Sigma e_i \log (1/e_i)$,

e_i : share of the ith occupation in the non-farm sector,
n: number of occupations.

Here; (a) DI= 0 for E = 0. It means that a particular occupation captures all workers: non-farm employment is not diversified occupationally. (b) DI = 1, when E = log n : maximum value. This happens when all workers are employed equally by different occupations: employment is completely diversified occupationally in the non-farm sector. So 0 £ DI £ 1. The higher (lower) the value of DI, the higher (lower) the degree of diversification of rural non-farm employment occupationally.

Estimates (Table 6) reveal that in rural India DI for male has increased from 0.834 in 1972-73 to 0.877 in 1993-94 and to 0.895 in 2004-05. The reverse trend is observed in case of female: 0.892, 0.684 and 0.785. DI for male exceeds that for female. Thus the occupational distribution of rural non-farm employment for male has been more and more diversified than that for female during the period under study.

Across the states we also observe (Tables 7 and 8) that DI for male has increased in all the constituent states in India during 1993/94-2004/05. The rising tendency is also observed in case of female excepting Haryana, H.P., Orissa, Punjab, U.P. and West Bengal. DI for male is lowest in A.P. (0.681 in 1993-

94) and Assam (0.815 in 2004-05) and highest in Punjab (0.897 in 1993-94) and H.P. (0.917 in 2004-05). Also, DI for female is lowest in Tamil Nadu (0.603 in 1993-94) and Punjab (0.584 in 2004-05); and highest in Rajasthan (0.913 in 1993-94) and Gujarat (0.946 in 2004-05). Thus among the states, Punjab and H.P. for male and Rajasthan and Gujarat for female have occupied more and more occupational distribution of RNFE. The RNFE for male and female are highly diversified in India and in almost all the states during the period under study. But the diversification for male is more than that for female in most of the states in Indian states during the period under study.

VI. GROWTH RATE OF RURAL NON-FARM EMPLOYMENT BY SEX

We have already discussed the changing structure of rural NFE in India and its constituent states during 1993/94-2004/05. Let us now examine the growth rate rural NFE in India and its constituent states during the pre-reform periods of 1983-1993/94 and the post-reform periods of 1993/94-2004/05.

Estimates (Table 9) reveal that the annual compound growth rate of rural NFE has varied in India and its constituent states during the pre-reform and the post-reform periods. At the all India level the growth rate of rural non-farm employment has increased from 1.46% during the pre-reform periods to 2% during the post-reform periods for male. But for female the growth rate has decreased from 2% to 0.62%. Sex-wise analysis reveals that the growth rate of female non-farm employment is higher than that of male one during the pre-reform periods while the reverse has happened during the post-reform periods in India and its constituent states.

Indian states have exhibited wide variations in the growth rate of NFE during the pre-reform and post-reform periods. The growth rate of male NFE has increased in almost all the states in India excepting Gujarat, Haryana, Karnataka, Maharastra, Rajasthan, T.N. and West Bengal from the pre-reform periods to the post-reform periods. While for female NFE the growth rate has also increased in almost all the states excepting Assam, Gujarat, Haryana, Karnataka, Kerala, Maharashtra, Punjab, T.N. and W.B. during the period under

TABLE 9
Annual Compound Growth Rate (%) of Rural Non-Farm Employment in India

States	*1983 to 1993/94*		*1993/94 to 2004/05*	
	Male	*Female*	*Male*	*Female*
A.P	0.32	-0.11	2.40	1.86
Assam	0.55	1.37	2.51	0.89
Bihar	-0.50	-2.59	2.47	5.16
Gujarat	2.95	4.46	0.58	0.57
Haryana	2.90	3.87	1.73	1.71
H.P.	2.29	4.87	2.50	5.26
Karnataka	0.89	3.12	0.55	-2.01
Kerala	0.72	4.66	2.32	1.12
M.P.	0.49	2.51	3.19	5.26
Maharashtra	1.71	1.66	1.11	0.53
Orissa	-0.60	2.22	3.84	4.76
Punjab	1.39	14.24	2.70	3.10
Rajasthan	3.56	1.76	2.28	3.97
Tamil Nadu	1.22	1.85	0.99	1.41
U.P.	1.25	0,68	2.74	4.04
W.B.	2.58	7.19	0.02	-1.35
India	1.46	2.98	2.00	0.62

Source : NSSO Data.

study. The growth rate is negative: (i) in Bihar and Orissa for male during the pre-reform periods, (ii) in A.P. and Bihar for female during the pre-reform periods, and (iii) in Karnataka and West Bengal for female during the post-reform periods. However, the growth rate has varied from: (a) –0.60% in Orissa to 3.56% in Rajasthan during the pre-reform periods and 0.02% in West Bengal to 3.84% in Orissa during the post-reform periods for male; (b) –2.59% in Bihar to 14.24% in Punjab during the pre-reform periods and –2.01% in Karnataka during the post-reform periods for female. Also, compared to male employment growth rate, female employment growth rate is higher in the states of Assam, Gujarat, Haryana, H.P., Karnataka, Kerala, M.P., Orissa, Punjab, T.N. and West Bengal during the pre-reform periods; and Bihar, H.P., M.P., Orissa, Rajasthan, T.N. and U.P. during the post-reform periods. Thus we observe that the growth rate of female employment is higher than male one in most of the states during the reform periods.

VII. CONCLUDING REMARKS

Labour force of male and female has declined in rural India during 1993/94-2004/05. But male labour force has declined more than female one. Male labour force participation rate (LFPR) is more than female one in India and its all the constituent states during this period. Indian states have exhibited wide variations in LFPR. This inter-state disparity has widened more for female LFPR than that for male one in rural India during the period under study.

Irrespective of sex, mode of employment in rural India has been changing during 1972/73-2004/05. Male self-employment is more than the female one. The incidence of causal labour has increased for both male and female. But female is casually employed more than male in India and its almost all the constituent states excepting Haryana, Kerela and West Bengal during the period under study. Also, irrespective of sex the rural employment in Non-Farm Sector has increased all through in India at the cost of Farm Sector during 1972/73-2004/05. But male non-farm employment is more than female ones. Among the states only Kerala has enjoyed significantly the absorption of the rural non-farm employment for both male and female due to the generation of the non-farm activities during the reform period of 1993/94-2004/05.

In rural India the occupational distribution of rural non-farm employment for male has been more and more diversified than that for female during 1993/94-2004/05. Thus among the states, Punjab and H.P. for male and Rajasthan and Gujarat for female have occupied more and more occupational distribution of RNFE. The RNFE for male and female are highly diversified in India and in almost all the states during the period under study. But the diversification for male is more than that for female in most of the states in Indian states during the period under study.

The annual compound growth rate of rural NFE has varied in India and its constituent states during the pre-reform and the post-reform periods. At the all-India level the growth rate of employment has increased from 1.46% during the pre-reform periods to 2% during the post-reform periods for male. But for female the growth rate has decreased from 2% to 0.62%. Sex-wise analysis reveals that the growth rate of female employment is higher than that of male one during the pre-reform periods while the reverse has happened during the post-reform periods in India and its constituent states.

References

Datt, R. (2001), "Economic Liberalization and its Implications for Employment in India", *The IEA Conf. Volume,* Vellore.

Dutta, R. (1995), "New Economic Reform—Need for Some Rethinking", *The Indian Economic Journal,* Vol. 42, No. 3.

NSSO (1997): Employment and Unemployment in India, 1993-94, Govt. of India.

NSSO (2000): Employment and Unemployment Situation in India, 1999-2000, Govt. of India.

NSSO (2003): Household Consumer Expenditure and Employment-Unemployment Situation in India, Govt. of India.

NSSO (2004-05): Employment and Unemployment Situation in India, Govt. of India.

Pal, P.K. and Pal, D.P. (2000), "Structural Reforms, Development and Employment Projection in India" in P.D. Hajela and M.P. Goswami (eds.) "Economic Reforms and Employment," Deep & Deep Pub. Pvt. Ltd., New Delhi.

Pal, P.K. (2001), "Development Pattern in India: An Inter-State Analysis", *Occasional Papers,* Department of Economics, Rabindra Bharati University, Vol. IX, 2001.

Pal, P.K. (2002), "Globalisation and Labour Utilization in Rural India: An Inter-State Analysis", *The Indian Economic Association Conf. Volume,* Trivandam.

Pal, P.K. (2003), "Rural Non-Farm Employment in India: An Inter-State Analysis", *Occasional Papers,* Department of Economics, Rabindra Bharati University, Vol. XI, 2003.

Pal, P.K. (2005), "Workforce in India: An Inter-state Analysis", *Artha Beekshan,* Vol. 13, No. 4, March, 2005.

Pal, P.K. (2006), "Pattern and Growth of India's Economy since 1980s", *Occasional Papers*, Department of Economics, Rabindra Bharati University, Vol. XIV, 2006.

Pal, P.K. (2009), "Labour in India during the Reform Period: An Inter-State Analysis", *Artha Beekshan*, Vol. 18, No. 3, March, 2009.

Pal, P.K. (2010), "Rural Poverty and Rural Non-Farm Employment in India: An Inter-State Analysis" in Komol Singha (ed.)," Rural Development in India : Retrospect and Prospects", Concept Publishing Company Pvt. Ltd., New Delhi, 2010.

Pal, P.K., Pal, G.P. and Pal, D.P. (1997), "The Indian Economy: A Structural Analysis: 1950-95" (Summary), *The Indian Economic Association Conf. Vol.*, 1997.

Panda, Manoj (2004-05): "Macroeconomic Sence: Growth and Equity Perspective" in Kirit, S. Parikh amd R. Radhakrishna (eds.): India Development Report, 2004-05, Indira Gandhi Institute of Development Research, Oxford.

Shetty, S.L. (1978), "Structural Retrogression in the Indian Economy since the Mid-Sixties", *EPW*, Special No.

Sundaram, K. (2001), "Employment-Unemployment Situation in the Nineties: Some Results from NSS 55th Round Survey, *EPW*, Vol. 36, No. 11.

Vaidyanathan, A. (1994), "Employment Situation: Some Emerging Perspectives", *EPW*, Vol. 29, No. 50.

Visaria, P. (1995), "Rural Non-Farm Employment in India: Trends and Issues for Research," *IJAE*, Vol. L, No. 3 (Conf. Number).

CHAPTER

3

Rural Development and Women Employment in the Indian Agriculture

S. Rengarajan

INTRODUCTION

Rural development is defined as 'improving living standards of the low-income population living in rural areas and making the process of their development sustaining'. It includes the programs of agriculture and related matters, irrigation, communication, education, health, supplementary employment, housing, training and social welfare. The initial strategy for agriculture and rural development in India focused attention on an area approach like Community Development Programme. Later the emphasis shifted to target group approach like Swarn Jayanti Gram Swarozgar Yojana.

Major areas of concern have to be addressed to make rural development more broad-based and balanced. Rural development necessitated rural investment, technology, and appropriate rural institutions. For this the policy attention needed are in the areas of employment, increase in public investment, agriculture, water management, rural institution reforms, rural non-farm sector, health and education, PURA Model and improving basic services, reduction in personal and gender disparities, and decentralization, and governance (Dhingra, 2006). Agriculture, employment and gender equality are of at most importance for rural development. Hence, this paper analyses rural development and women employment in the Indian Agriculture.

Women have key roles to play in the farming systems throughout the world. Women are involved not only household chores and child rearing in rural areas but are a major source of labour for food production, and account for a large portion of economic activity. In majority of the developing countries, women play two major roles in the rural areas. As the first role they have household responsibilities for child rearing, food preparation and other chores; and the other role, they are paid or unpaid workers in agriculture or off the farm. In many areas women manage the affairs of the household and the farm. When surveys are taken the true extent of involvement of women in agriculturally related activities is underestimated as women describe their principal occupation as housewife. They are then counted as economically inactive and this 'invisibility' of female employment ignores women and sometimes affects them when policies and programs are taken.

In African countries women play larger role in farming, nearly all the tasks connected with food production are left to women while men may tend livestock or produce cash crops and when men migrate to elsewhere the entire administration of the household is often left to women. In Malawi over two-thirds of those working full time in farming are women. Most of the areas in the world women are important to agriculture. Households headed by women make up to 20 to 25 per cent of rural households in developing countries, excluding China and Islamic countries. Women typically care for animals particularly chickens and pigs and tending garden vegetables and other food crops in Latin America. In the Caribbean countries women work as a cash labourers on plantation, sugar and fruit producing areas, earn a substantial portions of household income. In Asia, for example, Nepal female farming systems are known in which women have the 50 per cent household income in the subsistence farm.

Social, cultural and religious factors; population pressures; farming techniques; off-farm opportunities; colonial history; and many other factors determine the role of women in farming systems. As off-farm job opportunities, population pressures and farming technique change, women assume very different role even when the physical conditions of the area are same. Shifting cultivation with hand labour tend to lend itself more to female labour then does settled cultivation with a plough, but in settled agriculture and cash crops men typically do more work than previous, but often the role of women in the farm work still dominates.

In several Asian countries, in regions of intensive cultivation on small irrigated farms, men, women and children must work hard to generate enough production on a small piece of land to support themselves. Mechanisation in larger farms make women may devote a higher percentage of their time to housework, which is underappreciated, has tendered them to lower their status. Increased integration of present farmers into the labour market has increased the importance of women's role in agriculture, for example, in Lesotho 70 percentage of households are headed by women (Nortan and Alwang, 1993).

WOMEN AGRICULTURAL EMPLOYMENT

Women are active at every point in the food chain and are often responsible for protecting the integrity of food and ensuring its wholesomeness and safety. Apart from inadequate opportunities for wage labour, lack of command over productive resources act as a major constraint for those women who do undertake farming for the household. The shift in policy regarding agricultural production and land management, nationally and internationally, is based on the premise that since agriculture is increasingly becoming the domain of women. Enhancement of women's endowment through land will increase production, reduce poverty and ensure food security. As subsistence is the predominant role of women, diversification of employment is necessitated.

The contribution of women in agricultural work seems to be increasing judging from the number of women workers in crop production and if we include the work involved in livestock, poultry, fisheries, water conservation, forestry and work related to common property resources, the contribution of women to agriculture would far exceed that of men. Male migration from rural-to-urban, for women translate into a marked increase in agriculture work, including a wide range of farm tasks, a heavier workload, and less time for domestic tasks and child care (Krishnaraj, 2005).

The cultural and ecological diversity of the north-eastern region is apparent in the heterogeneity of its rice-farming system. Women in North-Eastern India generally do a greater amount of work than the men and also undertaking task that were traditionally in the men's domain. Because of the composite nature of agriculture, women also undertake multifarious activities, animal husbandry being one of the most time-consuming and labour-intensive. North-eastern women have a viable role in the economic life of their communities because they are responsible for producing and processing food for their families. Physical mobility, the freedom to work and to take certain kinds of decision seems to give women a sense of self-work and dignity. The ethnographic picture can be sustained by date on gender disparities in the region. Women's economic participation is known to be related to improved social status but women's production work continues to be underestimated despite effort to improve enumeration (Krishna, 2005).

Access to employment and by implication to earned incomes is a shaping condition of livelihood security for men and women. Kerala and West Bengal have registered very low female work participation rates, which mark significant caste-based differences, alongside substantial diversification of rural employment for both men and women. Women constitutes less than 20 per cent of all workers in agricultural sector in West Bengal as against 37 per cent in Kerala as shown by 55th round of the NSSO in 1999-2000. Kerala women in 'household duties' raised from 16 per cent of women non-worker in 1961 to 32 per cent in 1999. Migration of women is associated with the withdrawal of women from wage work. "Women is manual work in itself was not regarded as lowering the household rank—whereas seeing outside, under male gaze, or even worse under the employers gaze was". Low female work participation rates and greater diversification in women's employment in West Bengal and Kerala are intertwined with gender conservation. West Bengal and Kerala have been at the forefront of diversification of rural female employment since at least early 1970s. The proportion of female non-agricultural workers in workforce in 1999-2000 was in Kerala, 46.0 in West Bengal and 14.6 in all India level to male (Kodoth, 2005).

Notwithstanding in the expansion of irrigation in recent years, rainfed agriculture is still, of major importance in the regions of three districts of eastern Uttar Pradesh. In the vast rainfed areas of eastern Uttar Pradesh poor women provide 60 to 80 per cent of the total labour inputs in rice production and unpaid family and wage labourers in other farms, aside from their participation in non-rice crops and livestock care. Migration is a dynamic process and encompasses diversified forms of temporary, permanent, seasonal, year-long migrants and geographical mobility, and their role may shift from unpaid family helpers to decision-makers (Paris *et al.*, 2005).

Women work remains invisible in family enterprises. At present in rural India about 33 per cent of cultivators and about 47 per cent of agricultural labourers are women. Increased share by women in the agricultural workforce left them to accept low paid work in less prosperous areas. The workforce composition of the rural India shows a 4 per cent/shift of rural workforce in favour of women. The census 2001 showed that agriculture is the single most important contribution of employment in the rural sector so to the rural women.

Female agricultural labourer is high in some states compared to other. Agricultural labour is normally hired for crop production. The demand for hired is high in low mechanization, high crop intensity and high irrigated area. Female labour is important for rice cultivation due to limitations of migration. In the states of Andhra Pradesh, Gujarat, Karnataka, Madhya Pradesh, Rajasthan and Tamilnadu more than 50 per cent of agricultural labourers are women and the percentage of female cultivators is also high at about 30 to 40. Agriculture in more backward districts attracts larger percentage of women than men, and at the same time, more prospective districts have less number of women than men. Hence in backward districts agriculture has more than the average share of women than men. In Tamilnadu the percentage of women agricultural labour in the irrigated Thanjavur districts is only 44 while the backward districts Ramanathapuram is as high as 58 per cent and Namakkal, Pudukottai, Thiruvannamalai and Perambalur 55 to 60 per cent agricultural labour are women. This high percentage is due to the lower wage paid to women. The percentage of female agricultural labour is 60 to 70 and women cultivators is 55-68 percentage in the states of Chattishgarh, Himachal Pradesh, and Uttaranchal (Vepa, 2005).

In Tamilnadu agriculture continues to be the main stay for rural workers. The percentage share of male employment in crop production activities rose to 91 from 88, that of rural women 85.4 from 83 and in livestock activities 8.7 for women employment from pre-reform period. There was a relative stability at all India level. The share of obtaining for crop production activity was 84 per cent and livestock 11 per cent in respect of female employment. The employment in field crop production in Tamilnadu accounted for as high as 89.84 per cent and 89.52 per cent at all India. After 1990's, the restructuring of agricultural employment, the excessive dependence on field crop production sector for employment, overtook in number of states induced partly by shift in demand and partly by opportunities thrown upon by the open economic regime. For example, in Kerala there was a shift in employment from plantation to field crops products, and in Karnataka post-reform strategy seems to have favoured new crops like fruits, vegetables, flowers, etc. (T.N. Economic Appraisal, 2003-05).

FEMINISATION OF AGRICULTURE

Feminisation of agriculture is the increasing burden of work on them and lower compensations. Feminisation of

Indian agriculture has been emphasized by (i) increased involvement of women in agriculture at the beginning of this century compare to the earlier period; and (ii) underestimation of women contribution to agriculture due to faculty definition. Women continue to be immersed in agriculture and bear heavy burden of farming tasks with less access to non-form work than men. Feminisation of agriculture is specific not only to India, but also allover Asia, South-Asia, South-East Asia and in parts of West Asia. Work participation of women is mostly related to family enterprises and local work available, their upward mobility for employment is highly restricted (Vepa, 2005).

Women participate work in four ways, according to their caste and class position in paddy-rice markets at the end of twentieth century: (i) Women from pauperized, female, female-headed and/or low caste petty producing and trading households are confirmed to seasonal operations, to subsistence orientation, (ii) Female casual wage workers from the asset less class from the largest substratum of labour in grain milling and pre-milling process, (iii) in small family firms, unwaged family members have provided that part of the wage to labour in trading firms which takes the form of prepared food, and (iv) a typical milling and training firm is owned by the makes of a joint family and practices a diversity of trading and transformation activities women are used for the caste-based reproduction and expansion of firms first by means of their dowries on marriage and second through the practice of fictitious registration of a trading company in a women name (Harris-White, 2005).

MASCULINISATION

Three case studies will be used to interrogate the relationship between technical changes, prices and gendered institutional changes: Bangladesh in the late 1970s, Tamilnadu in the 1980s and West Bengal and Punjab in the early 21st Century (Harris-White 2005). Bangladesh in 1978, the long wave of labour displacement. As the late 1979s two-thirdsBangladesh paddy was milled by women. The modernization of rice milling replacing female labour by non-renewable energy priced at low level and disproportionately, biased against women. The structural concentration and ownership of assets are male-controlled. The last technological components hush-fired-mechanical driers (HFMID) have had a massive impact on the labour process, the burden of displacement being borne entirely by female. Casual labour wage in gender neutral, but the system is increasing male. Twenty-first century Punjab, an almost completely masculine labour process. The labour process has developed as almost entirely male. Women invisible in Punjab from the account of milling cost.

Male off-farm livelihoods are pulling men from agriculture. It is that the displacement of women from female off-farm livelihood is pushing them back to agriculture and helping to depress female wages there. Most definitely female marginalized from rice processing, they are not necessarily and always marginalized from productive activity, but their incorporation is on adverse terms.

GENDER AND EMPLOYMENT

The confliction of gender inequality with poverty is however analytically problematic. A disaggregated analysis of sex ratios (used as a proxy for gender quality) across districts in India, in fact, suggest that women's survival chances decline with prosperity. Given this disadvantage faced by women in terms of lower literacy levels, lower wage jobs in the labour market, restriction on mobility and responsibility for children and household maintenances, men have generally moved into non-farm activities, leaving women behind in rural areas to tend the land. In Zimbabwe 68 per cent, Tanzania 52 per cent and in India 85 per cent of women workers are dependent on agriculture (Rao, 2005). For forty years (1961-2001) the rice production grow world-wide at the rate greater than that of population and doubled. The division of tasks in rice production shows a structural diversity. The control over commercial assets is dominated by men and technological change displaces female labour disproportionately to that of men.

The implications of the gender relations have been examined (Rao, 2005).

(i) The responsibility of the women has been added by the development of ground water, intensified agriculture. But the specific cropping changes from cotton and vegetables to banana, coconut and grape were from crops using considerable female labour to those using little. If employment opportunities for women expand, accompanied by fair wages, it is assumed that the household income, hence consumption, will improve. Increasing work burden of women has negative consequences for children's well-being.

(ii) Taking wastelands grouping activities, the social and domestic constraints faced by women prevent them from operating freely in market, apart from the gendered nature of labour, credit and product markets themselves. This is to making the patriarchal institutions more gender-equitable.

(iii) Male assets are strongly correlated to the food expenditure especially rice. Considering male as the role of providers their responsibility is for the production and marketing of rice. When home production is insufficient for food security, women then irrigate in search of seasonal work; when men stay back to tend their fields and harvest paddy crops.

(iv) Changes in livelihood systems show women employment has increased marginally in the construction, manufacturing and services. NSSO data in the last two decades (Sectoral distribution of rural workers from 1977-78 and 1999-2000) reveal that while there has been considerable diversification, this has not only slowed down considerably in the 1990s, but has, mostly be confined to male workers.

(v) There have been changes in macro-policies. An analysis of agricultural wages across the same 15 states revealed that women's agricultural wages were consistently lower than the minimum wages, especially in Madhya Pradesh, Orissa, West Bengal, Uttar Pradesh and Haryana. Right to land may contribute the increasing access to other resources, whether this would necessarily lead to enhancing food security, or indeed more equitable gender relations, is not obvious.

(vi) The co-operation of male is needed from the lessons of land struggles. Land is an important social resource that mediates patronage and power relations at the local. The rise in the local women's group and the presence of supportive women within government and policy-making institutions, there has been enhanced pressure to recognize women in the process of land reform and titling. Women had come to the forefront, demanding a role is decision-making and control.

STATUS OF FEMALE WORKERS

In rural sector, women are likely to be employed more in agricultural activities such as farming, livestock, fisheries and forestry rather than in non-agricultural activities. Primarily, the work opportunities did not keep pace with the demand for work in the economy as a whole and particularly, in the rural sector. Female marginal workers are found both in crop and enterprises, mostly family enterprises as well as agricultural labour. There were 51.12 million female marginal workers and only about 30 million male marginal workers in rural India according to census 2001. The composition of main to marginal workers among female workers changed from 70:30 in 1991 to 54:46 in 2001. A decade ago all marginal workers were women. Women receive lower wages as labour, ownership control over resources limited as cultivators. The access to credit, technology and market information is highly restricted and shifting to better paid work is also narrow to women (Vepa, 2005).

Male out migration on the livelihood of rice farmers left their women with the following impacts (Paris *et al.*, 2005).

(i) The gender division of labour before and after migration there has been a slight change and increase in women's work load in addition to her own familial and domestic responsibilities, the role of sole bread winner and take over many male-specific activities.

(ii) The contribution of female family members among households with migrants are higher compare to men regardless of migration status to total labour inputs. This resulted high demand for unpaid female family and exchange labour.

(iii) In the migrant nuclear households wives bear the responsibility of budgeting, than the non-migrant household women. However major decision is postponed till the migrant visit. Decision on problems regarding the daily subsistence of the family, and keep it from slipping into deep poverty make the wives of left behind as decision-maker.

(iv) The women particularly from nuclear households of men out migrant face the constraints of lack of access to technical knowledge and new seeds; restriction of movement faraway from home make the land rent for others and loss of potential yield; labour scarcity, especially upper caste women, and insecurity if longer the distance and the duration of stay of the out migrant.

Hundreds of farmers in Vidharbha, Maharashtra state, have committed suicide due to unable to bear heavy burden of debt caused by declining prices of cotton. During 2001-02 and 2005-06 the number of farmers committed suicide in the four states of Karnataka, Andhra Pradesh, Maharashtra and Kerala were 5910, 1835, 981 and 201 respectively and in total 8927. Bt.cotton, expert studies have found out that it is not suitable for rainfed areas like Vidharba, but only suitable for irrigated areas. Over one-lakh farmers committed suicide in India between 1993-2003 (*Yojana*, 2006). Suicides continue unabated even now and effect of this increases the responsibility of the wives of these suicide farmers. The wives of these farmers should have the additional responsibilities of all farming activities and the burden of debt as majority/all farmers committed suicide may be male (data not available).

DEVELOPMENT PROGRAMMES IN INDIA

There was a stated commitment to social justice and redistribution through land reform recognizing that the inequalities in land holding were one of the primary forms of social and economic inequality. In national policies women were never mentioned separately, it is being assumed that as member of beneficiary households, they would automatically benefit. The Sixth Plan document (1980) recommended that all land distributed under the land reform programme should be registered jointly in the name of the both spouses. This was the first public document to recognize the importance of land to women in India. Special programme for women, namely, the Development of Women and Children in Rural Areas (DWCRA) was initiated.

During the seventh plan period (1985-90) the National Perspective Plan for Women (1998-2000) was drafted which stated "women's undiluted access to land.... would undoubtedly bestow on her necessary economic independence and power and would improve her social position in the family". The Eighth Plan (1990-94), while subsuming women once again within the chapter on social welfare, makes a link between the role of women in agricultural production and the need to increase women's control over economic resources and services. As part of the Tenth Plan, the New Agricultural Policy, 2000 emphasize social equity objective for recognizing women's rights to land.

There is a national argument for the need of land rights for women in terms of the welfare, efficiency and employment. Women's right in land will be recognized and women will be given preference in group activities for land conservation and improvement. Efforts will be made to grant property right in land to women where ever possible and self-help groups of women may be encouraged to take up activities like the generation of wasteland. Women will be given preferences in the allotment of ceiling surplus lands (Rao, 2005).

In order to mitigate the miserable conditions of farmers committed suicide in the four worst affected states and for easing the agrarian crisis and to prevent farmers from committing suicide the Government of India proposed a special package. This proposed package will cover six issues—credit, insurance, irrigation, agriculture productivity, lack of extension services and lack of marketing infrastructure (*Yojana*, 2006).

DEVELOPMENT OF DRYLAND AGRICULTURE

Dryland farming is broadly defined to cover rainfed agriculture dominated by low water requiring crops in the arid and semi-arid tropical region in India. Dryland agriculture is that not only rainfed agriculture but also of an arid and semi-arid environment. Probably the isohyet of 1000 mm. should be an appropriate limit to the arid and semi-arid zones of India. About 70 per cent of our agriculture is rainfed or dryland agriculture and even after utilization of all our water resources for irrigation, about half of the cultivated land will remain rainfed. Dryland farming contribute 41 per cent to the national food basket and support 40 per cent of human and 60 per cent livestock population.

While intensive irrigated farming is imperative for 'survival', dryland agriculture is necessary for 'equity'. The growing gap in incomes between farmers blessed with irrigation resources and those dependent on rainfed agriculture has to be considerably narrowed and that too soon (Singh, 1989).

A high agriculture growth of four per cent and industry growth for more than 10 per cent are needed for better structural change. Importance should have been first to agriculture, manufacturing, rural infrastructure, etc., in reforms for better employment, incomes and equity. Stepping up agricultural growth through biotechnology holds prospects for reducing regional and inter-persons disparities. The share of employment in manufacturing in Malaysia is 50 per cent, in Korea 62 per cent, in China 31 per cent and in India it is only 22 per cent, while share of agriculture is more than 60 per cent (Dev, 2006).

During mid-1960s the Green Revolution which was mainly concentrated on introduction of non-conventional inputs thus brought about what has been termed as seed-fertilizer revolution. The benefits of Green Revolution accentuated largely to wheat and rice grown more or less in same homogenous tracts both agro-climatically and socio-economically served with assured sources of irrigation and inhabited by resourceful farmers (Rao, 1986). In dryland areas where irrigation ratio is less than 10 per cent, the percentage of population below poverty was as high as 68.75, whereas in areas served with assured irrigation facilities (irrigation ratio is above 50 per cent) the percentage of poor was 26.46 (Jodha, 1986). The single neglect of dryland farming and cereals other than wheat and rice in policy has led to deterioration in the nutritional level of the poor in India (Krishnaraj, 2005).

Prime Minister Manmohan Singh wanted a second Green Revolution which would have special focus on dryland agriculture and would address the needs of the small and marginal farmers (*The Hindu*, 2005). Imports of pulses and oilseeds serve as indicators of our failure to launch a green revolution in dryland areas in spite of having the technologies and resources to do so. This is a human tragedy of vast dimensions where millions of children, women, and men are condemned to a life of malnutrition and poverty. Imports of crops of importance to the income security of farm families in rainfed areas implies generating more unemployment and misery in such areas. The mindset where the consumers applies only to the politically powerful urban population and ignores the population living in rural India, who are both farmers and consumers, will have to be destroyed if our country is to achieve Poorna Swaraj (Swaminathan, 2006).

Strengthening the conditions of female farmers and female labourers would also help in prove the food security at the household level. Generally women spend more of their income on household expenditure in turn help nutrition of their children (Rao, 2005). Women after grow food crops that are minor in terms of value of production but are important in the diets of families on small farms. Agricultural research often neglects these crops, and this neglect may have adverse effects on nutrition.

Development of dryland agriculture is important in the context of gender justice through equal wages, equal opportunity, upward mobility, right to land, avenues of acquiring skills which are as important to rural women as the community participation and gainful self-employment and credit in wage-based livelihood system. New technologies are relatively gender neutral. Female employment opportunities are ample in the dryland areas as in the case of post harvest processing. Crop diversification is vast in these areas as land factors are heterogeneous. Crop diversification leads to different processing enterprises are involved in dryland crops.

Farmers in Karnataka have been receptive to the operations of the agri-business firms and have been optimistic about their prospects in the liberalized economic regime. Modern floriculture has introduced a plantation type of agrarian system in several parts of rural Karnataka. Floriculture is highly labour-intensive activity. According to an official estimate, on an average, a modern, floriculture farm employs 50 workers per acre. Workers, especially women, have to do a variety of different tasks in the farms. Men are involved in maintenance of equipments and in climate control. In the villages around Bangalore people prefer to work in floriculture farm because such works command high prestige. Employment opportunities for both men and women seem to be increasing. Apart from floriculture farms, seed farms and horticulture nurseries prefer women workers (Panani, 1999). Women's group acquired private wastelands, which they own and work collectively as apart of the Women's Wasteland Development Project in Bankura (West Bengal). The project was initiated in 1980 as an experiment in employment generation through rural women's organization, the larger end being "fundamental redistribution of power and sources of power" (Kodoth, 2005).

There were some studies of the Drought Prone Area Programme Watersheds in different part of western India which was launched in 1996. These studies conducted after the completion of 80 to 95 per cent of the works under the programme by following 'before' and 'after' methods. In most cases the crop yields after the implementation of the programme were higher by over 60 per cent. The cropping intensity increased by 40 per cent to 60 per cent. A brief increase ranging from 150 per cent to 200 per cent in total output from crops, and two-thirds of this increase came from rabi crops, indicating improved stability in output and increase in employment. Dairying is another economic activity has also shown a marked improvement which is women oriented. Migration of labour has been reduced, enrolment of children in school, especially girls, has increased and there is increased participation of women in decision-making (Rao, 2000). These foregoing analyses point out that the development of dryland agriculture is imperative to promote women employment in the sphere of high income work opportunities, upward economic mobility, property right and right such as equal wages for equal work.

References

Dev, Mohendra (2006), "Ensuring Equitable Development through Reforms", *The Hindu*, January 21.

Dhingra, Ishwar C. (2006), The Indian Economy—Environment and Policy, Sultan Chand & Sons, New Delhi, pp. 367-78.

Harris, Barbara and White (2005), "Commercialisation, Commodification and Gender Relations in Post-Harvest Systems for Rice", *Economic and Political Weekly*, XL(25), June 18-24, 2530-42.

Jodha, N.S. (1986), "Research and Technology for Dryland Farming in India: Some Issues for Future Strategy", *Indian Journal Agricultural Economics*, 26(4), Oct.-Dec. : 287-94.

Kodoth, Praveena (2005), "Fostering Insecure Livelihood: Dowry and Female Seclusion in Left Development Contexts in West Bengal and Kerala", *Economic and Political Weekly*, XL (25), June 18-24, 2555-62.

Krishna, Sumi (2005), "Gendered Price of Rice in North-Eastern India", *Economic and Political Weekly*, XL (25), June 18-24, 2508-12.

Krishnaraj, Maithreyi (2005), "Food Security: How and For Whom?", *Economic and Political Weekly*, XL (25), June 18-24, 2508-12.

Norton, George and Jeffrey Alwang (1993), *Introduction to Economics of Agricultural Development*, McGraw-Hill, INC, Singapore.

Panani, M.N. (1999), "Trends in Cultural Globalisation From Agriculture to Agribusiness in Karnataka", *Economic and Political Weekly*, XXXIV (31), July 31, 2168-73.

Paris, Thelma, Abha Singh, Joyce Luis and Mahabub Hossain (2005), "Labour Out Migration, Livelihood of Rice Farming Households and Women Left Behind", *Economic and Political Weekly*, XL (25), June 18-24, 2522-29.

Rao, Hanumantha C.H. (1986), "Science and Technology Policy: An Overall View Broadest Implication", *Indian Journal of Agricultural Economics*, 41(3), July-Sept: 229-47.

Rao, Hanumantha C.H. (2000), "Watershed Development in India: Recent Experience and Emergency Issues", *Economic and Political Weekly,* XXXV(45), Nov. 4-10, 3943-47.

Rao, Nitya (2005), "Gender Equality, Land Rights and Household Food Security: Discussion of Rice Farming System", *Economic and Political Weekly,* XL(25): June 18-24, 2513-21.

Singh, R.P. (1989), "Technology Options for Small Dryland Farmers in India", in N.S. Jodha (ed): *Technology Options and Economic Policy for Dryland Agriculture: Potential and Challenge,* Concept, New Delhi, 57-78.

Swaminathan, M.S. (2006), "Avoiding an Unequal Social Bargain", *The Hindu,* July 6.

The Hindu (2005), "Work for Second Green Revolution Manmohan Singh Tells Scientists", *The Hindu,* January 4, p. 1.

Vepa, Swarna S. (2005), "Feminization of Agriculture and Marginalisation of Their Economic Stake", *Economic and Political Weekly,* XL (25), June 18-24, 2563-68.

Yojana (2006): "PM's Package", *Yojana,* Vol. 50, August, p. 24.

CHAPTER

4

Role of Non-Farm Sector in Rural Development

BIRENDRA KUMAR JHA AND RAM NARESH THAKUR

WHAT IS RURAL DEVELOPMENT?

The concept of rural development embraces a wider spectrum today and it is not only economic betterment which, of course, is the kingpin of development. There is no universally acceptable definition of rural development and the term is used in different ways and vastly in divergent contexts. Historically speaking, the term rural development was earlier known as community development which emerged during the period of Second World War as a technique for development of underdeveloped agrarian economy-based countries (S.N. Bhattacharya, 1970). As a concept, it means overall development is rural areas with a view to improve the quality of life of rural people. Rural development is comprehensive programme of activities which include agricultural growth, development of economic and social infrastructure, village planning, public health, education, functional literacy and communication, etc. (A.R. Tripathi, 1988). In this way it is clear that enrichment of quality and content of life of rural masses constitutes the real meaning of rural development. Thus, any progressive change of status of rural people is called rural development which cannot be said to come about unless each one in the village has the opportunity to improve his or her living conditions. Rural development no longer means agricultural development alone. It is also not a social welfare case of pumping money into rural areas to provide for basic human needs. It encompasses a spectrum of activities and human mobilisation to make people stand on their own feet and break away from all the structural disabilities which chain them to the condition in which they may live UNCRD, (1985, p. 2, Vol. 1). V.K.R.V. Rao has defined rural development as a process of optimum utilisation of the natural and human resources of a given rural area for the enrichment of the quality of life of the population (V.K.R.V. Rao, 1977). In fact, rural development is a concept aimed to provide all developmental potentialities in rural areas which could increase their standard of living. The concept of integrated rural development is also defined as a series of mutually supporting (inter-related) agricultural and non-agricultural activities oriented towards a stated objective which involves the progression of rural sub-system and their interaction leading to desired improvement in the rural system as whole (Y.L. Ahmed, 1975). The achievement of integrated rural development depends on inter-sector linkages, how one sector is related with an other sector and their functional linkages between different sectors, say agricultural sector and industrial sector and their dependence on infrastructural facilities which are combined to bring a desirable improvement in the rural areas.

WHY RURAL DEVELOPMENT?

Our government is committed to economic development which entails development of all segments of its population, whether living in metropolitan and urban areas or in semi-urban and rural areas. No doubt, urban and rural areas are two sides of the same coin of economic development. Whereas, the urban sector has witnessed both faster economic growth and development fuelled by the post-independence era of industrialization the rural sector is far behind in this race. Industries continue to get located in urban centres notwithstanding the tall talk of decentralization or giving of incentives for location in less developed area. Rural-urban divide is quite apparent in all spheres of human life. What worries most the policy planners is that the magnitude of disparity has accentuated during the post reform period due to change in the income growth trajectories of rural and urban areas. The growing rural-urban divide may have far-reaching economic, social and political ramifications. It is, therefore, essential to examine the extent of this divide and draw inferences for policy interventions for its rectification.

If we look at the rural share in the total population, we find that over 73% of Indian people live in rural India, while they get only 48% of national income. In 1999-2000, as against per capita income of Rs. 30217 in urban area, it was only Rs. 10606 in the rural area. In 1999-2000 the ratio was 2.85 indicating that per capita income in urban area is almost three times that of rural area.

Rural-urban income disparity can also be assessed on the basis of difference in level of household consumption expenditure that could serve as a proxy for household monthly income.

The 61th Round Survey of NSSO estimated that in 2004-05 5% of rural households spent less than Rs. 235 per person per month on consumption. Another 5% spent in the range Rs. 235-270 whereas the poorest 5% of urban households have MPCE less than Rs. 335 and another 5% spent in the range of Rs. 335-395.

In order to know the extent of disparity in rural and urban areas in case of basic amenities, data on five amenities, namely, Pucca house, piped drinking water, sanitary toilet, electricity and LPG cooking gas are shown in table. It is found in 2004 as against 74.1% urban households lived in Pacca houses, only 25.5% households in rural areas had this facility. In case of drinking water, it is observed that only about 28% rural households had tap water facility, whereas 71% urban

TABLE I

Trend in Rural-Urban Income Disparity in India.

Particulars	*1970-71*		*1980-81*		*1993-94*		*1999-00*	
	Rural	*Urban*	*Rural*	*Urban*	*Rural*	*Urban*	*Rural*	*Urban*
(1)	*(2)*	*(3)*	*(4)*	*(5)*	*(6)*	*(7)*	*(8)*	*(9)*
Share in NDP at Factor cost (%)	62.35	37.65	58.88	41.12	54.27	45.73	48.09	51.91
Share in Population (%)	80.22	19.78	76.88	23.12	73.51	26.49	72.53	27.47
Per capita income at current prices (Rs.)	529	1294	1245	2888	5783	13525	10606	30217
Ratio of urban income to rural income	2.45		2.32		2.34		2.85	

Source : GOL, NAS 2007, CSO, New Delhi.

Table 2
Monthly Per Capita Consumption Expenditure in Rural and Urban Areas

Year	MPCE at current prices	Rural		Urban		MPCE at constant price 1972-73	Urban Rural Ratio
		CPIAL (1972-73/100)	MPCE at constant price	MPCE at current Price	CP/UNME (1972-73/100)		
(1)	(2)	(3)	(4)	(5)	(6)	(7)	(8)
1973-74	44.17	100	44.17	63.33	100	63.33	1.43
1977-78	68.89	144	47.84	96.15	160	60.09	1.26
1983	112.31	227	49.48	165.80	258	64.26	1.30
1987-88	158.10	289	54.71	249.92	364	68.66	1.25
1993-94	286.10	520	55.02	464.30	618	75.13	1.37
1999-2000	486.16	833	58.36	854.92	998	85.66	1.47
2004-05	558.78	922	60.61	1052.36	1230	85.56	1.41

Source : NSS 61th Round Report on Level and Pattern of Consumption Expenditure, 2004-05.

households had this facility. Against 93.1% urban households having electricity, the corresponding percentage for rural households was only 55.7%. Only 25.9% rural households had toilet facility, while the corresponding percentage in urban areas was 83.1 LPG as a cooking gas is used by 48% households in urban areas whereas only 5.7% of rural households had this facility in 2004. Thus, the rural India is far behind the urban India in case of basic amenities.

TABLE 3

Disparity in the Access to Basic Amenities

Amenities	*1998*		*2004*	
	Rural	*Urban*	*Rural*	*Urban*
Pucca house (%)	19.0	66.0	25.5	74.1
Drinking water through Tap (%)	18.7	70.0	27.9	71.0
Toilet Facility (%)	17.50	74.5	25.9	83.1
Electricity (%)	48.4	89.1	55.7	93.1
LPG Cooking gas (%)	—	—	5.7	48.0

Source : GOI, CSO Selected Socio-economic statistics India, 2006.

All these are sufficient ground for a general and widespread feeling regarding superiority of urban in comparison to rural in terms of physical and material amenities. The picture of development which our leaders had in mind was that of Europe and America which were highly industrialized and urbanized and they succumbed to the wiles of industrialization, forgetting the traditional Indian masses living in villages (Jagannath Mishra, 1997). Seldom do we come across the recognition of the need for rural-urban balance as an important inter-sectoral balance in the process of development. As a result, the rural-urban disparity has been ever increasing over the plan period reaching an alarming dimension provoking disharmony between them. The first shot of urban bias was fired in the 1st Plan when it is said. "the necessary incentives for the producer cannot be created unless the prices of alternative crops are controlled. In the case of certain key commodities, it may be necessary to keep down their prices in order to obviate the need for price rises. in several industries which use these commodities." Here one finds a clue to the thesis that agricultural prices are to be depressed in order to promote industries and exports which is the main strategy of the planning process (D.M. Nanjundappa, 1981). Credit should be given to the 2nd plan for categorically stating that, "it seeks to rebuild Rural India to lay the foundations of industrial progress and to secure to the greatest extent feasible opportunities for weaker and underprivileged sections of our people and balanced development of all parts of the country." Thus, for the first time, it is found that something may be nearer a concept like balanced development of rural and urban parts of the country but unfortunately, policies required for establishing such a rural-urban balance were not pursued in consistent analytical frame. Instead, not with standing the objective of improvement of the living conditions of the masses, the pattern of development has moved on contrary lines (D.M. Nanjundappa, 1981). While it is true that some priority in resource allocation was given to housing in rural areas and the spread of primary education in the rural areas in earlier plans, the issue had to wait until the 5th plan for the introduction of the concept of the basic minimum needs of the rural economy which has been further extended in the 6th plan. A sizeable amount of more then Rs. 9074 crores (nearly 5% of the total plan outlay) has been allocated for rural development during the 7th plan as against Rs. 4762 crores spent during the 6th plan. Indeed, after the 2nd plan, it appears that the thrust has been in favour of regional considerations. Among the objectives, progressive reduction in regional inequalities has been mentioned in the later plan documents. To this date the central government has launched several countrywide programmes in the field of rural development but implementation of these programmes deserve much more resources and attention than what it has received so far. The development model that India adopted since independence has led to vast rural-urban disparities. Urban areas are highly developed and have all the modern amenities, whereas rural areas are grossly underdeveloped, dependent mainly on agriculture and lacking in even basic necessities. Agriculture; being subjected to vagaries of nature cannot sustain such high levels of population i.e., 70% who reside in rural areas. As a consequence large number of people are migrating from rural areas to urban areas in search for employment. This has put great strain on the civic infrastructure of the urban areas leading to problems like congestion, unauthorized construction, slums, waste disposal and the like and this has also adversely affected the law and order situation. Urban areas have now reached their saturation limit and are unable to absorb further migration. Quality of urban life has been degenerated and the entire urban system is on the verge of collapse. In such a scenario rural development is a must on urban lines. Rural-urban disparities cannot be allowed to continue further and it is heartening to take down that the Planning Commission has set ambitious targets during the 11th plan and the rural economy has to grow at double the present pace of growth of less than 3% . The Task is gigantic but not unachievable.

RURAL INDIA : THE ROAD TO PROGRESS

The real India does not live in Bombay, Delhi or Calcutta but in seven lakh villages (M.K. Gandhi, 1926). Gandhi realized that India lives in villages and without self-reliance or Gram Swaraj in villages, India's political independence too will remain incomplete. He perceived the tremendous natural and human potentialities of villages contributing to the country's development. Gandhiji saw that the only way to bringing hope and good living to the people in rural areas was by making the villages the central focus of economic development Programmes. Mahatma Gandhi understood India better than any other person in the 20th century. It is now being realized not only in India, but allover the world that Gandhi's vision of development and his simplistic approach to rural up-liftment is more relevant than any other model of development. Perhaps it is correct to use the parameters suggested by Dudley Seers about the stage of development of an economy viz. what has happened to poverty, unemployment and inequality in that particular economy. On the basis of these parameters, there is no doubt that the paradigm adopted so far needs to be changed for an alternative one in the near future for the sake of the huge majority of the suffering people living mostly in rural areas of the country. Facts sheet of our present status in rural development is stated to comprehend the challenges of India. Rural poverty in India has record of 27.1% i.e., 280 million rural poor make larger size than India had in 1947 constituting 13 Australia. Income poverty is 26% and nutritional poverty is 56%. According to Dr. Swaminathan's Food Insecurity Atlas, India has largest absolute number of undernourished people in the world. Rural India has an employment problem rather than an unemployment problem. Village workers earn very little per day or month despite working very long hours, and so on. According to National Social Watch 48% people in 13 states of India viz. Andhra Pradesh, Tamil Nadu, Maharashtra, Gujarat, Arunachal Pradesh, Assam, Jharkhand, Bihar, Chhattisgarh, Madhya Pradesh, Orissa, Rajsthan and Uttar Pradesh do not get two meals a day properly. No doubt, a survey conducted by the National Council of Applied Economic Research shows that our efforts for rural development have begun to bear fruits, and income levels and quality of life in rural India have registered in upward trend but much needs to be in this direction on Gandhian lines. One should not forget the stark reality that overall economic development of the country depends upon the development that takes place in the rural areas which are characterized by abject and widespread poverty, unemployment and inequality. Thus, it is rural development which opens the door of prosperity for the country as a whole Gandhi saw that the only way of bringing hope and good living to the people in rural areas was by making the villages the central focus of economic development programmes. The Gandhian strategy of rural development was based on his programmes of Village Swaraj and the Swadeshi movement. Under these two programmes he introduced many apparently very simple activities like charkha and Khadi, revival of household/cottage industries and village handicrafts, village sanitation and hygiene basic education for harmonious development of the whole personality, etc. It is widely accepted now that the era of rising expectations created by hard-earned independence and national planning is said to have given way to a period of rising frustrations in the Indian people. The modernization-industrialization-growth trickle down model of development has been found to be woefully inadequate to tackle the rural problems. Any development programme even if it is targeted towards the weaker sections, would automatically bring more benefits to those who have resources than to those who have none. Thus, our development efforts over the last sixty years and so, have reached a state where we have to think of a new and more realistic approach to development. It is in this respect that rural development for overall economic development needs to be emphasized.

For rural development our ex-President or A.P.J. Abdul Kalam has proposed the concept of PURA (Providing Urban Amenities in Rural Areas) in the vision-2020 Project initiated by him. Its objective is to make rural areas so attractive to investor as cities are then, rural areas too will generate urban-style employment to halt (if not reverse) rural-urban migration. In line with ex-President Kalama's vision of India's path to progress through the concept of PURA where he envisions independent cluster of villages in rural areas by building self-sustainable business enterprises, through four types of connectivity (Physical, electronic, knowledge and economic) and proper infrastructure facilities. The theme of PURA, apart from concentrating on reinforcing agriculture, will emphasize on agro-processing, development of rural craftsmanship, dairy, fishing, silk production, so that non-farm revenue for the rural sector is enhanced, based on the core competence of the region. The rural economy will also be driven by renewable energy such as solar wind, bio-fuel and conversion of municipal waste into power. In this approach, the aim is to provide sustainable development using the core competence of the rural sector. PURA is a vehicle for sustainable development. It is to improve the quality of life of over 220 million people living below poverty level in rural areas and makes special suggestions to remove urban congestion also. This sustainable rural development programme is for a nation of a billion plus people. In India the development of the rural sector is very important. The government, private and public sectors have been taking up rural developments in parts. During the last few decades, it is our experience that these initiatives start well just like heavy rains resulting into multiple streams of water flow. As soon as the rains stop, few days later all the streams become dry because there are no water bodies to collect surplus water and store it at the right place. For the fist time PURA envisages an integrated development plan with employment generation as the focus, driven by provision of habitat, health care, education, skill development, physical and electronic connectivity and marketing. Inspite of massive efforts and huge amount of resources pouring into multiplicity of programmes, rural development is not happening in the way and at the pace desired but proper rural development for overall sustainable economic development of our country is a must.

ROLE OF NON-FARM SECTOR IN RURAL DEVELOPMENT

Defining the RNFS (Rural Non-farm Sector) properly is extremely important as that will prescribe the contours of our discussion on its role in rural development. We define it as follows. "The RNFS comprises all non-agricultural activities—mining and quarrying, household and non-household manufacturing, processing, repairs, electricity, gas and water supply, construction, wholesale and retail trade, restaurants and hotels, transport, storage and warehousing, communication, financial and business services, community, social and personal services, and other services, in villages and rural towns of up to 50,000 population, undertaken by enterprises varying in size from household own account enterprises, all the way to factories (Fisher Thomas and Mahajan Vijay, 1996). As per GOI (1998), there were a total of 17.71 million rural enterprises of which 3.20 million or 18.1% were engaged in agricultural activities such as livestock rearing (dairy, poultry, piggery), forestry, fishing and agricultural services such as custom hiring of tractors, and managing of irrigation systems. The remaining 14.51 million or 81.9% of the rural enterprises were non-agricultural in nature. These comprised a wide range of activities including mining and quarrying, manufacturing both within and outside households, electricity, gas and water supply, construction, wholesale trade, retail trade, restaurants and hotels, transport, storage and warehousing, communications, financial, real estate and business services, and community, social and personal services.

The last category itself comprises a broad range all the way from school teaching to hair-cutting and laundry services. In terms of employment, we learn from GOI (1998) that nearly 40 million persons worked in RNFS enterprises. Of these, 73.9% were adult males, 22.5% were adult females while 3.6% were children. This amounted to 1.4 million children below the age of 15 working in RNFS enterprises. Over three quarters or 76.8% of the enterprises were own account enterprises, numbering 13.6 million, while 23.2% of the enterprises were establishments. Hereby it is worth-noting that enterprises operating with the help of household labour only are termed as own account enterprises, which are enterprises without any hired worker, in contrast to establishments which employ at least one hired worker. A vast majority of the RNFS enterprises are small and dispersed, and employ mainly self and family, there is a strong case for waiving labour legislation applicable to them. In terms of the number of persons employed per enterprise establishments in RNFS averaged 4.65 as compared to 1.5 persons in OAEs. Thus, establishments are a major source of wage employment and need to be encouraged. In terms of the nature of activities, the larger number of RNFS enterprises was in retail trade, totally 5.23 million. The next largest categories were community, social and personal services (3.82 million) and manufacturing (3.52 million) and transport and storage (0.48 million) were also significant (Vijay Mahajan and B.Y. Raju, 2005). The rural non-form sectors are widely dispersed and labour-intersive, require small amounts of initial capital and are ideally suited to provide jobs for the rural poor. However, the rural non-farm sector has not received as much policy and programme support as has been the case in other countries such as China (D. Rajasekhar, 2005) and moreover, there has been a definite decline in the quality of employment in rural India. The two prominent and inter-related trends are the decline in self employment and increasing casualisation.

Due to unavailability of estimates regarding growth in output or income of the rural non-farm sector, one can get an indication of the performance of this sector from the changes in its share in employment. There has been a diversification of employment in rural India with the share of rural non-farm employment having increased by 7.2 percentage points (or by

TABLE 4
Distribution of Workers by Employment Status in Rural India, 1972-2000

Employment Status	*1972-73*	*1977-78*	*1983*	*1987-88*	*1993-94*	*1999-2000*
Self-employed	65.3	62.9	61.0	59.4	58.0	55.6
Regular	9.3	7.7	7.5	7.7	6.4	6.7
Casual	25.4	29.7	31.5	32.9	35.6	37.7

Source : Reddy (2002), p. 63.

0.33 percentage points per annum) over 22 years between 1977-78 and 1999-2000. The annual rate of this shift is nearly the same between different periods during the pre-reform as well as post-reform periods, whereas one would expect the rate of shift towards rural non-farm activities to have increased in the post reform period. This could well be attributed to the slowing down of agricultural growth and insufficient infrastructural support to rural non-farm activities during the post-reform period. This is reflected in the slow pace of reduction in rural poverty during the post-reform period of 1990s when compared to the 1980s. This is not the surprising in view of the continued decline in public investment in agriculture and the squeeze in public expenditure on rural development, in general, following fiscal compression during the post-reform

TABLE 5
Broad Sectoral Distribution of Workers in Rural India (usual status PS+SS 1977-78 to 1999-2000)

Year	*Primary*	*Secondary*	*Tertiary*	*Rural non-farm (cols 3+4)*
1977-78	83.4	8.0	8.6	16.6
1983	81.5	9.0	9.4	18.4
1987-88	78.3	11.3	10.3	21.6
1993-94	78.2	10.2	11.5	21.7
1999-2000	76.1	11.3	12.5	23.8

Source : Dev (2001).

period. However, since productivity and wage rates in the rural non-farm sector are much higher than that in agriculture, the above facts do indicate that the shares of the rural non-farm sector in output and incomes over the last two decades have been significantly higher than its share in employment (Hanumantha Rao, 2005).

One of the most significant failures of economic development in post independence India is that the proportion of the work force dependent on agriculture has declined much less than expected. Despite a significant decline in the share of income of the primary sector and a substantial increase in that of tertiary activities, their employment shares have remained stable while the share of national income originating in agriculture has dropped from over 50% in the 1950s to less than 25% currently, the share of labour force engaged in agriculture, which was about 70% in 1951, still remains over 60%. The consequence is that the ratio of per worker domestic product in non-agriculture to that in agriculture, which was about 2 in the 1950s, to now well over 4. This widening gap between incomes in agriculture and non-agriculture is generally accepted to be a major factor underlying persistent poverty in the country (Abhijit Sen and Praveen Jha, 2005).

The role of the rural non-farm sector (RNFS) is crucial in generating productive employment and reducing poverty in rural areas because of the limited absorptive capacity of the urban sector and near saturation of the agricultural sector in further absorption of workers. The importance of the rural non-farm sector in reducing rural-urban differences in income levels, bringing greater integration between rural and urban economies and generally contributing rural and overall development is also now recognized. The theoretical proposition on the processes and patterns of development and the empirical observation of the experience of the past decades both point to the crucial role of the rural non-farm sector in accelerated and balanced development. The diversification of the demand basket in India from cereals to other food items and non-crop based agricultural products like animal husbandry and forest-based products, provides the demand base for the growth of RNFS. Opening up of the economy to international trade will further accentuate the diversification of the rural economy. The products of RNFS has also attracted increasing demand, although the trends are varied and mixed (T.S. Papola and A.N. Sharma, 2005). It is observed that there are strong linkages between agricultural development and growth of the RNFS as agricultural development leads to growth of non-farm activities not only through forward and backward linkages, but also through the utilization of invistible surplus generated by it. However, a large segment of RNFS is not directly related to agriculture. It consists of traditional skill-based crafts and local resource-based products. They are quite often concentrated in specific regions and locations with proper development of niche markets, they have significant potential to grow and provide employment and sustainable livelihoods to those engaged in their production and possibly for some others as well in the areas of product and market development and marketing.

In recent years, many studies in India and outside India on non-farm sector have come up and some of notable studies in this areas are Okafor (1983), Vijerbery (1990), Haggblade, Hazell and Brown (1987), Oshima (1986), Hearth (1996), for Schsdroll (1991), Ho (1986), Park (1980), Ksryo (1986), Cholamworng (1986) and Tslam (1991), whereas in India the studies by Krishnamurthy (1984), Vaidyanandan (1986), Basant and Rakesh (1989), Kashyap and Desai (1990, 1991), Basant and Parthasrthy (1991), Rayoprpa (1986), Unni (1991), Basu and Kashyap (1992), and Papola (1993), are noteworthy. The results of the above studies broadly reveal certain common factors which explain the growth of non-form sector. This sector faces a number of problems the most important among them being the lack of access to quality (adequate and timely) credit from the formal banks, and inadequate support of rural non-farm producers in the acquisition of skills and raw materials and the marketing of their products. Among other problems small size operations, backward technology, seasonal involvement, wages or salary, etc. are also important. Undoubtedly, poor-technology base of RNFS is one of the serious handicaps responsible for low productivity higher unit costs and sometimes poor quality of the end-product. As a result, they face severe competition from the well-organized, large and urban-based sector. In fact, with the introduction of eco reforms in 1991 and no real dynamic increase in the rural non-farm sector in early 1990s, this sector once again neglected. The nagging problem of how to absorb the large and growing workforce in the rural areas for rural development, perhaps, accounts for the re-emergence of interest in the non-farm sector. As the major factor responsible for the high level poverty level in rural India would be the lack of access to employment or economic substence. Hereby, it is worth-noting the fact that the RNFS employment increased from 16.6% in 1977-78 to 23.8% in 1999-2000 which cannot be considered sufficient for lessening the burden of population on farm sector. The employment of males increased by 9.4% while that of females registered a slower growth of 2.8% indicating a relatively less diversification in female employment. The problems of serious malnutrition and hunger in the rural areas can be attributed, in no small measure, to the worsening of the employment situation and lack of purchasing power. Notably, the incidence of poverty is the highest among casual workers who suffer most from the problem of underemployment and low-income employment. Even a near full employment situation would not be an entirely satisfactory state of affairs, unless the earnings of the workers are at a reasonable level. At present, due to underemployment and low productivity, a large proportion of working households belong to the category of those falling below the poverty line. In fact, open unemployment is increasingly becoming a characteristic of those above the poverty line. The poor cannot afford to remain unemployed. In 1993, nearly one-third among the employed were poor. But only 18-19 percent among the unemployed were poor (Hashim, 2009). Any strategy for improving the quality of employment on a sustained basis will require effective measures for raising productivity as earnings depend on the productivity of the worker, the important determinants of which are technology, infrastructure, capital formation, and skills. Thus, for overall economic development, rural development in India is a must and for rural development rural non-farm sector may play a crucial role through alleviating poverty for which employment at reasonable wage level is a must.

CONCLUDING REMARKS

As expected non-farm sector in India is related to the process of economic development in general and rural development in particular. If improving the living standard of the rural masses is the objective, as it emphatically is, of planning in India, the rural non-form sector has to be preferred. If rural development is our target without which overall development is not possible, the RNFS needs to be given due attention. This requires a heroic step of imposing a moratorium at current level on urban facilities so as to release more resources, energy and administration for the uplift of the rural economy with the help of the RNFS. If not, the rural-urban conundrum is bound to lead to further inequalities and class conflicts which dimension it is difficult to fathom at this stage. It is the fore most task of economists today to create a climate of opinion which can provoke such an anti-urban bias movement at all levels. Thus, there is no scope for urban-bias for overall economic development which can be possible only when rural India becomes prosperous. Really a focused policy and programme for establishing rural-urban linkages is vital to facilitate the growth of the RNFS. Linking rural producers with the urban market would help the farmer in getting fair prices for their products and adjusting to changing market conditions.

For increasing the productivity of the RNFS technology is a crucial factor and so the use of both incremental and leap-forgging technologies is required to give a boost to the RNFS.

In the changed competitive environment as a result of LPG policy in the global era the creation and upgradation of skills for workers as well as for potential entrepreneurs. For acquiring some special technical skills or knowledge to be transferred to local enterprises, local institutions need to be encouraged to enter into collaboration with national level technical research institutions and large companies. For the growth of units in the RNFS, there is need to provide business development services (BDS), that might make value addition in terms of providing strategic information on markets, the prevailing demand-supply situation, availability of raw materials, technology sourcing and other related matters.

Basic infrastructure facilities are a must for augmenting opportunites for income generation activities and to facilitate and create forward and backward economic linkages and so provision of basic physical infrastructure in rural areas must receive the highest priority.

There is a need for working out an integrated strategy to finance the rural non-form sector to enable this sector to graduate from the subsistence to a commercial activity and so a special RNFS fund on the line of Venture Capital Fund (VCF) needs to be set-up to look into the diverse needs of the RNFS such as technology upgradation, quality standardization, market support including development of brands, etc. And this fund can be set-up with contributions from Government of India, National Bank for Agriculture and Rural Development (NABARD), Small Industries Development Bank of India (SIDBI) and through private sector participation.

Besides all those, for healthy growth of the RNFS a need to rationalize the working of existing agencies and regulations in the emerging environment, with a view to reducing regulatory burden on enterprises and providing incentives to grow, diversify and employ more workers of course with adequate measures to protect them from exploitation and hazardous working conditions cannot be ignored.

And above all what is highly desirable is to recognize the importance of the rural non-farm sector in faster growth not only for rural development but also for overall economic development of the country.

References

Mishra, Jegannath (1997), My vision for India's Rural Development, Vikas Publishing House, New Delhi.

Hashim, S.R. (2009), Employment and Unemployment in a Society in Transition in Growth, Employment and Labour Markets (ed.) by Krishnamurty and R.P. Mamgain, Danish Book, Gazipur.

Nanjundappa (1981), The Presidential Address in IEA from Diamond to Platinum (ed.) by R.K. Sen.

Gandhi, M.K. (1926), *Young India*, Oct. 7.

Papola, T.S. and Sharma, A.K. (2005), Towards a Policy Agenda for Rural Non-farm Sector in Rural Transformation in India (ed.) by R.

Nayyar and A.K. Sharma, Institute for Human Development, New Delhi.

Sen, Abhijit and Jha, Praveen (2005), Rural Employment : Patterns and Trends from National Sample Survey, *Ibid*.

Rajasekhar, D. (2005), Micro-finance and Rural Non-form Sector: Some NGO Experiences, *Ibid*.

Mahajan, Vajay and Raju, B. Yerram (2005), Institutional Framework and Enabling Environment for Rural Non-farm Sector, *Ibid*.

Fisher, Thomes and Mahajan, Vijay (1996) The Forgotten Sector : Rural Non-farm Employment and Enterprises in Rural India, Oxford and IBH, New Delhi, and ITDG, U.K.

Tripathi, A.R., (1988), Can Bank Alone Achieve Rural Development, *Yojana*, July 16-31.

GOI (1998), Economic Census, CSO.

Rao, V.K.R.V. (1977), Integrated Rural Development.

Ahmed, Y.L. (1945), Administration of Integrated Rural Development, International Labour Revolution.

Different Governmental Documents.

Bhattacharya, S.N. (1970), Community Development and Analysis of the Programme.

CHAPTER

5

Role of Non-Farm Sector in Rural Development

PRASHANTRAO M. YADWAD AND
ANUPAMA C. MAREGUDDIKAR

INTRODUCTION

The development process of present developed countries shows a certain pattern of economic growth over a long period of time. *Simon Kuznets (1979)* in his path breaking empirical work on modern economic growth of present developed countries points out that as economic growth takes place, the primary sector shows a decline in its relative contribution to national output, even though its absolute share to national output increases. Then the growth of primary sector leads to an increase in the relative share of secondary or industrial sector to national output over a long period of time. Once the economy becomes industrialized, the share of secondary sector starts declining in its relative importance and the tertiary sector consisting largely of Services of all kinds begins to increase till it reaches the major proportion of national output *(Simon Kuznets, 1979)*.

However, this pattern of economic growth is not occurring in India, if one looks at sectoral diversification of Indian economy. Although there has been a decline in the share of the primary sector in national output from about two-thirds to one-third in the post-independence period, the share of the secondary sector has increased at a much slower pace than could have been expected. Moreover, before the country reached a stage at which it can be called an industrialized economy", the share of the service sector in national output has begun to increase. This is some thing which is contrary to the experience of the present developed countries like the USA or UK *(Vyasulu 1990)*. It could be noted that even though there was a decline in the share of contribution of agriculture to national income, there was no decline in the share of labour force in agriculture until 1970. Even after two decades of Independence, the share of agriculture in labour force confined to be around 70 per cent as in 1971. There was a decline in the 1970s by 3 per cent; there was further decline in the 1980s by another 2 per cent, resulting in the percentage share of agriculture being reduced to 65 percent. Such a decline in the share of agriculture and the corresponding increase in the share of non-agriculture inspired hopes of sustained growth of rural non-farm sector (RNFS). Since the 1970s, when the share of primary sector in terms of both employment and national output started declining, the growth of rural non-farm sector has begun to be seen as an alternative solution to the problem of rural unemployment and poverty *(Parthsarathy, Shameen and Reddy 1998)*.

MEANING OF RURAL NON-FARM ACTIVITES

The term 'non-farm' encompasses all the non-agricultural activities; it includes manufacturing activities, trades and services in rural areas. And also, the seasonal and contractual jobs, unconnected with farming as such, available within the village or in the nearby town, are a part of non-farm employment *(Chadda, 1986; 141: Edgreen and Muqtada, 1989; 38:*

Rajshekhar, 1995; 279). The rural non-farm activities constitute an important category of income for the poor in developing countries which are characterized by problems such as mounting population pressure, diminishing land frontiers, small and fragmented landholdings due to declining land-man ratio and a high incidence of unemployment. The non-farm activities provide supplementary to the small and marginal rural household, especially during the slack season.

TRENDS AND PATTERN OF RNFA

TABLE I

Percentage of Workers in RNFAs in India

Year	*Male*	*Female*	*Persons*	*AGR-RNFS*	*AGR-RFS*
1972-73	16.7	7.3	14.4	—	—
1977-78	19.3	10.3	16.6	50.7	1.64
1983	22.2	12.2	18.5	4.32	1.62
1987-88	25.4	15.8	21.7	4.39	0.28
1993-94	25.9	13.8	21.6	0.33 Neg	—
1983 to 1993-94	—	—	—	2.83	0.17

Source : Visaria, 1995, p. 402; NSSO, 1990; Government of India, 1996.

Table 1 presents the percentage of workers involved in rural non-farm activities in India. It is clear from the table that proportion of rural labour in non-farm activities increased significantly from 14.4 per cent in 1972-73 to 21.7 per cent in 1987-88 and it marginally declined to 21.6 per cent in 1993-94 and thereby showing the inability to absorb labour force in RNFS. The similar trend, by and large, is also observed for male and female non-farm activities until 1987-88 and virtual stagnation in 1993-94 *(Unni 1996)*. The proportion of rural female workers engaged in secondary sector declined from 10 per cent in 1987-88 to 8.3 per cent in 1993-94. The percentage male workers in rural non-farm employment increased from 16.7 per cent in 1972-73 to 25.4 per cent in 1987-88 and slightly went up to 25.9 per cent in 1993-94, whereas in the case of female non-farm activities increased from 7.3 per cent to 15.8

per cent and it markedly went down to 13.8 per cent during the same period. The proportion of female non-farm activities uneven spatially as well as gender-wise. It was not only lower than that of males, but also registered a declining trend in the recent past (Table 1). Available literature suggests that lower female participation in non-agriculture could be due to "limited diversifications of women's work" *(Unni 1989)*.

Table 2 shows that the percentage of workers involved in rural non-farm activities in major States of the country. It is clear from the table that the states like Kerala, West Bengal, Tamil Nadu, Punjab, Haryana, etc., have witnessed a higher proportion of diversification into non-farm activities, while the states like UP, Karnataka etc. have shown a moderate diversification between 1987-88 and 1993-94. Moreover, there

TABLE 2

Percentage of Workers in RNFAs By Major States in India

State	*1977-78*	*1987-88*	*1993-94*
Kerala	34.6	41.5	43.6
W.B.	24.4	28.2	36.7
Tamil Nadu	22.0	29.6	29.5
Punjab	18.5	23.5	25.3
Haryana	18.3	21.4	28.1
Andhra Pradesh	17.6	22.4	20.7
UP	17.3	17.7	20.0
Bihar	15.3	17.2	15.7
Karnataka	15.2	18.1	18.8
Orissa	15.0	24.1	19.1
Maharashtra	14.2	17.1	17.4
Gujarat	13.0	28.5	21.3
Himachal Pradesh	12.2	18.0	19.7
Rajastan	11.6	26.6	20.1
Madhya Pradesh	8.6	12.3	10.2
All India	16.7	21.7	21.6

Source : Unni, 1991, p. 110; NSSO, 1992; Government of India, 1996.

are some other states such as Gujarat, Madhya Pradesh, Orissa, Andhra Pradesh, Bihar etc., which have experienced negative growth in RNFS particularly between 1993-94. Hence, India as a whole, the proportion of rural labour in non-farm activities marginally declined with strong regional variations in the early 1990s.

TABLE 3

Percentage Distribution of Workers by different RNFAs in India

Activities	*1972-73*	*1977-78*	*1983*	*1987-88*	*Growth Rate*
Agriculture	85.5	83.4	81.5	78.3	1.2
Non-agriculture	14.4	16.6	18.4	21.7	—
Mining and Qurrying	0.3	0.4	0.5	0.6	6.4
Manufacturing	5.4	6.2	6.8	7.2	3.9
Electricity, Gas and water	0.1	0.1	0.1	0.2	11.5
Construction	1.4	1.3	1.6	3.3	7.9
Trade, Hotel etc.	2.5	3.3	3.4	4	5
Transport and Storage	0.8	0.8	1.1	1.3	8.7
Services	4.1	4.5	4.9	5.1	3.3
All	100	100	100	100	1.8

Source : Unni, 1991, p. 111; NSSO, 1992; Government of India, 1996.

Moreover, the growth in the percentage of rural non-agricultural workers is not uniform across the organization of work (Table 3). We have just noted that the basement of rural non-farm sector seems to be widening in most parts of India. In the recent past, within the rural non-farm activities such as mining and quarrying, household and non-household manufacturing activities, electricity, gas, water, construction, trade and hotel, transport and storage and services have been showing varying degree of increasing trends over a period of time. This trend is quite discernible in the case of service related activities than that of manufacturing activities. Further, it could be noted that within the manufacturing sector, the household manufacturing activities have been losing its relative importance due to competition with the modern, small scale industry. On the other hand, the rising importance of the non-household sector for manufacturing processing, servicing and repairs, has been most evident in all parts of rural India *(Chadha 1986)*. Therefore, non-household manufacturing sector is gaining much importance, while household manufacturing tending to be deteriorated.

DETERMINANTS OF RURAL NON-FARM SECTOR

The rural non-farm activities are being influenced by an improved socio-economic infrastructure, state incentives, higher degree of profit expectation in non-farm activities and degree of urbanization. The above determinants of rural non-farm sector have led to constitute three important hypothesis.

(A) Linkage Hypothesis

Growth of Rural Non-agriculture employment depends on agricultural prosperity through its production and consumption linkages. *Vaidyanathan (1986)* found that a significant and positive relationship between level of non-agricultural employment and crop output per head of agriculture population and a negative relationship with the size of operational landholdings.

(1) Production Linkages

The growth of rural non-agricultural activities could be brought about by backward and forward linkages that the rural non-agricultural sector has with agriculture.

(2) Consumption Linkages

The consumption linkages operated through an increase income of rural farmers has more effective impact on the growth of non-farm sector than the production linkages. However, in developing countries like India, backward linkage of production is not quite operative but forward linkages are relatively more effective to stimulate the growth of rural non-farm sector.

(B) Residual Sector Hypothesis

The second dimension to the growth of rural non-farm activities can be termed as "distress diversification" into unproductive or low paid non-agricultural jobs. Such a distress diversification occurs especially when the workers are underemployed in agriculture. This distress-induced non-farm sector has been put forward as the residual sector hypothesis" by Vaidyanathan (1986). He found a strong positive relation between unemployment rate and percentage of non-agricultural workers.

Level of Urbanisation Hypothesis

The growing urbanisation may also facilitate the growth of non-agricultural employment in rural areas in the following ways. The areas may cater to the demand for non-agriculture products and services in the nearby urban area. Secondly, some of the residents from rural areas may engaging non-agricultural occupations in the nearby town and commute their work place regularly. *Jeemol Unni* found that the percentage of urban population positive impact only on male non-agricultural workers. Thus the debate on whether the growth of non-farm employment is "distress diversification" or not remains inconclusive *(Unni 1998)*.

IMPORTANCE OF RURAL NON-FARM SECTOR

In view of ever growing labour force, diminishing land frontiers, declining employment elasticities in agriculture and urban organized manufacturing sector, ever increasing problems of urbanistion etc., the rural non-farm sector plays an important role in the following ways.

(I) Income from Non-farm Activities

The non-farm are second in importance to agriculture in several countries of the world in terms of their contribution to GNP. For the smallest landholding categories in each country, non-farm income sources accounted for more than 50 per cent of annual income. Therefore, non-farm activities are important as far as enhancing the income of small and marginal rural households is concerned *(Samuel 1979)*.

(2) Non-farm Activities Reduce Income Inequalities

There is also an evidence to show that non-farm activities reduce income inequalities in rural areas (Samuel 1979). The evidence from some Asian and African countries suggests that Gini coefficient tended to decline when income from non-farm activities is included. The income from non-farm sources is found to be more in the case of small and marginal farmers as against the large farmers.

(3) RNFAs Reduces Income Instability

The evidence also shows that the rural non-farm activities act as a means to reduce the income instability of the households over a year. The empirical studies show that small and marginal farmers have more tendency to participate in rural non-farm activities especially during the off-season to supplement their annual household income and reduce the income instability. Therefore, farm and non-farm activities tended to move in opposite direction *(Liendholm and Kilby 1989; Anderson and Leisonson 1980)*.

(4) RNFAs Reduce Rural Urban Migration

The rural non-farm activities also reduce rural-urban migration and thereby minimizing the problems of growing urbanisation in most of the developing countries. The experience of Taiwan country shows that lots of rural employment opportunities were created and enabled a large number of Taiwan's population to participate in rural industries without having to leave the countryside.

PROBLEMS OF RURAL NON-FARM SECTOR

Notwithstanding the important role assigned to the non-farm sector in the development process of the country, this sector is facing several problems. Some of them are:

(1) Dominance of Traditional Units

The predominant of traditional small-scale sector, handicrafts and other industries has not resulted in much dynamic in the sector.

(2) RNFS has Less Production Linkage Compared to Consumption

Infact, the sustainable growth or rural non-farm sector largely dependence on its production and consumption linkages. As the production linkages are not adequately operative in rural India, consumption linkages would lead to increase the demand for non-food produced in the urban areas. So that industrial basement in urban informal sector would grow much faster at the cost of rural non-farm sector.

(3) Lack of Social Overhead Capital

The growth of rural non-farm sector is limited by infrastructural facilities such as all weather road transport, marketing, electricity etc.

(4) Problem of Marketability

In most of rural areas, several manufacturing units are not equipped with up-to-date technology and as result of which the quality of the products is quite poor as compared to the goods which produced in urban manufacturing sector.

(5) Problem of Institutional Credit to RNFAs

Though the NABARD has initiated a host of packages including credit programmes towards the growth of rural non-farm sector, many of these rural non-farm units have not been able to obtain required quantum of financial assistance.

(6) Non-farm Sector is not Homogeneous

In fact, rural non-farm sector includes several activities which are not homogeneous in nature. So that it is difficult for the authority to formulate uniform policy packages under which rural non-farm units could make much progress. Thus, the rural non-farm sector has facing several problems which are handicapping the progress of it.

CONCLUSION

The employment elasticities in relation to output in agriculture seem to be declining. On the other hand, capital intensive technology in manufacturing sector led to a decline in employment elasticities and resulted in a virtual stagnation in the growth of employment in this sector. Therefore, small and marginal farmers are finding it difficult to derive wage employment either in agriculture or in the secondary sector. In these circumstances, rural non-farm sector seems to be one of the best options available for small and marginal farmers to enhance their employment and to stabilize incomes.

These activities provide supplementary employment to small and marginal farmers households especially during the slack season. Consequently, incomes of these households tend to be smooth in a year. These activities are widely dispersed, labour intensive, require little capital and ideally suited to provide jobs for the poor. RNFAs also have potential to reduce income inequalities and rural-urban migration. The non-farm sector is, therefore, seen as a method by which the problems of unemployment, particularly rural employment, can be reduced. Hence, rural non-farm activities (RNFAs) play an important role in developing countries such as India. Therefore the rural non-farm sector should be developed in a manner as to create adequate employment opportunities to absorb a large number of rural unemployed labour forces in the Next Millennium.

CHAPTER

6

Rural Development Non-Farm Sector : A Need for Sustainable Development in India

P. SUBRAMANYACHARY AND M. REDDI RAMU

"The prosperity of India lies in the prosperity of Villages".
—Mahatma Gandhi

BRIEF HISTORY OF RURAL DEVELOPMENT

The concept of Rural Development in India is not new concept, From British onwards efforts have been taking by the government for the development of rural areas. Mahatma Ganghi also proposed some schemes for rural development. Earlier five year plans had been prepared based on the theme of rural development along with priority to rural development. The expectations of economists collapsed as the failure of trickle down effect which refers that the fruitful results of economic development reach to bottom from top. So, in 1974 department of Rural development came into force as a part of ministry of Food and Agriculture. On the 18th August, 1979 new Ministry of Rural reconstruction was emerged. It was renamed as Ministry of Rural development on 23rd of January 1982. Later Rural Development Ministry was converted into a part of Agriculture Ministry in 1985. In 1991 on July 5th Rural development was announced as separate Ministry. Again in 1995, it was formed as Ministry of Rural Areas Employment and Poverty Alleviation. In 1999 finally it was formed as Ministry of Rural development. For the development of rural areas the Ministry of Rural Development has been implementing number of programmes focusing on the elements like health, education, safe drinking water, housing and roads.

MEANING OF RURAL DEVELOPMENT

There is no comprehensive definition for rural development. "Overall development of villages is called Rural Development. This refers development of agriculture and allied sectors, rural industries and rural infrastructure. As majority of population lives in villages, their backwardness stands for backwardness of economy as whole. So, rural development has emerged as "a strategy designed to improve the economic and social life of a specific group of a people i.e. rural poor". It involves extending the benefits of development to the rural population who seeks a livelihood in the rural areas.

OBJECTIVES OF RURAL DEVELOPMENT

In the Indian context rural development programmes have aimed at achieving the following objectives:

1. To change the attitude of rural people towards development.
2. To promote democratic leadership by setting up local self-government.
3. To provide basic needs such as drinking water, health care, sanitation, housing and employment, i.e. to improve standard of living

4. To improve communal harmony and unity among rural people
5. To develop farming and non-farming activities without adverse affecting the environment.

PROGRAMMES INTRODUCED FOR RURAL DEVELOPMENT

For the development of any country proper utilization of natural and human resources is inevitable. The developing countries not in apposition to utilize natural resources as they have lack of advanced technology. India is also not except to this as developing economy. The land mark in the history of rural development is introduction of Community Development Programme on October 2nd, 1952. After that for the elimination of poverty and unemployment in the rural areas, the programmes like Small Farmers Development Agency (SFDA), Marginal Farmers and Agriculture Labour Development Agency (MFADA), Minimum Needs Programme (MNP), Food for Work Programme (FWP), Development of Women and Children in Rural Areas (DWCRA), Training for Rural Youth for Self-Employment (TRYSEM), Ganga Kalyan Yojana (GKY), Jawahar Rojgar Yojana (JRY) Swarna Jayanthi Swagram Yojana (SJSY), Anthyodaya Programme etc. Besides this, NABARD, RRB, SBI and other commercial banks are helping for the development of rural areas. During 1995-96 Rural Infrastructure Development Fund was established and maintained by NABARD. The budget expenditure for Rural Development during 2007-08 is 32,508 crores. In the plans Expenditure on Rural Development has been increasing. In the 11th plan the government has target that an increase in forests and cover tree by 5 percent and provision of safe drinking water by 2009. But in the programmes there was no much more priority for the protection of environment.

IMPORTANCE OF NON-FARM SECTOR : RURAL DEVELOPMENT

The Rural Non-Farm Sector (RNFS) refers all non-agricultural activities like mining and quarrying, household and non-household manufacturing, processing, repair, construction, trade and commerce, transport and other services in villages and rural towns undertaken by enterprises varying in size from household own account enterprises to huge industries. The level and growth in urbanization is expected to have a positive impact on the level of non-farm employment in rural areas. In many areas, agriculture alone cannot provide sufficient livelihood opportunities. Migration is not an option for everyone and where possible, policy-makers may in any case prefer to limit the worst excess of urbanization with its associated social and environmental problems. Rural Non-farm employment can play a potentially significant role in reducing rural poverty and numerous studies indicate the importance of non-farm enterprise to rural incomes. Livelihood diversification is often characterized as being driven by two processes. 1. Distress push, where the poor are driven to seek non-farm employment for want of adequate on-farm opportunities. 2. Demand-pull where rural people are able to respond to new opportunities. In the former situation large numbers may be drawn into poorly remunerated low entry barrier activities, while the later are more likely to offer a route to improved livelihoods.

Rural Non-farm activities may absorb surplus labour in rural areas, help farm-based households spread risks, offer more remunerative activities to supplement or replace agricultural income, offer income potential during the agricultural off-season and provide a means to cope or survive when farming falls. So the policies must aim to improve the assets held by the poor or increase their productivity. The following different types of assets provide an appropriate way in which to structure the evidence on livelihood choices and outcomes.

Natural Capital

The rural resource stocks from which resource flows useful for livelihoods are derived (e.g. land, water, wildlife, bio diversity environmental resources).

Social Capital

The social resources upon which people draw in pursuit

of livelihood (e.g. networks, membership of groups, relationship of trusts across to wider institutions of society).

Human Capital

The skills, knowledge, ability to labour and good health importance to the ability to pursue different livelihood strategies.

Physical capital

The basic infrastructure and production equipment and means which enable people to pursue their livelihoods. (e.g. transport, shelter, water, energy and communications).

Financial Capital

The financial resources which are available to the people and which provide them with different livelihood options (e.g. savings, supply of credit, regular remittances, pensions).

Under Globalization the following measures can be suggested regarding Rural-Non-Farm activities:

1. Provision of adequate and suitable infrastructure support
2. Availability of fiscal and financial incentives
3. Provision of Marketing outlets
4. Information about improved technology
5. Maintenance of harmonious relationship between farm and non-farm activities.
6. Policy review and changes in must be made as per requirement.

DEGRADATION OF ENVIRONMENT

At present, the most dangerous common problem to mankind in the world is degradation of environment. In order to make comfortable life, human beings are going to exploit natural resources by applying modern technology and on the other side, totally neglecting future generation's needs. Industrialization, massive deforestation and urbanization result in downfall of quality in land, water, air and depletion of ozone layer, acid rains, green house effects, global warming, etc.

Environmental Degradation is decrease in the quality of environment. In other words it refers a visible reduction in the availability of goods and services from the physical environment and the renewable natural resource base. The degradation of environment affects badly on every one on the planet and different fields. Poverty still remains a basic problem at the root of several environmental problems. So there is necessity to protect environmental resources in rural areas for future generation's needs.

SUSTAINABLE DEVELOPMENT

The present world countries both developed and under developed countries realized the danger of environmental degradation and spending more funds to prevent and abatement of environmental pollution. The concept of Sustainable Development conveys the message that environment and economic development are closely interdependent and mutually supportive. It deals with quite simply the largest and most extreme problem even faced by humanity. The term Sustainable Development was probably coined by Barbara Ward (Lady Jackson), the founder of the International Institute for Environment and Development, who pointed out that Socio-Economic Development and environmental protection, must be linked. The World Commission on Environment and Development (WCED), 1987, headed by Brundtland defined Sustainable Development as follows :

> "Development that meets the needs of the present without compromising the ability of future generations to meet their own needs."

In economic terms Sustainable Development is defined as "Economic system in which the number of people and the quantity of goods are maintained at same constant level, which is ecologically sustainable overtime and meet at least the basic needs of total population." So in the name of Sustainable Development world countries are trying to protect environment.

RURAL INDIA : ENVIRONMENTAL PROBLEMS

Rural India is facing the severe problem of degradation of environment. The problems like lack of sanitation, shortage of safe drinking water, ill-designed habitats and kitchens, living of animals and human in close proximity etc. are very common. Besides this, human beings are misusing natural resources like soil, water, forests etc. indiscriminately. The farmers and agricultural labours are suffering from soil erosion, chemical seeding, chemical fertilizers, pesticides spraying, etc. Unscientific agriculture is also causing for environment problems in rural areas.

India is sixth largest and second fastest growing producer of Green House Gases (GHG) in the world. According to Tata Energy Research Institute, India is losing 10% of national income due to environmental degradation and the availability of fresh water declined by two-thirds. The water requirement of major water consuming industries such as agro-based industries, refineries petro-chemicals, and fertilizers has grown 40 times, but these are not yet treating the huge waste water generated. 2.5 million Pre-mature deaths have been recording due to indoor and outdoor air pollution. Regarding soil pollution in the country, out of 329 million hectares of total geographical area, 144 million hectares are subject to water and wind erosion and 30 million hectares through salinity, water logging, etc. Recently, noise pollution is also becoming as major environmental hazard in rural areas. There is a forecast that there will be more pressure on water and other natural resources in India by 2020. Apart from this improper implementation of Land reforms and Green revolution which is benefited for large farmers badly affected rural environment scenario.

There are some hurdles to the Sustainable rural development. The factors like improper participation of people in the programmes, too much political interference, much dependency on executive officials, corruption, wrong records of rural particulars, lack of integration among the rural poor, etc. show their adverse impact on rural scenario.

POLICY MEASURES FOR SUSTAINABLE RURAL DEVELOPMENT

For the overall development of the country in the long run, protection of environmental resources, particularly in the rural areas is needed. India has two-tier system for combating pollution, i.e. Central Pollution Control Board (1974) and State Pollution Control Board In this regard, the Indian Constitution adopted the articles like 48A, 51A(g) and 253. On the basis of these articles, the Indian Parliament enacted Water Act of 1974, the Prevention and Control of Pollution Act (Air Act), 1981, the Environmental Protection Act, 1986, The Public Liability Insurance Act 1991 have been made for the protection of environment resources. But the performance of Boards and the implementation of acts should be improved. The following measures are useful to protect environment especially in rural areas, while performing non-farm activities.

1. To educate and train the politicians, bureaucrats, NGOs and policy-makers who are responsible for making and implementation of laws over environmental degradation in rural areas when undertaking the non-farm activities.
2. *By keeping non-farm activities in mind Vanasamrakshna Samithis* are to be upgraded with knowledge, mission and vision.
3. The sarpanches or administrators of villages and their ward members are given exposure to environmental protection and their merits for farm as well as non-farm activities.
4. At the cost of non-farm activities, a strong dose of advertisements would be given in press and electronic media about degradation and result of sustainability in rural.
5. Environmental clubs in schools, colleges, universities are to be prompted and make it mandatory in villages by giving priority to the non-farm activities.
6. In connection with non-farm activities, Environment budgeting and auditing is to be introduced in village Panchayats.

7. Promote the good habits among people to grow a tree, to nurture a pet animal which are suitable to the non-farm activities.
8. Public participation is needed for sustainable development. For this motivation of the people about the importance of sustainable development is needed while introducing non-farm activities in rural areas.
9. Wind, solar, bio-gas, hydel power are to be promoted than fossil fuels and they should be used in non-farm activities as much as possible
10. World environment day, earth day, etc. are to be celebrated seriously and sincerely which reminds the importance of environmental resources among rural people for proper understanding the non-farm activities.
11. The literacy rate should be improved as it is key factor to change their attitude and control migration to urban areas and also awareness about rural sustainable development.
12. Brining awareness among the rural people regarding non-farm activities and undertake them at suitable environment by using the services of NGO and other voluntary organizations.

CONCLUSION

For the prosperity of any economy either in Developed or Underdeveloped country sustainable rural development is required by giving priority to non-farm activities. Though governments are implementing number of programmes there is more focus on elimination of poverty and unemployment and less priority to protection of environment. It is necessary that by taking the strategies of developed countries, India has to go to achieve sustainable rural development.

REFERENCES

Haggbladge, S., Hazell, P.B. and Brown, J. (1989), Farm/Non-farm Linkages in Rural sub-urban Africa, *World Development*, 17(8) 1173-1201.

Smith, D. (2000): The Spatial Dimension of Access to the Rural Non-farm Economy Draft paper, Chatttam, UK, Natural Resource Institute.

Earth Summit-2002, *Udyoga Samacharam*; Sept. 1-15, p. 12.

Aarthi Dhar (2004): Impact of Pesticides: Poisoned Land/water, *The Hindu Survey of Environment*, 2004, pp. 13-17.

Report of World Commission on Environment and Development, 1987.

P.M. Prasad (2006): 'Environment Protection: Role of Regulatory System in India', *Economic and Political Weekly,* April 1-15, pp. 1278-86.

Dr. I. Satya Sundaram (2002): "Rural Development", Himalaya Publications, Mumbai.

Y.V. Krishna Rao (2004): Rural Development—Problems and Challenges.

Report of Ministry of Environment and Forest.

www.indiabudget.nic.in

www.unep.org

www.ecologic.de

CHAPTER

7

The Growth and Employment in Non-Farm Sector

B. ESWAR RAO PATNAIK

INTRODUCTION

The present paper is a modest attempt to analyze the contours of change in the growth and employment in Non-Farm Sector in Orissa. The Present paper is based on secondary data and the materials of the current study are drawn from Plan documents of government of Orissa, Economic Survey, 2007, Government of Orissa, estimates by N.S.S.O. several Indian Economic Journals, Standard texts and several dailies of Times of India. The academic exercise spans over a periods of two decades, from 1980 to 2000.

The economy of Orissa is primarily agricultural and rural oriented, with concentration of 8.5% of its population in rural areas. In contrast, 74 percent of population dwell in rural areas in India. Orissa is the poorest state of the country in which 39.9% of population live below poverty line in 2004-05, above the national average of 21.8%. Materials deprivation, social deprivation and social deprivation of the masses are manifestations of grinding poverty of the masses. Generation of large scale employment opportunities is a pre-requisite to contain poverty and inject more money into the body of the economy.

The agricultural sector is over saturated and hence the rural non-form sector marched forward to absorb a large number of rural workers in the state of Orissa. The Rural Non-Farm sector has occupied the center stage in the development of the economy in our efforts to achieve an all inclusive growth, check migration of human capital from villages to cities and broaden the employment base of the economy.

CONCEPT OF RNFs (RURAL NON-FARM SECTOR)

The world Development Report, 1995 has defined the rural non-farm activities, as all economic activities in rural areas, except agriculture, live-stock, fishing and hunting. A constellation of activities are included in RNFS, (i) Activities undertaken by farm households as independent producers in their homes, (ii) the sub-contracting of work to farm families by urban-based firms, (iii) non-farm activities in villages and urban enterprises, (iv) commuting between rural residents and urban non-farm jobs (Lanjouw and Lanjoun : 1995). Broadly speaking, RNFS comprises all non-agricultural activities, handicrafts, mining and quarrying, manufacturing, processing repairs, construction, trade and hotelling, transport and communications, personal and other services in rural areas.

Reasons for the Emergence of RNFs

(i) It is well known that, the structural transformation of every economy is the outcome of development process. So, a dynamic economy, that is passing through structural transformation must undergo a diversification in the occupational pattern of living of the work force. The economy of Orissa has registered a mere 1 percent to 2.5 percent growth in employment in farm sector. So, stage is set for the play of the non-farm sector in the feeble economy of Orissa, as labor absorption can not be accomplished in farm sector alone.

(ii) Development economists at country level argue that, the subjects of Orissa eke out a living from multiple sources of livelihood, the prominent among which is the non-farm rural activities. Approximately, 30 percent of rural incomes accrue from RNFS.

(iii) The case for rural non-farm activities in Orissa emerge from the fact that, the extension of RNFs by public authorities may reduce rural urban disparities in development.

(iv) In the recent part, there has been migration of rural workers to cities in quest of livelihood. The protagonists of rural non-farm activities contend that, the launching of a number of rural industries in villages may defer the current tendency of migration of human resources from rural areas to urban areas.

(v) Rural industries are labour-intensive and capital light. Hence, rural industries may widen the employment base of the economy.

(vi) There prevails yawing gap between male workers and female workers in the pace of development. The employment of female workers in handicrafts, and manufacturing may bridge the pace of gender gap in development.

(vii) One flower to the garland of RNS is its cost effectiveness. The creation of jobs in RNS costs only 10% to 20% of what it costs in large scale industries.

(viii) The feather to the cap of RNFS is its capacity to utilise local raw-materials and spur enterprises of the locality.

It follows that, there exists vast potential for the development of RNFS in the poverty-stricken economy of Orissa.

INTER-LINKAGE HYPOTHESIS AND ORISSA

The seminal contribution of John F. Mellor has been, the focus, he has laid on the capacity of agriculture to stimulate a swarm of new economic activities in rural non-farm sector. What Mellor has highlighted is : (i) Consumption-expenditure linkages and labour flow linkages of agriculture, Conventional wisdom shows that, there are close links between the growth of farm sector and non-farm sectors of the economy.

The line of reasoning developed by Mellor has suggested that, the geographical isolation and tastes of rural population combined to create demand for locally produced goods with increase in incomes of farmers. The spurt in productivity in agriculture has the impact of enhancing farmers incomes which gets manifested in expanding the consumption demand for labor-intensive goods, which pushes the wheels of RNFs forward.

Mellor has hinted at the possible effect of technological innovation with expansion food supply and distribution of initial gains of development to already wealthy class. This initial increase in rural incomes sets in motion a sequence of multiplier effects and paves the way for expansion in production and employment in other sectors of the economy, namely, consumers goods industries and small scale units. The injection of more income into the body of low income group, via more employment levels brings about a spurt in the demands for foodgrains production.

TABLE I

Expansion of Non-farm Sector in Orissa

Annual Growth rate of Rural-Non-farm employment in Orissa (US – PS + SS Basis)

State/Country	*Growth during pre-reforms period*	*Growth during post-reforms period*	*Difference in growth rate between two periods*
Orissa	0.86	6.20	3.33
India	3.05	3.81	0.75

Source : N.S.S.O reports on Employment and Unemployment for 38th, 50th and 61st rounds.

TABLE 2

Sectoral Composition of the Rural Non-farm Sector in Orissa

State	Year	Mining and Quarrying	Manufacturing	Electricity, gas and water	Construction	Secondary	Wholesale and retail trade	Transport storage etc.	Services	Tertiary
(1)	(2)	(3)	(4)	(5)	(6)	(7)	(8)	(9)	(10)	(11)
Andhra Pradesh	1993-94	3.47	33.21	0.27	8.20	45.15	21.67	5.05	28.13	54.85
	2004-05	4.19	30.38	0.39	11.74	46.71	23.34	9.21	20.74	53.29
Assam	1993-94	0.75	17.70	1.13	3.11	22.69	32.86	6.12	38.33	77.31
	2004-05	0.87	12.29	0.58	9.57	23.32	34.92	9.53	32.24	76.68
Bihar	1993-94	2.90	22.56	1.47	8.12	35.05	28.79	6.35	29.80	64.95
	2004-05	1.59	26.10	0.31	19.28	47.29	27.72	9.27	15.73	52.71
Gujarat	1993-94	2.19	42.83	0.76	11.19	56.97	13.87	7.14	22.01	43.03
	2004-05	2.10	34.26	0.52	13.07	49.96	20.72	11.77	17.55	50.04
Haryana	1993-94	1.89	14.17	1.42	15.34	32.82	19.01	12.51	35.67	67.18
	2004-05	0.36	24.66	1.53	23.15	49.71	21.02	11.35	17.92	50.29
Karnataka	1993-94	4.03	35.40	0.95	7.84	48.21	19.95	3.78	28.06	51.79
	2004-05	2.20	33.06	0.31	10.83	46.39	23.62	10.56	19.43	53.61
Kerala	1993-94	3.41	29.02	0.70	13.22	46.34	20.70	9.34	23.63	53.66
	2004-05	2.31	23.48	0.46	18.42	44.67	20.97	12.24	22.12	55.33

Madhya Pradesh	1993-94	12.08	31.75	1.58	8.87	54.12	15.83	4.54	25.51	45.88
	2004-05	3.87	27.66	0.63	21.15	53.31	22.31	4.45	19.93	46.69
Maharashtra	1993-94	2.08	28.40	0.93	13.73	45.15	18.72	6.18	29.95	54.85
	2004-05	1.33	27.91	0.55	14.51	44.30	23.37	11.30	21.04	43.88
Orissa	1993-94	5.59	33.32	0.68	9.50	49.09	23.03	3.72	24.16	50.91
	2004-05	2.46	35.44	0.42	17.81	56.12	21.91	6.37	15.59	43.88
Punjab	1993-94	0.00	19.33	4.56	13.63	37.52	19.31	10.44	32.72	62.48
	2004-05	0.20	22.42	2.18	26.83	51.63	19.44	10.60	18.32	48.37
Rajasthan	1993-94	9.15	17.84	0.84	34.74	62.56	11.85	5.30	20.29	37.44
	2004-05	4.31	21.43	0.63	35.63	62.00	14.93	7.57	15.50	38.00
Tamil Nadu	1993-94	1.24	43.25	0.76	7.83	53.08	16.22	6.94	23.76	46.92
	2004-05	0.81	40.07	0.32	16.43	57.64	18.45	8.12	15.79	42.36
Uttar Pradesh	1993-94	0.73	32.04	0.73	9.77	43.24	21.51	7.67	27.58	56.76
	2004-05	0.70	31.85	0.29	19.81	52.65	22.67	7.79	16.89	47.35
West Bengal	1993-94	0.55	43.84	0.21	6.66	51.25	20.47	7.47	20.80	48.75
	2004-05	0.68	36.04	0.21	10.63	47.56	25.62	9.56	17.26	52.44
All India	1993-94	2.74	32.40	0.89	11.00	47.04	19.82	6.70	26.44	52.96
	2004-05	1.79	29.44	0.46	17.78	49.47	22.58	9.09	18.85	50.53

Source : *The Indian Economic Journal*, Vol. 55, 2007, p. 55.

The Arithmetics of the foregoing Table 2, reveal that, the annual growth rate of non-farm employment in the backward state of Orissa during the pre-reforms period was 0.86 percent and 6.20 percent during the post-reforms periods. The corresponding average at national level was 3.05 percent.

An enquiry into the above table depicts that, the secondary sector claims the lion's share in provision rural non-farm employment for rural works namely, 63.62%, while tertiary sector provides employment to rural workers by 36.32. In secondary sector, manufacturing, construction, mining and quarrying, electricity, gas and water supply are employment providers for rural workers in descending order. In tertiary sector, wholesale and retail trade, services and transport and storage have absorbed transport and storage have absorbed work force in the state of Orissa.

Reflection reveals that, the focus of mining and quarrying, electricity, gas and water supply as job providers to workers is receding back in the state. The shares of transport and storage in rural non-farm employment have marginally improved from 3.72% in 1993-94 to 6.37% in 2004-05. Further, wholesale and retail trade and services have a key role to play in the state of Orissa in providing employment to human resources, not withstanding a decline in their percentage share to rural non-farm employment in 2004-05, compared with 1993-94.

For the rural females, manufacturing sectors holds the key for employment in Orissa in 2004-05. Next to manufacturing, construction and quarrying deserve mention, in respect of providing gainful employment to rural females in secondary sector. The key areas of employment for rural females in NFRS in tertiary sector are services wholesale and retail trade, transport and storage (Tertiary sector accounts for 22.66% of total employment for rural females of Orissa in RNFS (2004-05)).

The story is slightly different with regard to rural male workers in Orissa. In the rural non-farm sector manufacturing, construction, mining and quarrying have provided employment to rural male workers in descending order (secondary sector). In the tertiary sector, wholesale and retail trade services transport and storages are employment providers to rural male workers in the state of Orissa in descending order. To conclude, the recent spurt in rural non-farm employment seems to have occurred due to slack in farm employment. The recent initiatives taken by NABARD for developing the rural non-farm sector are rural entreprenurship development programme, refinance facilities, i.e. investment credit and production credit the provision of Swaraj credit card scheme for rural artisans up to Rs. 25000, creation of awareness among entrepreneurs and infrastructure development.

Small and medium industry is a vibrant source of Industrial dynamism and growth of employment. Khadi and Village industries need focussed attention of public authorities. Efforts may be made to implement the recommendations of national commission for enterprises in the unorganized sector i.e. provision of old age pensions and life insurance for BPL workers (in the post-reforms period). It can be summed up that, the State of Orissa has enjoyed not only a high rate of growth of non-farm employment but also a low rate of growth of farm employment. The rural-non-farm sector seems to have been the sector of last resort and it would continue to remain so in the ensuing years unless there is impressive improvement in employment opportunities in the farm sector.

References

Bagchi, K.K., 2007, Employment and Poverty Alleviation Programmes in India": An Appraisal, Abhijeet Publishers, New Delhi, 2008.

Bhaumik, Sankar Kumar "Growth and Composition of Rural Non-Farm Employment in India in the Era of Globalization", *The Indian Economic Journal*, Volume 55, Oct.-Dec. 2007.

Several Dailies of *Times of India*.

Orissa Economic Journal, 2009, Volume 43.

Economic Survey, Govt. of Orissa, 2007.

Samal Kishore, C., Rural Non-Farm Sector Appropriateness of Equity Hypothesis in Orissa, Orissa Jan.-June-July-Dec. 2009, Volume 43.

Patnaik, B. Eswar Rao, Employment Generation in Orissa: Challenges and Policy Options, *Orissa Eco. Journal*, 2009, Volume 43, Jan.-July and Oct.-Dec. 2009.

Behera, Surendranath "Rural Non-Farm Sector in Orissa; Opportunities and Options, *Orissa Economic Journal*, 2009, Volume 43.

CHAPTER

8

The Role of Non-Farm Sector in Rural Development

A. KURUSWAMY AND E. NEELA PERUMAL

India is one of the fast growing economies in the world. Agriculture happens to be India's major occupation. It provides livelihood to more than 64 per cent of the labour force, contributes nearly 26 per cent of the Gross Domestic Product (GDP) and accounts for about 18 per cent share of the total value of the country's exports.[1] Despite its great importance, agriculture today failed to absorb the rapidly growing rural population due to continuous failure of monsoon in major part of our country. Further, there has been gradual decrease in the productivity from the farm sector. This situation has paved the way for a shift in the preference of the rural mass to non-farm sector to earn their daily bread since agriculture is less productive and that too seasonal in nature. In this study, an attempt has been made to bring out the role of non-farm sector in rural development.

STRUCTURAL CHANGES IN THE INDIAN ECONOMY

The structural changes in the Indian economy are visible in a slow transfer of labour away from agricultural and towards non-agricultural employment in rural areas. One of the four characteristics of Simon Kuznet's,[2] "Modern Economic Growth" relates to a shift of the labour force from the agricultural sector to the non-agricultural sector during the process of economic development. Following this insight, the Indian Five Year Plans have aimed at rapid structural changes in the country's GDP and work force.

SECTORAL SHARE OF GDP

The sectoral share of GDP in India is given in Table 1. It shows the declining nature of share of GDP by agriculture and allied sector.

TABLE I

Sectoral Share of GDP

(in percentage)

Year	*Sectors*			*Total*
	Agriculture and Allied	*Industry*	*Services*	
1950-51	55.4	16.1	28.5	100
1996-97	31.5	24.9	43.6	100
2000-01	23.4	26.2	50.4	100
2001-02	23.2	25.3	51.5	100
2002-03	20.9	26.5	52.6	100
2003-04	21.0	26.2	52.8	100
2004-05	19.2	28.2	52.6	100
2005-06	18.8	28.8	52.4	100
2006-07	18.3	29.3	52.4	100
2007-08	17.8	29.8	52.4	100

Source : Compiled from *RBI monthly Bulletin*, Feb. 2007 and 13 (i-cube) at Planning Commission, New Delhi, Centre for Monitoring Indian Economy (CMIE).

RURAL NON-FARM SECTOR

The importance of the rural non-farm sector is well recognised in modern times. The necessity for expanding the network of non-farm activities for rural development, improvement of employment, productivity and earnings, poverty reduction and reducing inequalities of income and wealth has been gaining significance in terms of policy in the world.[3] Since agriculture is a seasonal activity, their income generating activity also seasonal. Non-agricultural activities in rural areas play a crucial role in providing simple consumer goods and services to the rural household.[4] Agricultural sector alone cannot provide the ultimate solution for rural poverty, unemployment and underemployment.[5] Colin Clark and Simon Kuznet persuasively remind us of the declining share of agriculture in national income and employment. The important point to be noted here is that a developing economy must steadily reduce its dependence on agriculture and expand its non-farm sectors to facilitate the transfer of workforce from agriculture.

NON-FARM SECTOR IN INDIA

The role of non-farm sector is crucial both in generating productive employment and alleviating poverty in rural areas because of the limited absorptive capacity of the urban sector and near saturation of the agricultural sector in further absorption of workers. Policy-makers are increasingly recognising the importance of rural non-farm sector in providing sustainable livelihoods to a large number of people in rural areas. India's Tenth Five Year Plan has placed high expectation on the potential of this sector in generating productive employment.[6]

In India, agriculture and non-farm sector together accounted for close to 70 per cent of all employment in India in 1999-2000. In recent years, the burden of providing additional employment to the growing labour force has fallen upon the unorganised non-farm sector which accounts for only 30 per cent of employment in rural and urban areas combined.

According to Lanjouw and Shariff[7] (1999), in India, not only is the non-farm sector an important source of income of the rural households, but different types of activities appear to be of different relevance to the poor. Such trends and patterns of diversification promote the modernisation of the structure of rural economies.

IMPORTANCE OF RURAL NON-FARM ECONOMIC ACTIVITIES

In India, majority of the people live in rural areas and depend on agricultural activities. However, the share of household income from rural non-farm activities is growing. In the words of Berdegue *et al.* (2000),[8] Rural Non-Farm Employment Activities (RNFEA) are a part of the solution to at least three major problems in rural areas in developing countries, namely, modernization, environment, and alleviation of poverty. However, the three major problems are interrelated. The modernization of farming sector helps to reduce poverty in rual areas. The development of RNFEA helps to modernize the rural environment, through the development of industry and services. Finally, the development of RNFEA helps to solve one of the major problems in rural areas, which is poverty.

Non-farm rural employment has many advantages. Development of Non-farm sector in rural areas enables us to use the unutilized labour force productively in the rural areas, most of which cannot be moved away to urban areas.[9] Non-farm activities also give fillip to other activities including agriculture through their forward and backward linkages, promoting a clear integration of agriculture and industry in the rural economy. Diversification of agriculture not only provides more employment to human labour but also increased productivity of land and labour.[10]

The rural non-farm sector in most developing countries can :

1. Reduce unemployment by absorbing growing labour force;
2. Arrest rural-urban migration;

3. Employ women and provide seasonal employment or residual employment for those left out of agriculture;
4. Pave way for the use of more appropriate technology;
5. Improve household security through diversification;
6. Produce commonly produced goods at lower prices, and
7. Promote a more equitable distribution of income.[11]

The rural non-farm sector alone can reduce pressure on land which is responsible for low agricultural productivity. In the absence of non-farm income opportunities, the rural poor would be even worse-off, and the income distribution even more skewed.[12] Expansion of non-farm activities in rural areas alone can arrest the large scale rural-urban migration. By localizing employment, they can bring down pressure on urban infrasfructure. [13]

There is a growing recognition that non-agricultural activities in rural areas play a crucial role in providing simple consumer goods and services to the rural households. Such activities also provide a humble but critical income to the landless labour.[14]

The development of non-farm sector helps in diversifying the rural economy and reducing the rural-urban distortions. The rural non-farm activities have potentials to reduce income inequalities, mitigate fluctuations in incomes of the poor and arrest rural-urban migration. Allied activities and other business activities are undertaken, since agriculture is a seasonal activity.[15]

MAJOR PROBLEMS IN RURAL NON-FARM SECTOR

While diversification of agriculture is welcome, there may arise the phenomenon of agricultural distress diversification. The agriculturally backward region may depend on some form of non-agricultural activity for their livelihood. There is evidence from many countries that the extent of secondary employment in non-farm work is also extensive and important for small and landless farm families.[16]

Not a few factors have contributed to the poor performance of RNFS. There is stiff competition from the large and medium industries. Other problems relate to product technology, product acceptability and financial viability. One thing is very clear that the non-farm employment should lead to agricultural prosperity. A rise in non-farm employment because of distress diversification does not deliver the goods.

The globalization process may adversely affect RNFS. Even family-based industrial enterprises in a village may face a decline in its economic fortune just because the demand for its product is declining with the availability of cheaper substitutes through imports.[17]

STRENGTHENING RURAL NON-FARM SECTOR

One must be very careful in the selection of non-farm enterprises in rural areas. The management of these units deserves proper attention. As far as possible, rural industries should heavily depend on total resources and meet local needs. Finance has to play an important role in strengthening rural non-farm sector. Banks should think that small loans to poor are no longer a social obligation, but potential business opportunities.[18] The modernization process of rural economy would be incomplete without sustainable growth of non-farm sector in rural India. This modernization process requires huge amount of capital.

Finance is an important input for economic growth. It is, therefore, obvious that institutions like banks, which command huge financial resources, can play a decisive role in shaping the economy of a country by judiciously deploying their funds in such important activities which lead to over all economic growth.

Gandhiji once observed that the solution to India's problems lies in her five lakh villages. But, most of the farmers, who live in the rural areas, suffer a lot because of absence of adequate supply of institutional finance and their indebtedness to the informal credit agencies.[19] A French proverb states that "credit supports the farmers as the hangman's rope supports the hanged". Institutional as well as non-institutional agencies provide credit to the rural poor.

NON-FARM SECTOR IN KANYAKUMARI DISTRICT

Kanyakumari district, named after the Virgin Goddess "Kanyakumari" lies at the southern most tip of Indian peninsula, where three seas, namely, Indian ocean, Bay of Bengal and Arabian Sea confluence, is predominantly an agro-based district in Tamil Nadu. In this district, 60 per cent of the population in the rural areas mainly depend upon agriculture for their livelihood. Most of the farmers, that is, 99 per cent have less than two hectares of land in this district and their income from the land is quite insufficient to maintain their family.

NON-FARM ACTIVITIES

Non-Farm activities are undertaken by the poor people in the rural areas because of the seasonal nature of agricultural employment, failure of monsoons, high cost of cultivation, etc.,

TABLE 2
Occupational Pattern in 2008-09

Sl. No.	*Category*	*Number*	*% of Workers to Total Workers*
1.	Allied Agro-Activity	9871	2.14
2.	Small Farmers/Marginal workers	60,195	13.07
3.	Artisans	59,854	12.99
4.	Cultivators	61,567	13.37
5.	Agricultural labourers	1,77,410	38.52
6.	Household industry manufacturing, processing, servicing and repairs	38,514	8.36
7.	Other works	53,133	11.55
8.	Total workers	4,60,544	100.00
9.	Non-workers	12,15,490	
10.	Total population	16,76,034	

Source : Annual Credit Plan, 2009-10, Kanyakumari District.

and these activities provide additional employment and income for their family throughout the year. Table 2 explains the occupational pattern in Kanyakumari district.

Table 2 indicates that more than 80 per cent of the working population depends on agricultural sector for their livelihood. This sector provides seasonal employment only. Hence, the non-farm activities play a significant role in providing employment and income of the rural people in this district.

FINANCIAL INSTITUTIONS IN KANYAKUMARI DISTRICT

Financial institutions in Kanyakumari district play a very important role in the process of rural development. "Banks are the financial pillars of an economy".

TABLE 3
Financial Institutions in Kanyakumari District

Sl. No.	Banks	No. of Branches			Total
		Urban	S. Urban	Rural	
1.	Public Sector Banks	33	71	13	117
2.	Private Sector Banks	12	15	7	34
3.	Co-operative Banks	7	16	116	139
4.	Other Financial Institutions	1	—	—	1
	Total	53	102	136	291

Source : Annual Credit Plan, Kanyakumari District, 2009-10.

Table 3 shows that co-operative banks concentrated more in rural areas. Hence, co-operative banks are 'rural-friendly'.

NON-FARM ACTIVITIES IN KANYAKUMARI DISTRICT

Kanyakumari district has not registered any appreciable advancement in the industrial field despite its resources like high literacy rate, abundant supply of natural resources, wide network of transport and communication, a good number of financial institutions, etc. The non-farm activities spreaded in all the nine blocks are listed below. In an over-populated agricultural-oriented economy, 70-80 per cent of the population is engaged in agricultural sector, structural changes involve the expansion of the non-agricultural sector, so that, the proportion of population in the agricultural sector is progressively reduced.

TABLE 4

Non-Farm Activity Spread in the District

Sl. No.	Name of the Block	Name of RNF Activities
1.	Agasteeswaram	Artistic pottery, coir/coconut-based products, seashell products, palm products, marine products, bricks, tiles, Xerox, centring, rice mill, bakery.
2.	Rajakkamangalam	Artistic pottery, seashell products, marine products, coir/coconut-based products, printing.
3.	Kurunthencode	Artistic pottery, banana fibre products, seashell products, coir/coconut-based products.
4.	Munchirai	Coir/coconut-based products, marine products.
5.	Melpuram	Herbal products.
6.	Thiruvattar	Rubber products, herbal products, banana products.
7.	Thuckalai	Coir/coconut-based products.
8.	Killiyoor	Hallow bricks, palm products, marine products.
9.	Thovalai	Banana fibre, bricks, coir, tiles

Source : Annual Credit Plan, Kanyakumari District, 2009-10.

Employment

People in Kanyakumari district are more educated than the people in other districts of Tamil Nadu. 9 per cent of the population is literate and has a large number of unemployed population. Agricultural sector alone cannot provide the ultimate solution for rural poverty, unemployment and

TABLE 5
Other Major Sector Activities

Sl. No.	*Major Sector*	*Activities*
1.	Agro-processing	Rice and flour mill, coconut oil extraction, Tamarind products, flower-based, coir, banana powder, palm products, etc.
2.	Food products	Jaggery, coconut, bakery products, masala powder, sweets, palmyra products, pickle-making, etc.
3.	Fish and meat	Fish processing, dry fish processing, meat processing.

underemployment. Employment status is an important factor, which influences the income of the households and thereby determines their standard of living. Self-employment opportunities would contribute to the growth of this district.

The occupational distribution of working population in Kanyakumari district reveals that only 15.44 per cent people derive employment from industry. Nearly 25.73 per cent depend on non-farm sector for their livelihood. Majority of the labour force (3.5 lakhs) depend on the industrial and tertiary sector for their future employment opportunities. Hence, the only meaningful way in which the occupational structure can be changed is to faster the growth of non-farm employment. These structural changes require huge amount of investment.

Table 6 states that, the target and achievement are much more lower in non-farm sector than the other sectors. This shows the declining trend in non-farm activities in this district.

CONCLUSION

Out of the total population, 34.90 per cent live in rural villages in Kanyakumari district. But, the number of cottage industries to the total rural population is only 0.79 per cent and agro-based industries 0.19 per cent which is very much inadequate. Hence, the non-farm sector, which can play a vital role in the district in view of their capacity to generate employment opportunities in the rural sector at a low capital

TABLE 6
Sector-wise Credit Flow

(Amount Rs. in '000)

Sector	*March 2007*		*March 2008*		*March 2009*	
	Target	*Achievement*	*Target*	*Achievement*	*Target*	*Achievement*
(1)	*(2)*	*(3)*	*(4)*	*(5)*	*(6)*	*(7)*
Agriculture and Allied Sector	7606852	8173639	7022401	8574324	8512005	2102193
Non-Farm Sector	707605	582126	893898	427724	962345	378614
Other Primary Sector	1925562	1997512	2523940	1844316	2963995	2835049
Grand Total	10240019	10753277	10440239	10846364	12438345	5315856

cost, needs to be developed. Neglect of non-farm sector will lead to a slack condition in the development of rural villages.

Notes and References

1. India 2001: A Reference Annual, Research, Reference and Training Division, Publication Division, Ministry of Information and Broadcasting, Government of India, New Delhi, p. 370.
2. Kuznet, Simon (1979), "Modern Economic Growth-Rate, Structure and Spread", New Delhi, Oxford and IBH Publishing Company.
3. Samuel, P.S. (1986); "The Asian Experience in Rural Non-Agricultural Development and its Relevance for China", World Bank Staff Discussion Papers, No. 757.
4. Jeemol Unni (1991), "Regional Variations in Rural Non-Agricultural Employment—An Exploratory Analysis", *EPW*, January 19, p. 109.
5. Ravindra Pirabu, R. and Thangamuthu, C. (2003): "Non-Farm Employment among scheduled Caste Households in Tamil Nadu; A Study", *Southern Economist*, Vol. 42, No. 9, September 1, p. 13.
6. Robini Nayyar and Alakh N. Sharma (2005); "Rural Transformation in India : The Role of Non-Farm Sector", Published by Institute for Human Development, New Delhi, p. 9.
7. Lanjouw, P. and Shariff, A. (1999), "Rural Poverty and Non-Farm Employment in India, Evidence from Survey Data", Department of Development Economics, Free University of Amsterdam, Amsterdam, Netherland.
8. Berdegue, J.A., Reardon, T., Escobar, G., and R. Echeverri (2000), "Policies to promote Non-Farm Rural Employment in Latin America, Working Papers No. 75, Overseas Development Institute, London, U.K.
9. T.S. Papola and V.R. Misra (1980); "Some Aspects of Rural Industrialisation", *EPW*, Special Number, October, p. 1735.
10. D.C. Dhawan and A., S. Kahion (1975), "Diversification of Agriculture offers more efficient use of Labour Resources in Punjab", Agricultural Situation in India, October, p. 505.
11. Jean, O. Lanjouw and Peter Lanjouw (1995): "Rural Non-Farm Employment : A study", Policy Research Working Paper, World Bank, Washington, p. 1463.
12. Aswini Saith (1992): "The Rural Non-Farm Economy : Processes and Policies", ILO, Geneva, pp. 27-32.
13. Islam, R. (1987), "Rural Industrialisation and Employment in Asia", ILO-ARTEP, New Delhi, p. 2.

14. Kilby, P. and C. Liedholm (1986): "The Role of Non-Farm Activities in the Rural Economy", invited paper prepared for 8th World Congress of the International Economics Association, December 1-5, New Delhi.
15. T.K. Senthil Kumar and S.M. Chockalingam (2009), "Non-Farming Activities—A Permanent Solution to the Problem of Rural Indebtedness", *Southern Economist*, March 1, 2009.
16. World Bank (1978), "Rural Enterprise and Non-Farm Employment". A World Bank Paper, Washington.
17. Chandha, G.K. (2002); "WTO and Employment for Rural Households—Challenges from within and without; in Reform and Employment", Concept Publishing Company, New Delhi, pp. 349-50.
18. I.S. Sundaram (1984); "Anti-Poverty Rural Development in India", B.R. Publishing Corporation, New Delhi, p. 220.
19. Pazhani, K. and Jesi Isbella, S. (2006): "Long-term Finance and Development", Mohit Publications, New Delhi, p. 31.

CHAPTER

9

Role of Non-Farm Sector in Rural Development

D.R. Bachhav and Dilip Arjune

The Government of India set-up a committee with M. Bhagvati as Chairman to suggest measures to solve the employment problem. The committee suggested schemes like rural electrification, rural housing and Non-firm sector in Rural Development Schemes so the publication of the Bhagwati committee report in 1973. The Government took the measures to Rural Development Programmes.

The problem of rural poverty is old and massive one. A number of experiments have been made by India Government and Planning Commission of India. After Independence to alleviate rural poverty. Now-a-day the rural poverty is the major issue in the official level. Since the beginning of the planning era in India in 1951 a number of development schemes and programmes have been formulated and implemented from time to time by state and Central Governments for Non-farm sector in Rural Development on a experimental basis for the benefit of Rural peoples. At the very beginning of the planning on 2nd October 1952 in 165 blocks of the country was implemented 64 National Extension Services (NES) in 1953.

After Independence Government of India introduce a number of schemes in the practice. The schemes were Intensive Agricultural Development (IADA) in 1966, Intensive Agricultural Area Programme (IAAP) in 1969-70, Drought Prone Area Programme (DPAP) in 1971-72, Hill Area Development Programme (HADP) in 1974-75. Command Area Development Programme (CADP) in 1974-75, Integrated Tribal Development Programme (ITDP—1976-77), Food For Work Programme (FFWP—1977), High Yielding Variety Programme (HYVP), Small Farmers Development Agency (SFDA), Marginal Farmers and Agricultural Laboures (MFAL), Minimum Need Programme (MNP), National Rural Employment Programme (NREP), Rural Landless Employment Gurantee Programme (RLEGP), Training of Rural Youth for Self-Employment (TRYSEM), Sampoorna Grameen Rozgar Yojana (SGRY), Rural Water Supply Programme, Pradhan Mantri Gram Sadak Yojana (PMGSY), Rural Housing, Sawarna Jayanti Gram Sadak Yojana (SGSY), Provision of Urban Amenities in Rural Areas (PURA), Watershed Development Programme (WDP), Hariyali Programme and Rural Water Supply Programme, Jawahar Rozgar Yojana (JRY) and Employment Guarantee Scheme of Maharashra (EGS), all above programmes were formulated by the Government as a measure to attack rural poverty at its very root level, these programmes were benefited the people who are economically weaker section of the society but the Government of India being very careful about introduced a approach called Integrated Rural Development Programme (IRDP) in a few selected blocks at that time. It was objective of twenty point programme. IRDPs major objectives was poverty alleviation in the field of rural development and to help in identifying the rural poor families to enable it to cross the poverty line, the objective is achieved by providing productive assets and inputs to the target group. The assets which could be in primary and secondary sectors are provided through financial assistance in the form of subsidy by the Government and term credit advanced by financial Institutions to upliftment of poverty line from rural peoples.

CONCLUSION

All above several Schemes which aims at generating employment and income of below the poverty lice peoples. The impact of the non-firm sector in Rural Development Programme in terms of retention of assets, continuing with some repayment and no overdue, significant number (More than 40%) beneficiaries utilize above Rural Development could be a better explained with the significant contribution made by the poorest among the poor even among SC/STs and open categories in the country. In addition, the schemes like bullocks, bullock-cart, dairying, other animal husbandry and trading generated higher income in general in all over the India. But the proper conclusion is that emerges from the above experience show that Land-linked schemes had better performance in the various categories of beneficiaries. That means Land seems to be the primary requisite for the truthful result of the anti-poverty programmes are wed with the implementation of land reforms perhaps the upliftment of poverty scenario could be better.

REFERENCES

Gram Vikas Programmes at a Glance, Ministry of Rural Development.

Indian Economy, Ruddar Datt and K.P.M. Sundharam, p. 400.

'Kurukshetra', pp. 22-23, and 28, *Journal of Ministry of Rural Development.*

CHAPTER

10

Role of Processed Food Sector in Rural Development

Deepshikha Singhal and Ashish Kumar

INTRODUCTION

India is one of the strongest and fastest-growing economies in the South-East Asian region, and has become an important export-oriented destination. For India, growth is an imperative to be counted as a major economic powerhouse by the end of this century's first quarter. India needs to accelerate its economic growth beyond the existing rates of 7 percent per annum. India has transformed itself from a manufacturing economy to a food and agriculture-based economy with one of the highest GDP in the South-East Asian region. The food and agricultural contribution in the GDP is 17% and is anticipated to increase manifolds. Agricultural and Processed Food Products Export Development Authority (APEDA) is a nodal and governing body working under the Ministry of Commerce and is responsible for implementing and enforcing all the agricultural and processed food exports from India to numerous destinations. The export basket consists of wide range of products containing fresh fruits and vegetables, poultry, dairy and processed products to name a few.

There has been a structural change in the composition of food exports from India over the past couple of decades. Traditional exports have continuously declined and have been replaced by agricultural and processed food exports. Expansion of agriculture food production and exports has become an important trend for India since such exports have intrinsic characteristics of manufactured goods. The structural change in food exports is believed to generate superior growth performance. Reasons for this belief include knowledge spillovers, and terms of trade gains. In particular, food production, especially in the final stage, appears to be labor-intensive.

Agro-Food exports from India and other developing countries are expected to continuously expand in the future since world demand for agricultural and processed food continues to grow in response to diet upgrades resulting from rising incomes, growing health consciousness, and urbanization. Factors such as international migration, communication revolutions, and international tourism and trade liberalization initiatives under various rounds of world trade negotiations also contribute to an expansion of demand for this sector. Thus, ability to expand agriculture and processed food production and exports into the world market would provide an important impetus for their economic development.

The outward orientation of the Indian economy after the commencement of liberalization and globalization started after the introduction of reforms export sector have helped in generating employment and raising income levels of the people. However, India is a nation of more than one billion people and the policy of inclusive growth also calls for exploiting food and agriculture industry competitiveness in the global market. This needs intervention at the level of research and development and appropriate policy-making in this sector.

This paper illustrates the key facts, challenges and outcomes of the Indian export portfolio. A lot more opportunities and key recommendations are rolled out with the commencement of different phases in the paper. Also an attempt has been made to devise strategies for further agro-exports growth in this sector. India's exports should grow exponentially in the years to come to achieve the goal of inclusive growth and long-term development.

SECTORAL ANALYSIS OF THE AGRO-PROCESSED FOOD INDUSTRY

Background

Processed food products are penetrating the large potential in a rapid motion presented by Indian population. The demand has been rising at accelerating pace and there is enough latent market potential waiting to be exploited through developmental efforts. This demand for these foods has captured a large amount of the food retail market in India. Major factors driving the change are changing lifestyle, eating and cooking habits, increasing young population.

The policy of liberalization and globalization has been the primary catalyst for the growth of this section in Indian food which is set to witness many more changes with newer offerings being available. The economic reforms in this sector have provided impetus to the demand (domestic as well as foreign) for such products. The changes in the lifestyles of the people owing to increased level of income have been another catalyst for the demand for the agro-processed food. Opening up of retail chains by various players is futher pushing up the demand for such products. The international markets have shown a positive response to Indian cuisine with Non-Resident Indian (NRI) communities and foreigners expecting to grow this market in the years to come. Another observation is that consumers are on the lookout for processed foods that are offered to them in hygienic, nutritional and attractive packaging at an affordable cost. Great care is needed during their packaging, so that they remain prevented from impurities. Among the emerging trends is that the consumer is looking for innovative flavors and new tastes, which is a big challenge for the food industry.

The advantages that the processed foods offers to the domestic and international markets has resulted into the increase in the export of processed food products from Rs. 6674 crores in the year 2006-07 to Rs. 9065 crores in the year 2007-08.

From the export perspective the sector can be branched in to two major categories. Firstly the processed fruit and vegetables and secondary the other processed foods including Alcoholic beverages, Aerated drinks, Snacks and Confectionery, etc.

Processed foods are evolving as a vast upcoming sector in the domestic market and also for the extensive market overseas. The sector has shown a linear growth in the past years and with the new efforts this is presumes to be expanding exponentially. Indian export market on the basis of value generated has gained a percentage change of approx 34% in 2007-08 from its previous year (2006-07) which was approx 21% when compared to 2005-06. If these growth rates are valued in terms of quantity then there is a percentage increase of approx 96% in 2007-08 from its previous year (2006-07) which was approx 92% when compared to 2005-06.

The major categories in this sector which are at the apex growth are confectioneries. Majority of categories have shown a vast growth in 2006-07 compared to 2005-06 but the same growth has not been achieved in 2007-08. The same can be inferred from the exhibit.

The decline can be because of different reasons but the concentration should be on overcoming them. There should be a holistic approach which will prune the laggard in an effective manner. In the same line of thought a framework is composed to increase the export rate of this sector manifold. The framework consist of the following extensions.

Potato Production and Export Potential with Special Reference to AEZ Agra

India is one of the largest producers of potato production in the world. It accounts about 25 percent of all the vegetables. Potato being a nutritious product provides food security. It is the world's third most important food crop after wheat and rice. Its production is around 300 Mn MT globally per annum.

EXHIBIT I

	2005-06		% Change		2006-07		% Change		2007-08	
	Qty.	*Value*	*Qty.*	*Value*	*Qty.*	*Value*	*Qty.*	*Value*	*Qty.*	*Value*
(1)	(2)	(3)	(4)	(5)	(6)	(7)	(8)	(9)	(10)	(11)
Dried and Preserved Vegetables	133787.98	39699.87	-10.85	7.69	119270.43	42754.17	5.41	0.56	125726.28	42993.81
Mango Pulp	134613.2	36424.12	16.51	38.87	156835.52	50582.79	6.32	0.76	166752.17	50968.51
Pickles and Chutineys	135381.85	26098.14	7.26	12.50	145216	29359.48	-12.16	-14.64	127558.47	25061.71
Other Processed Fruits and Vegetables	137146.82	47991.97	26.03	37.92	172851.58	66191.34	6.56	7.60	184197.82	71219.94
Confectionery	112644.06	26497.09	304.76	229.64	455935.85	87346.11	428.17	221.79	2408129.7	281068.2
Alcoholic Beverages	66051.95	20901.64	-23.13	6.18	50771.79	22194.22	17.42	52.55	59614.33	33856.2

Source : Technopak Analysis.

EXHIBIT A

Proposed Framework

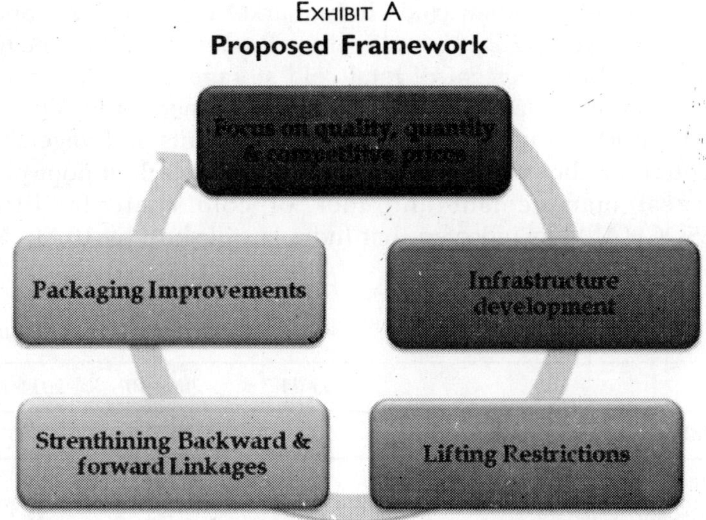

India account for about 0.5 percent of total potato exports in the world. The major importers are Pak, SL, UAE, Mauritius, Nepal, Singapore and Kuwait. The major exporters of potato are Netherland, France and Germany. Uttar Pradesh accounts for about 30 percent of total potato production and about 51 percent of cold storage capacity of India. Realizing this potential the GOI has declared Agra region as Agri-export zone at Agra.

Share of States in Potato Production

Sl. No.	*Name of State*	*Percent*
1.	Uttar Pradesh	30
2.	Bihar	15
3.	West Bengal	20
4.	Punjab	07
5.	Gujarat	04
6.	Madhya Pradesh	03
7.	Rest of India	21
	All India	100

At present Indian cold chain market is worth $ 2.6 Bn. It is expected to $ 12 Bn by 2015. Uttar Pradesh and West Bengal have around 65 percent of total cold storage installed capacity of the country. National Commission of farmers in its Vth and final report has reported that wastage of fruits and vegetables account for about Rs. 500 Bn annually due to lack of poor post-harvest management and lack of cold chain facilities. ASSOCHAM has reported that India is still short by 10 Mn MT of cold storage capacity.

(Market in USD billion)

	2003-04	*2009-10*	*2014-15*
Total	102	182	300
Select Sectors			
Fruits and Vegetables	1.1	6.4	12.2
Dairy	25.8	55.6	97.8
Flour milling	11.6	14.4	16.9
Breads	0.8	2.1	4.4
Biscuits	1.5	4.0	8.7
Snack and ready to eat foods	0.6	3.3	10.8
Pasta-based foods	0.2	0.8	2.3
Marine Products	1.6	3.8	8.4

The Competitive Edge

It's an established fact that the processing level in India in comparison from other developing countries is very minute. Within the country, there are wide variations in productivity levels. Punjab, Haryana, Andhra Pradesh, Tamil Nadu and Kerala may have attained productivity levels of a world standard. But other regions are way behind. The production potential reveals that the processing can be a vital option to bring the exponential growth in reality. India ranks first in global fruit production and second in case of vegetables. The ranking on processing level need not to be commented. There are various interventions initiated by the government to increase the level of processing but the adoption level is at nascent stage.

On ground it has been observed that there is a shift in cropping pattern in favor of horticulture in India in the past couple of years as it supports food processing. Analysis of the economic and processing feasibility of this shift away from cereals to fruits and vegetable shows that it's economically viable and beneficial to shift towards horticulture production, but this diversification needs to be planned in a systematic manner in the same line of value addition and sustainable processing. The Indian food-processing industry is primarily export oriented. India's geographical situation gives it the unique advantage of connectivity to Europe, the Middle-East, Japan, Singapore, Thailand, Malaysia and Korea. One such example indicating India's location advantage is the value of trade in agriculture and processed food between India and Gulf region

Certain strategies and policies are also adopted by the government in this regard. The main steps adopted by the government to boost Indian food processing at global prospect are :

- Food processing industries have been listed as one of priority sectors for bank lending.
- Zero excise duty on fruit and vegetable processing units
- Foreign equity up to 100% is permitted for most of the processed food items
- Zero excise duty on items like fruits and vegetables products, condensed milk, ice cream, meat production
- The excise duty on ready to eat packaged foods and instant food mixes has been brought down to 8% from 16%
- Excise duty on aerated drinks has been brought down to 16% from 24%
- India is globally competitive and has comparative advantage in production. These major production states should be targeted for enhancing the export potential of the country. Lessons from other developing countries focus on the growth of the processing industry through increased participation

of small and marginal entrepreneurs in a potent manner to facilitate a sustainable up-gradation for the industry.

Industry Interface

The food processing industry is emerging in India at a fast pace. At present it is ranked fifth among different industries of India. Nearly, one-third of the entire Indian food market share comprises of processed food. With policy measures from the government, the food processing Industry accounts for 13% of the country's exports. The food processing industry is presently growing at 14% against 6-7% growth of last couple of years. The food processing industry received foreign direct investments (FDI) a sum of Rs. 7190 crores in 2007-08 against Rs. 285 crore in the previous fiscal. The cumulative FDI received by the industry from April 2000-January 2009 stood at Rs. 38016 crores.

However, India's share in exports of processed food in global trade is only 1.5%; whereas the size of the global processed-food market is estimated at Rs. 160 trillion and nearly 80% of agricultural products in the developed countries get processed and packaged.

The cheap raw material availability and efficient labor connection is attracting plenty of global players in Indian food processing industry. Other than Indian products these processors are manufacturing various other international foods. For these investors India is an efficient food processing base for their products demanded worldwide. Some of the new investor, landed to India for food processing interventions covers Del Monte, Danon and other majors.

It is estimated that by 2012, India's processed food output is likely to grow by 45% to touch Rs. 4,50,500 crores, while packaged food sales will increase by 65% to reach Rs. 1,08,500 Crores. On a per capita year basis, per capita packaged food spending is expected to grow by 55% to rupees ninety thousand by 2012.

Besides foreign investors, food processing sector is a major attraction for Indian corporate houses to invest. Reliance, Godrej, Bharti, ITC, DSCL, Tata and Mahindra and Mahindra are prominent corporate houses with end-to-end integrated operations in the food processing. The major destinations for Indian processed food are U.S.A., U.A.E., Iran, U.K. and Indonesia.

RECOMMENDING THE WAY FORWARD

Despite abundant production and increasing global demand, India contributes to about 2% in the total agro export. The major challenges plaguing the agro export remain low and inconsistent product quality, poor packaging, poor infrastructure support, constraints in investment in agriculture, lack of synergy among various stakeholders, and inadequate market promotion. Hence, there is an urgent need to build on India's competitiveness and identify critical success factors.

Infrastructure Development

A major impediment in promoting exports is the lack of adequate infrastructure, particularly cold storage facilities and transportation, road networks, freight and cargo facilities. There is a need to encourage public-private partnership in building such facilities and ensuring their proper maintenance. There is no dearth of financial assistance as there are several incentives being provided by Government of India under its capital investment subsidy scheme as well as those available under the schemes envisaged by APEDA. Concerted efforts need to be made in this direction in collaboration with commercial banks.

Marketing Strategy

In the current scenario where all the quantitative restrictions have now been removed and there is an increased opportunity for the developing countries to have access to global markets, it is imperative that a marketing strategy is worked out, focusing on major items of import by countries and to concentrate on such products using the comparative advantage. The countries in the European Union, African countries and the CIS countries need to be given greater attention. Proper marketing strategy needs to be geared up to meet the raw material requirement of processing units and ensure a sustainable export market for the processed products.

Contract Farming

Contract farming needs to be encouraged not only to provide a broad base for raw materials for processing but also for the supply of the right type of inputs and other linkages necessary for the acceptability of the quality standards for competitive exports.

Human Resources Development

There is also a vital need for human resources development and to train the exporters about the quality standards and the sanitary and phytosanitary measures that need to be complied with. More "Cluster-based" training programs for the organized and unorganized sector need to be organized to upgrade their products and process to comply with the stipulated quality/hygiene standard.

Market Access and Information

There is a need to provide continuous updating of data on market information, market access, procedures and processed, etc.

Biotechnology

India has been recognized as one of the five top biotechnology leaders in the Asia Pacific region. In terms of number of patents filing, India ranks third in Asia. Biotechnology, leads to reduction in cost and improvement in productivity. Given the low-cost but high caliber work force, there is a need for optimal utilization of intellectual and biological resources with a view to bring cost effectiveness in production.

Trade Related Intellectual Property Rights

India should launch genetic and legal literacy movements immediately to sensitize panchayats and rural families on the implications of the protection of plant verities and Farmers Rights Act 2001 and Biodiversity Act 2002, since they contain provisions for recognizing and rewarding the contributions of the primary conservers of biodiversity and holders of traditional knowledge.

Credit Facilities

The EXIM Bank, in consultation with APEDA and the Ministry of Agriculture, may set-up Farm Export Promotion Cells in each AEZ and provide necessary technical support and guidance to the exporters. It can also open offices in each state in order to promote agri-export and also establish overseas branches in countries where Indian exports are favorite destinations.

Economies of Scale

Economies of scale and brand-banding can only happen when large and big companies enter the sector. In this respect, contract farming and corporate farming should be extended credit facilities with liberal terms and making storage, movement, processing, marketing and trade of farm commodities free from regulations and controls. It is necessary to streamline the procedure for export financing of agricultural products which are perishable in nature and making it entrepreneur-friendly. Likewise, since perishable products are traded on commodity basis and export is generally on consignment sale basis, the procedure for obtaining export credit guarantee cover should be streamlined and made exporter-friendly and in this respect a comprehensive insurance cover right from the stage of production to export can also be considered.

Policy Interventions

The following policy options should receive our attention at the earliest with a view to prepare ourselves in meeting the full impact of WTO:

- Increased investment in agriculture
- Increased flow of institutional credit to facilitate agricultural exports
- Diversification from cereals to high-value crops such as fruits and vegetables, floriculture, spices, animal husbandry, fisheries, medicinal and herbal crops etc.
- Promoting and encouraging public-private partnership to facilitate investment in infrastructure such as in irrigation, agriculture research, electricity,

roads, rural markets, cold storage and transportation etc in an endeavour to reduce transportations costs

- Organizing farmers into associations that would jointly produce and process commodities for international markets at both the regional and global levels including formation of, and motivation to, SHGs for cultivation, processing, marketing, nurseries, seeds production etc. and linking such initiatives through contract farming and corporate farming
- Increased investment on developing viable and cost effective seeds industry
- Developing institutions and providing support to them for the vertical integration of production, processing, packaging and marketing of agricultural produce with public-private partnership
- Policy framework for the contract and corporate farming should be streamlined
- Improving sanitary and phytosanitary measures as well as the adoption of Codex Alimentations standards of food safety and simultaneously evolving SPS standards for out domestic products as well as imports including strengthening the capacity of the state government institutions for educating the farmers with regard to SPS requirements
- More advanced and convenient packaging should be introduced. Internationally there is an increasing emphasis towards eco- friendly packaging using biodegradable materials. Efforts need to be made for reduction of excise duty on packaging material
- Government should consider modifying and rationalizing the taxation framework for the agro export sector, scrap excise duty on packaging material for basic foods and reduce custom duty on food processing and packaging machinery.

Therefore, there is a need to sensitize the stakeholders to work in collaboration and to deliberate on the issues and challenges, in order to offer wider export opportunities and necessary infrastructure support to Indian exporters.

CONCLUSION

India with enviable wealth of resources of fruits and vegetables, dairy, marine and agricultural products is set to emerge as a global leader in the quality processed food business, given of course the right direction and inputs. The government too has recognized this potential and is pro-actively prioritizing the food processing industry for growth. The success of this industry, besides other factors, depends largely on packaging and processing technology. Global marketing of our food products will to a large extent depend on our ability to deliver these products in forms that will not only ensure freshness of the edibles but also meet consumer convenience and growing environmental concerns.

CHAPTER

11

Role of Rural Non-Farm Sector in Rural Development : Problems and Prospects

MOHAN C. PATEL AND SAJENA M. PILLAI

INTRODUCTION

India's quest for equality and social justice had a long journey. It commenced with the First Five Year Plan in 1951 and ended as a faded dream in 1991. When India celebrated her golden jubilee of freedom, the richest 10 per cent of her people were consuming ten times more than the poorest 10 per cent (33.5 per cent of national income compared to 3.5 per cent of the latter) and the richest 20 per cent five and a half times more than the poorest 20 per cent (46.1 per cent of the national income against 8.1 per cent of the latter). Even sixty years after independence, agriculture still remains the main source of livelihood of rural population, 70 per cent of the population still depends on agriculture (51.7 per cent are self-employed in agriculture and allied activities and 25.8 per cent are agriculture laborers). Non-agriculture provides livelihood to 39.77 per cent (11.6 per cent are non-agricultural laborers and 15.8 are self employed in non-agriculture, However, there are inter-state variations in the distribution of population across occupation groups. Self-employment in agriculture and allied activities are dominant occupation in all the states. In the states of Andhra Pradesh, Himachal Pradesh, Kerala, Tamilnadu, and West Bengal and North-eastern region less than 40 per cent of rural population depends on agriculture and activities as their principal source of income. In the states of Andhra Pradesh and Tamilnadu difference between self-employment in agriculture and allied activities and agricultural wages as principal source of income is less as compared to other states. In the states of Himachal Pradesh and Rajasthan reliance on agriculture wages as principal source of income is quite small (0.66 per cent and 1.06 per cent respectively). Share of self-employed in non-agriculture, as principal source of income is high among the states of Bihar, West Bengal and North-eastern region. Salaries as principal source of income are quite significant in the states of Haryana (21.93 per cent), Himachal Pradesh (25.56 per cent), Punjab (20.69 per cent), Uttar Pradesh (18.98 per cent) and North-eastern region (22.53 per cent). States of Himachal Pradesh (9.62 per cent) and Kerala (8.90 per cent) have sizeable proportion of population getting principal source of income from other sources.

The issues concerning rural development are largely centered on the unfair income, opportunities and access of its populace. These inequities assume accentuated proportions when compared with urban segments. There is fundamental structural differentiation between rural and urban segments in terms of respective factors of production due to the distinct characteristic of rural economies. On account of relatively much intense and basic relationship with natural endowments, the rural economies are generally oriented to production of primary goods. There is a fair generalization in stating that aggregated income accrual to the rural households from production of such primary goods is higher than the urban households. The rural sectors, in turn are net suppliers of primary produce and generally, the net consumers of secondary and tertiary goods and services. The demographics, human and natural resource endowments and their linkages lead to varying discrepancies of the dichotomy of economic activities and income generation of people and the resultant inter and intra-regional differentiations in livelihood and well-being.

DEMOGRAPHIC INDICATORS OF RURAL POPULATION

TABLE I
Percentage Distribution of Households by Household Type (Rural)

Household Type	*% households*
1. Self-employed in Agriculture	35.9
2. Self-employed in Non-Agriculture	15.8
3. Total self-employed (1+2)	51.7
4. Agricultural labour	25.8
5. Other labour	10.9
6. Total Rural labour (4+5)	36.7
7. Others	11.6
8. All	100.0
9. Agricultural Households (1+4)	61.7

Source : NSS 61st Round Report no. 515

0.437 0.475 0.535 0.250 0.263 0.297

RURAL WORKFORCE COMPOSITION OVER TIME

In the wake of globalization, the real task before the economy was to usher the people centered development process that is equitable and sustainable. The process of development involves not only the locational shift of the workforce but also the occupational diversification. Given the sluggish urban employment growth and binding land constraints rural occupational diversification seems to be the only escape route to relative prosperity. Over the time the changes in the rural occupational diversification has been slow and unsteady. The different rounds of NSS survey show the signs of rural occupational diversification.

Percent Distribution of Usually Employed Persons by Broad Industry Division

Broad Industry Division	*Male*			*Female*		
	50th Rd. (1993-94)	*55th Rd. (1999-2000)*	*61st Rd. (2004-05)*	*50th Rd. (1993-94)*	*55th Rd. (1999-2000)*	*61st Rd. (2004-05)*
Agriculture	74.1	71.4	66.5	86.2	85.4	83.3
Mining and Quarrying	0.6	0.6	0.7	0.4	0.3	0.3
Manufacturing	7.0	7.3	7.9	7.0	7.6	8.4
Electricity, Water, etc.	0.3	0.2	0.2			
Construction	3.2	4.5	6.8	0.9	1.1	1.5
Trade, Hotel and Restaurant	5.5	5.8	8.3	2.1	2.0	2.5
Transport, Storage and Communication	2.2	3.2	3.8	0.1	0.1	0.2
Other Services	7.0	6.1	5.9	3.4	3.7	3.9
All	100	100	100	100	100	100

NEED FOR THE RURAL DIVERSIFICATION

The demographic pressure and socio-economic inequalities in rural domains of developing countries further complexes the relationship between humans and endowment. For instance, about 30% of world population is in the developing countries of South and South-East Asia with less than 7% world landmass. As derived from FAO Statistics (FAO, 2005) this region has almost 40% of world's agricultural dependent population with less than 20% global arable land resources. With such uneven distribution of production assets, low levels of literacy, skills, awareness and connectivity and limitations of alternative options for livelihood, the high incidence of poverty in these regions becomes the structural outcome.

Against this background, the scope of increasing real income of farmers and bringing sustained improvement in their well-being, solely through farming operations, is seriously constrained. There is concern on the incidence of deep rooting of poverty amongst the households depending on single income from farm activities (UN-Wye Group, 2007). The rural economies in developed countries are relatively more diversified and majority of their rural households have larger share of non-farm income accrual. The empirical evidences of change in rural economic activity composition of developing economies are being documented (FAO-RIGA 2007) and more rural families are earning from non-farm work, the process is slow in several regions due to limited skills and opportunities. The earnings from agriculture continue to be a fundamental source of livelihood for 90 percent of rural households, particularly the poor.

While emphasizing the catalytic role of accelerated agricultural growth for development and overall economic growth, the planning processes have also viewed (World Bank, 2008, Planning Commission of India, 2007) that such agricultural growth may not to be the source of increasing direct employment and earning per head. Considering the negligible employment elasticity to agricultural growth, creation of non-agricultural opportunities, diversification of rural economy and expansion of Rural Non-Farm Employment are addition to the strategies of managing vulnerabilities associated to the farm sector and bringing meaningful structural change in the rural socio-economic conditions.

Any strategy towards the development and improvement of well-being of population therefore needs to take into account these fundamental issues in relation to agrarian structures. The scientific understanding of these differentiations becomes a prerequisite for evolving and implementing development agenda. For sustainable improvement in the rural livelihood, particularly in developing economies, studies on various aspects of the rural economic diversification are the contemporary policy requirements. Since this subject takes into account a wider perspective of economic activities in rural domain, the data profile required to examine its varying dimensions is also expected to be much larger and complex. Often the statistics and indicators are not available from single source in desired format and confirming to the conceptual requirements. It necessitates the mining of rural development statistics for deriving relevant indicators needed for synthesizing rural economy, rural household livelihood and their well-being.

There is general acknowledgement that not only the economic condition of rural household improves with the blending of non-farm economic activities with farm activities; it has positive impact on efficiency of their farm enterprises. It integrates with the multi-pronged strategy in the framework of action against poverty, stimulating enhancement of entitlement and access. The opportunities, empowerment and security are the three factors that have complimentary and supplementary role in neutralization of economic deprivation. These three factors are also closely associated with the process of economic diversification. If the opportunity of doing multiple activities enhances returns and exposure and thereby empowers the economic and social well-being, the empowerment through literacy, skill, knowledge, awareness, resources and connectivity improves the capacity and scope of harnessing the opportunities. The resultant derivatives are augmented remuneration and returns from diverse sources, contributing to stability of economic condition, security, reduction in vulnerability and risk mitigation. Therefore, studies on different dimensions of diversification of rural economy, improvement in the measurement, factorization and impact and exploration of its indicators are needed for furthering rural livelihood development and well-being.

The experience of economic growth of the countries observes that "high agricultural growth is generally accompanied by the growth in the rural non-farm sector as it has forward and backward linkages with agriculture.

Non-agriculture sector is principal source of income of nearly 40 per cent of rural population. Expansion of non-agriculture activities in the rural areas will reduce rural poverty. However, infrastructure constraints and poor endowment of human capital (level of literacy and skill) are main hindrance in expansion of non-agricultural activities in the rural areas. Available NSS data on rural labour workforce shows that casualisation of labour force is increasing. This increasing trend of casualisation of labour force is a serious threat to the eradication of poverty

RURAL DIVERSIFICATION (VARIOUS OPPORTUNITIES)

Crop Diversification

One of the basic forms of rural economic diversification is the crop diversification. The diversified cropping pattern in a region emerges due to allocation of arable land resources for cultivation of number of alternative crops. The Indian agrarian space is endowed with diversity of agro-climatic conditions and varying degree of augmentation of farming resource through irrigation infrastructure, crop specific farming technologies, diversified demand and post-harvest linkages. Such as farm diversification helps in reducing farming risk due to climatic, market and other such aberrations and often improves resource use efficiency. However, the crop diversification has been subjected to resource endowment of farmers in terms of land, water, technology, seeds and soil besides externalities such as agro climatic conditions, sustainability and the response to market. The skewed distribution of infrastructure such as road, transportation, market, post-harvest handling, irrigation and power are found to be the impediments for both horizontal and vertical diversification. Nevertheless, the crop diversification not only indicates the options and opportunities of cropping, it also harmonizes the supply to demand of diverse commodities and in the process diffuses the price volatility in the market.

Inter-state analysis of indicators of depth and intensity of poverty shows that, these goes hand to hand with incidence of poverty. Any policy formulated towards reducing incidence of poverty will reduce depth and intensity of poverty also.

TAPPING THE UNUTILIZED IRRIGATION RESOURCES

Secondly, tapping the unutilized irrigation resources and adoption of certified/quality seeds and high yielding variety technology can reduce rural poverty. A rise in agricultural output will have impact on rural poverty in two ways, first, directly reducing rural poverty through rise in income levels of cultivators and agricultural labors.

INCREASING DEMAND FOR NON-AGRICULTURAL PRODUCTS

By increasing demand for non-agricultural products which will lead to growth of non-agriculture employment in rural areas. Government's role in provision of irrigation facilities is important, as a rise in Government investment in farm sector will attract private investment in farm sector. Provision and better utilisation of economic and social infrastructure in rural areas will act as a catalyst to growth in rural areas and will have impact on rural poverty alleviation.

ANIMAL HUSBANDRY, POULTRY AND FISHERIES

An extension of the same to more meaningful form of farm sector diversification is through animal husbandry, poultry and fisheries and its measurement in terms of value of outputs. It has been widely acknowledged that in semi-arid central and western India having lesser scope of multiple cropping, animal husbandry reduces the vulnerability of farmers. In the regions where forward integration of small cattle holders has been strengthened by institutions such as cooperatives, the economic conditions of farmers have improved. The cooperatives, self help groups and other institutions of marketing etc. have stimulated the process of on-farm and off-farm diversification by putting the opportunity, empowerment and security as the rural development package. The extension of farming activities to certain on-farm post-harvest operations not only adds to the farm gate value creation, it also expands the production entrepreneurship of the farmers to services.

EXISTENCE OF ENTERPRISES IN THE RURAL AREAS

From the point of view of diversification of economy in

the production boundary, one may also look into the existence of enterprises in the rural areas and producing non-agricultural goods and services. In India, there is a significant presence of small and tiny non-agricultural enterprises in the rural areas. There is preponderance of informal and unorganized enterprises in the rural economy, both in terms of their number as well as workforce. Out of total own account enterprises (without hired workers), 11.1 million (92%) non-agricultural manufacturing and 9 million (91%) of service sector (excluding domestic trading) enterprises are located in rural sector (NSS 62nd and 63rd Round, reference period 2005-06 and 2006-07 respectively). However, in terms of GDP, these rural enterprises have much smaller share.

The rural non-agricultural entrepreneurial diversification may not be simply assessable in terms of their number and GDP share. There are aspects of economy of scale, operating efficiency and technology used in the corresponding large enterprises located in industrial hubs, which are not easily measurable but impinge on efficacy of rural non-farm diversification.

The diversification through crops and on and off-farm production offers limited perspective of rural economic diversification. It is confined to the production boundary of agriculture and allied sector and producing entrepreneurial units of the farms. Without undermining the significance of such diversification, that eventually strengthens integration of farming with post-farming and off-farming activities, the economic gain to its stakeholders will be restricted to the growth potential of farm sector. For the rural economy to sustain in the long-run, the scope of its diversification would necessitate expansion to the wider dimensions of livelihood diversification.

LIVELIHOOD DIVERSIFICATION

The exposure of livelihood in rural agrarian segments of developing countries has been acknowledged and the livelihood security is one of the central theme needing attention in the liberalized and market reformed agricultural trade regime. The rural livelihood diversification therefore is an integral dimension of development agenda for strengthening rural livelihood and sustaining livelihood security.

There are two ways to look into livelihood diversification. One, the individuals and/or their groups perform different activities. In other words, the individuals are capable to engage in the alternative choices in the labour market and undertake different forms of rural employment; both farm as well as non-farm. From the point of view of rural development, the rural employment diversification is considered to be driving force (UN-Wye Group, 2007). Two, the rural income diversification enabling individuals or households to have income sourced from the diversified sources. There is differentiation in employment diversification and income diversification as both are broadly complementary but may not necessarily be synonymous. The employment diversification is measured in terms of labor force participation in diverse industries and occupation. The wages and remunerations from different employment would add up to income. However, the income diversification is more comprehensive, since it would also account for transfer payments (rents, interests, dividends, etc.) to individuals.

As stated above, the crop and farm diversification have potential to augment income and strengthen livelihood. But due to its confinement to labour participation in the farm related activity, it remains diversified in the limited sense. Further, the domain of crop and on farm diversification is the production boundary of primary goods, hence the stability and security of livelihood remains vulnerable despite such diversification. The domain of rural livelihood may extend beyond the rural production boundary. The commutation of rural people to urban neighborhood for their work and jobs as well as income transfers from urban to rural add to the wider dimensions of livelihood diversification. Following sections explore on measurement issue concerning rural livelihood diversification with specific reference to India.

POLICY IMPLICATIONS

Thus, the importance of the rural non-farm sector in generating productive employment and reducing poverty in rural areas, reducing rural-urban differences in income levels, bringing in greater integration between rural and urban economies generally contributing to the rural development. In this context, there are various policies and programmes and schemes that have relevance for the rural non-farm sectors. They are very few but are not found to be very effective. Therefore, the policy measures for the rural non-farm sector require a major reorientation in order to meet the new challenges.

POLICY FOR SMALL-SCALE INDUSTRIES

Most of the rural enterprises are in the form of small scale industries and cottage industries, a separate policy is required for their development. For this the subsidies and protection policy should be once again be maintained. There should be proper information on avalibility of markets, information technology and skills.

INFRASTRUCTURE

The infrastructural gaps at the regional level should be reduced. Appropriate infrastructure should be developed for the promotion of rural non-farm sector. Provision of basic physical infrastructure like roads, power, communication in rural areas must be given high priority. Private sector should be particularly involved in building industry and product specific infrastructure.

FINANCIAL SERVICES

There is a need to work out an integrated strategy to finance the rural non-farm sector. A special Venture Capital Fund should be set-up to meet the diverse needs of the non-farm sector such a technology upgradation, quality standardization, market support including development brands, etc.

RURAL-URBAN LINKAGE

Rural-urban Linkages should be developed to enable the growth of the rural non-farm sector. Such linkages should be developed to enable rural manufacturing units to obtain critical raw materials and access government facilities such as subsidized credit, marketing support, facilities for training, etc. With the strengthening of these linkages rural producers can easily go to nearby urban centres and establish channels for the flow of commodities and information.

TECHNOLOGY

Technology is a crucial factor in increasing the productivity of rural non-farm sectors. The use of both incremental and leap-frogging technology is required for the growth of rural non-farm sectors.

SKILL DEVELOPMENT AND TRAINING

The creation and upgradation of skills for workers and enterprises have become extremely important in the competitive environment. Training for rural labors should be organized on a decentralized basis, based on sector or area specific pririties and coordinated by local institutions. The ITIS and private technical institutions should be reframed to provide appropriate training to the rural entrepreneurs. Business development services should also be strengthen to make value addition in terms of providing strategic information on markets, the prevailing demand and supply situation, availibilty of raw materials, technology sourcing and other related matters.

REFERENCES

Ahluwalia, Montek (1978), *"Rural Poverty and Agricultural Performance in India"*, *Journal of Development Studies*, April 1978, pp. 298-323, 1978.

Bourguignon, Francois and Christian, Morrisson (1992), *"Adjustment and Equity in Developing Countries : A New Approach"*, Development Centre of OECD, Paris, 1992.

Centre for Monitoring Indian Economy, (1997), "*Profile of States*", March 1997.

Chadha, G.K. and Alakh N. Sharma (Ed.), (1997), "*Growth, Employment and Poverty: Change and Continuity in Rural India*", Vikas Publishing House Pvt. Ltd., Delhi, 1997.

Chandrasekhar, C.P. and A. Sen, (1996), "*Has Poverty Really Come Down Following the Reforms?*", Macroscam, Business Line, January 15, 1996.

Government of India, (Various Issues), "*Annual Plans*", Planning Commission, New Delhi.

Government of India, (Various Issues), "*Indian Labour Journal*", Labour Bureau, Shimla.

Government of India, (Various Issues), "*Agricultural Statistics at a Glance*", Directorate of Economics and Statistics, Department of Agriculture and Cooperation, Ministry of Agriculture, New Delhi.

Government of India, (1997a), "*Estimates of Poverty*", Press Information Bureau (Press Release), New Delhi, March 11, 1997.

Government of India, (1997b), "*Approach Paper to the Ninth Five Year Plan (1997-2002)*", Planning Commission, New Delhi, 1997.

Gupta, S. P., (1995), "*Economic Reform and Its Impact on Poor*", Economic and Political Weekly, pp. 1295-1313, June 3, 1995.

Pant, Devendra Kumar, (1997), "*Human Development, Demographic Transition and Economic Growth Linkage—An Econometric Analysis with India Data*", NCAER Working Paper Number 67, July 1997.

Sen, Abhijit, (1997), "*Structural Adjustment and Rural Poverty*", in Growth, Employment and Poverty: Change and Continuity in Rural India by Chadha, G.K. and Alakh N. Sharma (Ed.), Vikas Publishing House Pvt. Ltd., Delhi, 1997.

Sen, Amartya K., (1976), "*Poverty—An Ordinal Approach to Measurement*", Econometrica, Vol. 44, March 1976, pp. 219-31, 1976.

World Bank, (1990), "*World Development Report: Poverty*", Oxford University Press, 1990.

Nayyar, Rohini and Sharma N. Alakh (2005), Rural Transformation In India—The Role of Non-Farm sector; Institute for Human Development, 2005.

Rajiv Mehta (2009), Wye City Group On Statistics On Rural Development And Agriculture Household Income—Rural Livelihood Diversification and its Measurement Issues: Focus India.

CHAPTER

12

Rural Development in India : Issues and Challenges

S.M. Jawed Akhtar and Nahid Akhtar Siddiqi

1. INTRODUCTION

"India lives in its villages".

—Mahatma Gandhi

Rural India is real India. Literally and from the social, economic and political perspectives the statement is valid even today. Over seventy-four per cent of the total population of India lives in villages. Rural India still contributes about half of the national income. People in rural areas should have the same quality of life as is enjoyed by people living in sub-urban and urban areas. Further, there are cascading effects of poverty, unemployment, poor and inadequate infrastructure in rural areas on urban centres causing slums and consequential social and economic tensions manifesting in economic deprivation and urban poverty. Hence, Rural Development which is concerned with economic growth and social justice, improvement in the living standard of the rural people by providing adequate and quality social services and minimum basic needs becomes essential. Therefore after independence, the ultimate objectives of all planned development efforts are for the well-being of the masses. Since the very beginning of economic planning (1950), rural development has been the main concern of our country. Development is no longer identified with a mere increase in national income or even per capita national income. The increased income should be so distributed that as to result in a significant diminution of inequalities of income and wealth. The absolute poverty must be abolished with the guarantee of minimum level of employment. Public distribution of goods and services are to be provided at lower rates for the poor. Development is also expected to include the areas of health, education, roads, water supply, electricity, cultural values and welfare. With all these aspects, the standard of living improves. Thus, the purpose of development has to enrich total quality of life. Now it is widely accepted that the development is not only the provision of opportunities for development but also their actual utilisation by the people for whom they are intended and involves the creation of the facilities necessary for such utilisation.

Rural development is not simply an economic proposition; it has social, psychological, and cultural dimensions as well. It is a multidimensional and multidirectional concept. To be precise, rural development is a programme designed to improve the socio-economic living conditions of the rural poor. It also aims at raising their cultural level and reorienting their rich traditions. It seeks to achieve increased rural production and productivity, greater socio-economic equity and a higher standard of living for the rural poor. It is partly ameliorative and partly development-oriented. Development is interlinked with motivation, innovation and the active participation of the beneficiaries, *inter-alia;* this calls for organisation and management. Rural development recognises the importance of improved food supplies and nutrition, as well as the importance of such basic services as health, housing, education and expanded communications, which will go a long way in enhancing the productivity of the rural poor. Moreover, it aims at providing gainful employment, so that the rural people too may contribute their mite to the national product. Rural development implies a fuller development of existing resources, including the construction of infrastructure, such as roads and irrigation works, the introduction of new production technology, the revival of traditional arts and crafts, and the creation of new types of institutions and organisations.

2. ISSUES PERTAINING TO RURAL DEVELOPMENT

(i) Literacy and Education

The simple literacy (basic reading, writing and numeracy) and the functional literacy had been presented as panaceas in the 1950s and the 1970s, respectively, with a view that if everybody learned how to read and write or learned this would enhance development. On the other hand, education plays an important role in the progress of an individual's mind and country. Ignorance and poverty, the two major speed-breakers in the swift developing country, can be overcome easily through education.

(ii) Construction of Link-Roads

Though unknown, a potentially important share of the benefits to the poor from rural roads cannot be measured in monetary terms. This is extremely important if the benefits of development have to move beyond the limited confines of cities to the vast hinterland so that the millions of toiling farmers can also become partners in progress. Urban-rural road links play a vital role for carrying urban prosperity into the heartland of a developing economy.

(iii) Self-employment Generation

Employment generation, especially the self-employment generation in rural sector attach great importance to poverty alleviation and mitigation specifically of the wide variations across states and the rural-urban division. A set of prudentially selected programmes of self-employment in rural sector plays as the panacea to remove multi-dimensional nature of poverty through helping a lot. The anti-poverty strategy has three broad components—promotion of economic growth, promotion of human development and targeted programmes of poverty alleviation.

(iv) Health Awareness

The goal of health awareness programme of rural development is to create awareness, a stirring of both heart and mind, about health care conditions, challenges, and solutions among the rural people.

(v) Awareness of Existing Opportunities

Awareness of opportunities available in rural sector means to provide relevant and usable information to the rural youth regarding (both the skilled and the unskilled) labour markets and access to relevant training to help them make decisions about the labour market options available to them.

(vi) Family Planning

Population growth is though a global problem but in heavily populated developing countries the problem is rather acute. An effective family planning programme is necessary so as to curb high population growth that erodes the increase in employment opportunities.

(vii) Land Reforms

Land reforms mean deliberate change in the way agricultural land is held or owned, the methods of its cultivation, or the relation of agriculture to the rest of the economy. The most common objective of land reform is to abolish feudal or colonial forms of landownership, often by taking land away from large landowners and redistributing it to landless peasants.

(viii) Extension of Rural or Agricultural Credit

An Integrated Rural Development Programme (IRDP) needs to identify the poor, give them credit and subsidy to purchase a productive asset to raise their earnings, recover loans, and recycle loans progressively to help poor villagers. There is a need to popularise banking facilities and make villagers account-holders. New type of accounts to suit their requirements should be considered. Moreover, the main problem in rural credit is not merely of credit but the way in which it is spent. Therefore, the agencies providing rural credit should aim to promote sustainable and equitable prosperity of the villagers through effective credit support, related services, institutional development and other innovative initiatives.

(ix) Extension of Canal Irrigation

Irrigation is an essential component of progressive agriculture. In fact, development of irrigation has become synonymous with agricultural development and rural prosperity. Canal irrigation system is much more challenging than the groundwater irrigation systems. It is the dominant water transfer technique in developing countries and is performing well in many parts of the world.

(x) Rural Electrification

Rural electrification means to facilitate availability of electricity for accelerated growth and for enrichment of quality of life of rural population. It is argued that rural electrification is essential in the longer term perspective of rural development since electricity is considered as a prerequisite for economic development and improvement of overall standard of living of the rural inhabitants. Electricity is also considered as a potent force capable of elevating and providing the much needed dynamism into the rural economy.

3. CHALLENGES OF RURAL DEVELOPMENT

(i) Slow Down in Agricultural and Rural Non-Farm Growth

Both the poorest as well as the more prosperous 'Green Revolution' states of Punjab, Haryana and Uttar Pradesh have recently witnessed a slow-down in agricultural growth. Some of the factors hampering the revival of growth are:

- *Poor composition of public expenditures*: Public expenditure on agricultural subsidies is crowding out productivity-enhancing investments such as agricultural research and extension, as well as investments in rural infrastructure, and the health and education of the rural people. In 1999-00, agricultural subsidies amounted to 3 percent of GDP and were over 7 times the public investments in the sector.

- *Over-regulation of domestic agricultural trade*: While economic and trade reforms in the 1990s helped to improve the incentive framework, over-regulation of domestic trade has increased costs, price risks and uncertainty, undermining the sector's competitiveness.
- *Government interventions in land, labour, and credit markets*: More rapid growth of the rural non-farm sector is constrained by government interventions in factor markets—land, labour, and credit—and in output markets, such as the small-scale reservation of enterprises.
- *Inadequate infrastructure and services in rural areas*: Due to less returns from rural areas, public and private sectors are not much interested in investment for infrastructure and services provision in rural areas.

(ii) Weak Framework for Sustainable Water Management and Irrigation

- *Inequitable allocation of water*: Many states lack the incentives, policy, regulatory, and institutional framework for the efficient, sustainable, and equitable allocation of water.
- *Deteriorating irrigation infrastructure*: Public expenditure in irrigation is spread over many uncompleted projects. In addition, existing infrastructure has rapidly deteriorated as operations and maintenance is given lower priority.

(iii) Inadequate Access to Land and Finance

- *Stringent land regulations discourage rural investments*: While land distribution has become less skewed, land policy and regulations to increase security of tenure (including restrictions or bans on renting land or converting it to other uses) have had the unintended effect of reducing access by the landless and discouraging rural investments.

- *Computerization of land records has brought to light institutional weaknesses*: State government initiatives to computerize land records have reduced transaction costs and increased transparency. But this has brought to light institutional weaknesses.
- *Rural poor have little access to credit*: While India has a wide network of rural finance institutions, many of the rural poor remain excluded, due to inefficiencies in the formal finance institutions, the weak regulatory framework, high transaction costs, and risks associated with lending to agriculture.

(iv) Weak Natural Resources Management

- *A purely conservation approach to forests is ineffective*: Experience in India shows that a purely conservation approach to natural resources management does not work effectively and does little to reduce poverty.
- *Weak resource rights for forest communities*: The forest sector is also faced with weak resource rights and economic incentives for communities, an inefficient legal framework and participatory management, and poor access to markets.

(v) Weak Delivery of Basic Services in Rural Areas

- *Low bureaucratic accountability and inefficient use of public funds*: Despite large expenditures in rural development, a highly centralized bureaucracy with low accountability and inefficient use of public funds limit their impact on poverty. In 1992, India amended its Constitution to create three tiers of democratically elected rural local governments bringing governance down to the villages. However, the transfer of authority, funds, and functionaries to these local bodies is progressing slowly, in part due to political vested interests. The poor are not empowered to contribute to shaping public programs or to hold local governments accountable.

4. SUGGESTIONS

(i) Enhancing Agricultural Productivity, Competitiveness, and Rural Growth

- *Enhancing productivity*: Creating a more productive, internationally competitive and diversified agricultural sector would require a shift in public expenditures away from subsidies towards productivity enhancing investments. Second, it will require removing the restrictions on domestic private trade to improve the investment environment and meet expanding market opportunities. Third, the agricultural research and extension systems need to be strengthened to improve access to productivity enhancing technologies. The diverse conditions across India suggest the importance of regionally differentiated strategies with a strong focus on the lagging states.
- *Improving water resource and irrigation/drainage management*: Increase in multi-sectoral competition for water highlights the need to formulate water policies and unbundle water resources management from irrigation service delivery. Other key priorities include: (i) modernizing Irrigation and Drainage Departments to integrate the participation of farmers and other agencies in irrigation management; (ii) improving cost recovery; (iii) rationalizing public expenditures, with priority to completing schemes with the highest returns; and (iv) allocating sufficient resources for operations and maintenance for the sustainability of investments.
- *Strengthening rural non-farm sector growth*: Rising incomes are fuelling demand for higher-value fresh

and processed agricultural products in domestic markets and globally, which open new opportunities for agricultural diversification to higher value products (e.g. horticulture, livestock), agro-processing and related services. The government needs to shift its role from direct intervention and overregulation to creating the enabling environment for private sector participation and competition for agribusiness and more broadly, the rural non-farm sector growth. Improving the rural investment environment includes removing trade controls, rationalizing labour regulations and the tax regime (i.e. adoption of the value-added tax system), and improving access to credit and key infrastructure (e.g. roads, electricity, ports, markets).

(ii) Improving Access to Assets and Sustainable Natural Resource Use

- *Balancing poverty reduction and conservation priorities*: Finding superlative combinations for conservation and poverty reduction will be critical to sustainable natural resource management. This will involve addressing legal, policy and institutional constraints to devolving resource rights, and transferring responsibilities to local communities.
- *Improving access to land*: States can build on the growing consensus to reform land policy, particularly land tenancy policy and land administration system. States that do not have tenancy restrictions can provide useful lessons in this regard. Over the longer term, a more holistic approach to land administration policies, regulations and institutions is necessary to ensure tenure security, reduce costs, and ensure fairness and sustainability of the system.
- *Improving access to rural finance*: It would require improving the performance of regional rural banks and rural credit cooperatives by enhancing regulatory

oversight, removing Government control and ownership, and strengthening the legal framework for loan recovery and the use of land as collateral. It would also involve creating an enabling environment for the development of micro-finance institutions in rural areas.

(iii) Strengthening Institutions for the Poor and Promoting Rural Livelihood

- *Promoting community-based rural development*: State Government efforts in scaling up livelihood and community-driven development approaches will be critical to build social capital in the poorest areas as well as to expand savings mobilization, promote productive investments, income generating opportunities and sustainable natural resource management. Direct support to self-help groups, village committees, user's associations, savings and loans groups and others can provide the initial 'push' to move organizations to higher level and access to new economic opportunities. Moreover, social mobilization and particularly the empowerment of women's groups, through increased capacity for collective action will provide communities with greater "voice" and bargaining power in dealing with the private sector, markets and financial services.
- *Strengthening accountability for service delivery*: As decentralization efforts are pursued and local governments are given more prominence in the basic service delivery, the establishment of accountability mechanisms becomes critical. Local governments' capacity to identify local priorities through participatory budgeting and planning needs to be strengthened. This, in turn, would improve the rural investment environment, facilitating the involvement of the private sector, creating employment opportunities and linkages between farm and non-farm sectors.

5. CONCLUSION

Rural development is vital for India because of its demographic and geographic characteristics. All of these characteristics can affect the performance of the country, as well as the different strategies and programmes that will be implemented by the government. No doubt, villages are in a state of neglect and under-development, with impoverished people, as result of past legacies and defects in our planning process and investment pattern. But the potential in rural India is immense. What if every village in the country is provided with basic amenities, like drinking water, electricity, health care, education, transport, communication and other facilities, with only a smaller population of the village engaged in agriculture and the remaining in other gainful occupations? When this happens India will turn into mighty country. The purchasing power of the rural population throwing enormous demand for goods and services will boost the national economy tremendously. The day will see the reverse migration of people from the urban slums back to the villages.

Rural Development is the subject to come to the forefront and rural banking will serve the backbone of this development. Rural development can be more meaningful only through the participation of clienteles of development. Just as implementation is the touchstone for planning, people's participation is the centre-piece in rural development. People's participation is one of the foremost pre-requisites of development process both from procedural and philosophical perspectives. For the development planners and administrators it is important to solicit the participation of different group of rural people, to make the plans participatory.

It is important to consider the different aspects that are connected with standard of living in the rural areas, together with the different economic activities that are being performed in that area. Rural circumstances are changing; development thinking is changing; and rural development policy needs to keep up: these are the core messages of the review. Finally, it is worth pointing out that rural development is not exactly a single sector, so that some kind of higher level planning device will be needed.

References

Agrawal, A.N. (2006), *"Problems of Development and Planning"*, Wishwa Prakashan, New Delhi.

Aslam, M., (2004), "Rural Development Experiences: An Asian Perspective", *Journal of Rural Development*, Vol. 37, No. 2.

Caroline, A. and M. Simon (2001), "Rethinking Rural Development", *Development Policy Review*, Vol. 19, No. 4.

Chandra, Garg Lakshmi and Jindal, Sadhna (1989), *Rural Development—A Critical Appraisal*, Nitisha Publications.

Farrington, J. *et. al.* (2001), "Rural Development and the New Architecture of Aid: Convergence and Constraints", *Development Policy Review.*

Frank, E. and Stephen, B. (2001), Evolving Themes in Rural Development 1950s-2000s" *Development Policy Review.*

Prasad, B.K. (2003), *Rural Development: Concept, Approach and Strategy,* Sarup and Sons Publication, New Delhi.

Sahu, B.K. (2003), *Rural Development in India,* Anmol Publications Pvt., Ltd.

Singh K.K. and S. Ali (2001), *Role of Panchayati Raj Institutions for Rural Development,* Sarup and Sons Publication, New Delhi.

Singh, R.R. (1999), "Rural Development: Concept and Significance of the Issue" in Devendra, K. Das (ed.), *Rural Sector and Development: Experiences and Challenges,* Deep and Deep Publications, New Delhi.

World Bank (1997), "Rural Development—From Vision to Action: A Sector Strategy Paper", Washington, D.C.

PART II

NON-FARM SECTOR AND EMPLOYMENT OPPORTUNITIES

CHAPTER

13

Rural Non-Farm Sector : Opportunities and Options

SUVASH CHANDRA PRADHAN

In the World Development Report, 1995, RNFS (rural non-farm sector) is defined as the sector which includes all economic activities in rural areas except agriculture, livestock, fishing and hunting. It is not in any sense a homogeneous sector. It may include :

(a) Activities undertaken by farm households as independent producers in their home.
(b) The sub-contracting of work to farm families by urban bases firms.
(c) Non-farm activities in villages and rural town enterprises.
(d) Commuting between rural residents and urban non-farm jobs.

The RNFS includes a wide range of economic activities whose composition may vary from country to country. The most common convention is to include animal husbandry, hunting and taping, forestry, logging, fishing, etc. Thus, the RNFS has positive role to play in promoting growth and welfare.

The RNFS can be classified as follows :

1. Informal.
2. Formal.

Informal RNFS can be classified into:

(i) Modern.
(ii) Traditional.

Modern informal RNFS may be classified into:

(i) Establishment Sector.
(ii) Non-establishment Sector.

Traditional informal RNFS may be classified into:

(i) Artisans (Caste-based)
(ii) Service Workers (Caste-based)

Case study—The village Thakurpatna Dist. Kendrapara having concentration of artisans of most developed Panchayat of Kendrapara district of Odisha. It is observed that there is a fall in the sale of products of traditional RNFS. It is consisting of three categories of artisans.

(a) Blacksmith
(b) Carpenter
(c) Potter

The fall in the sale is due to the availability of substitutes produced by modern formal industry located in urban areas. There may be rise in the number of people engaged in traditional sector, but the sale and income of this sector have declined owing to the competition of the formal urban-based Industries.

Growth of Non-farm Sectors in Odisha (Growth Rates of NSDP in Non-farm Sector of Odisha at 1980-81 Prices)

Sectors	*1980-1990*	*1990-2000*	*1980-2000*
Mining and Quarrying	8.72	12.70	11.95
Manufacturing	15.16	-19.15	-0.46
Construction	3.55	0.49	4.27
Electricity, Gas and Water supply	5.17	-2.92	3.52
Transport	11.90	6.71	7.8
Communication	7.89	13.69	8.82
Tourism and Trade	5.66	6.75	5.48
Banking and Insurance	14.54	9.78	11.26
Railways	12.81	4.69	10.63

Source : Orissa Human Development Report, 2005.

The table reveals that the performance was better during 1980-90 compared to 1990-2000.

PROMOTION OF RURAL NON-FARM SECTOR BY NABARD

The National Bank for agriculture and rural development (NABARD) has taken various promotional measures for expansion and modernization of the rural non-farm sector through different loan schemes for promoting rural entrepreneurship and rural Industrialization. Recognizing the importance of RNFS for faster economic development of the rural areas, NABARD has taken a number of initiatives with the refinance support and promotional assistance for development of RNFS.

(1) Rural Entrepreneurship Development Programmes (REDP)

It aims at developing entrepreneurial activity and skill to those who are willing to set-up micro-enterprises for creation of sustainable employment and income opportunities in rural area in a cost-effective manner.

(2) Refinance Facilities through Different Loan Schemes

The refinance facilities are broadly classifies into two categories : (i) Investment credit, (ii) Production credit. Under investment credit NABARD provides refinance to commercial Banks, Regional Rural Banks (RRBs) and Co-operative banks for a wide spectrum of manufacturing, processing and service sector activities under RNFS. Various refinance schemes formulated are :

(i) Composite Loan Scheme
(ii) Integrated Loan Scheme
(iii) Small Road and Water Transport Operators Scheme.(SRWTOS)
(iv) Schemes based on pre-sanction procedure.
(v) Soft Loan Assistant for Margin Money.

Rural artisans, craftsmen, small entrepreneurs, and NGOs are covered under the refinance scheme for expansion, diversification, modernization, under renovation through SHG—Bank linkage through Informal Credit Delivery System.

(3) Swarojgar Credit Card Scheme (SC Scheme)

NABARD is committed to rural prosperity. Its Swarojgar Credit Card Scheme was introduced by the Govt. of India in 2003 in consultation with NABARD and RBI. It was formulated as a special Credit Card Scheme benefiting the rural artisans and small entrepreneurs. The Scheme is normally valid for 5 years subject to satisfactory operation and renewed on a yearly basis through simple review process. The normal limit under the skill is 25,000. The Scheme is to be implemented by all Commercial Banks, RRBs, State Cooperative Banks, DDCBs, PACS, SCARDBs, PCARDBs and Scheduled Primary Co-operative Banks. Beneficiaries under the scheme would automatically be covered under the group insurance scheme and the premium would be shared by bank and the borrower equally.

POLICY FOCUS

Keeping in view the crucial role played by the Rural Non-farm sector. Following issues deserve priority attention calling for thrust areas of policy measures needed for a radical transformation of the rural economy.

(i) A dynamic small enterprise promotion policy for the state is a crucial need for future industrial development. Focusing on industrial clusters, especially supporting their technology upgradation and promoting external orientation shall be a potential area of intervention.

(ii) Necessary support need to be extended and encouragement should be provided to the potential entrepreneurs and rural artisans for skill upgradation, and acquisition, capacity building, design development of handicrafts and marketing of products.

(iii) Prospective entrepreneurs should be encouraged through periodic awareness campaigns of NFS potential in the area.

(iv) Entrepreneurial involvement would be more attractive and viable by encouraging local demand base. Dismal performance of the industrial structure can be avoided by local entrepreneurial drive.

(v) Lack of proper infrastructure has severely impaired the growth and diversification of industries in the state. Emphasis need to be given on infrastructure development, especially transport and power. The road condition and power supply need to be improved substantially. The railway network need to be fully operationalised for both goods and passenger traffic connecting depressed regions.

(vi) There is a need to build up entrepreneurial and managerial skill and encouragement to higher skilled entrepreneur venture to utilize the huge potential in different districts in industries, services and business sector. Skill development is the key to quality and productivity and employment of local people in various industries.

(vii) Marketing support can be strengthened through construction of retail outlets for handicrafts at tourist places productivity.

(viii) Training programmes on processing technology, quality control accounting pricing and market support are necessary for village industries. Setting up vocational training institutes may be promoted in Government and private sector to cater to the manpower requirement of the industrial sector. Investment of NGOs in this respect will be more fruitful.

(ix) Large number of Orientation meets and gender sensitization needs by NABARD to promote financing under NFS will improve the ground level credit flow to women.

(x) Since a sizeable population are earning their livelihood from handicraft and cottage industries focused attention for their upliftment is to be made adequate and consistent efforts of various agencies like KVIC/KVIB and the Government deserve special attention. The Industrial, Handloom, Weavers and Handicraft cooperative societies need to be activated to facilitate procurement of raw materials, access to credit and marketing.

(xi) Adequate coordination between the Government agencies and banker will avoid delay in processing of proposals for the potential entrepreneurs. Govt. agencies like ITDA, DRDA, DICs, should have greater coordination among them.

CONCLUSION

There is a sharp positive linkage between the Rural and Non-form sector and development of the rural economy. Necessary mechanism need to be developed to put great emphasis on institutional finance, training, and marketing support, rural entrepreneurship development and greater integration between the people, Government and NGOs for a sustainable development. Organized efforts by the Government, Educational institutions, PRIs and NGO, should be expedited to raise the level awareness and motivate rural youth towards non-farm employment especially in the informal sector. Rural growth centres and special economic zones would promote employment and income of workers in RNFS. In view of low quality education of the rural unemployed youth, there is need for initiating short-term skill-oriented relevant training programmes also.

References

Chadha, G.K., "Non-farm Employment for Rural Households", *Indian Journal of Labour Economics*.

Dev, S. Mahendra, "Non-Agricultural Employment in Rural India", *EPW*.

Papola, T.S., "Rural Non-farm Employment", *Indian Journal of Labour Economics*.

NABARD : The Report of Study Group on RNFS.

Samal, K.C. (2008), "Informal Sector, Concept, Dynamics and Linkages".

CHAPTER

14

Growth and Pattern of Rural Employment in India : In Post-Liberalisation Period

RANA NASEER

INTRODUCTION

Employment structure that existed in India before the economic reforms of 1991 has the genesis in the economic and political events took place in colonial period and goes back to the second half of the eighteenth century. These developments are East India Company's acquiring control over the large part of the Indian Territory and subsequently effecting changes of far reaching consequences in the system of land revenue collection, which in turn altered the entire landscape of agrarian relations in India.

With the decline in the urban handicrafts in the 19th century there was a huge migration of labour from urban to rural areas. The agriculture being the only source of employment such migration augmented the demand pressure on land resulting in the increase in rent and worsening the condition of farmers as a result of increase in the agriculture dependent labour force.

The net result of such developments was agriculture to remain the only sector of any consequence. It could be gauged from the fact that agricultural sector accounted for about half of the country's national income and about two-third of employment of work force at the time of independence. Agriculture, as a result, remained backward, non-diversified activity where labour of very low or even zero marginal productivity could be found.

With economic development, the employment pattern of a country undergoes significant changes. An increasing number of new jobs opportunities are created in the non-farm sector. In farm sector, development largely takes place in the form of increased productivity and thereby releasing many workers from this sector. As a result of which labour force shifts from farm sectors to non-farm sectors. Besides, the factors such as high wages, fixed working hour, better working conditions, availability of modern facilities, etc. in NFS induce the workers to migrate from farm sector to non-farm sector. Economic liberalization and the consequent structural adjustment programmes introduced during 1990s, the shift of employment opportunities from primary sector to other sectors is increasingly viewed as a positive indicator for better growth and performance of the economy.

CHANGES IN THE STRUCTURE OF EMPLOYMENT

Economic reforms initiated in 1991 were of far reaching consequences for the economy as they attempted the wide ranging restructuring. One of the significant moves was to limit the public sector investment in many areas as it sought to correct the fiscal imbalances. But simultaneously it had opened up the economy considerably to facilitate private sector investment. Such things in turn created opportunities for the employment avenues in non-farm sector too. It is in this context that an examination of employment scenario in post-reform period becomes significant.

Thus it can be argued the pattern of employment as a whole underwent significant changes as a result of economic development which was accelerated in post-reform period. Economic liberalization and the structural adjustment programmes introduced during 1990s and consequent shift in employment opportunities from primary sector to other sectors is increasingly viewed as a positive indicator for better growth and performance of the economy. For assessing the impact of economic reforms on employment with respect of age group, sex, location, industry, level of education etc. are assumed to be of great significance in the pattern of employment and the sectoral distribution of the workforce provide a meaningful insight as to the sufferings and gains of the new economic policy.

The changes observed in the employment structure are summarized in the Table 1.

TABLE I

All India Usual (PS+SS) Status Workforce Participation Rates by Gender and Rural-urban Location (1972-73 to 2004-05)

(*Percent of Pop.*)

NSS Round	*R_M*	*R_F*	*U_M*	*U_F*
27(1972-73)	53.6	31.4	49.4	13.2
32(1977-78)	54.4	32.6	50.0	15.3
38 (1983)	54.3	33.0	51.0	14.6
43 (1987-88)	53.9	32.3	50.6	15.2
50 (1993-94)	55.3	32.8	52.1	15.5
55 (1999-00)	53.1	29.9	51.8	13.9
61 (2004-05)	54.6	32.7	54.9	16.6

Notes : PS—Principal Status, SS—Subsidiary Status, R_M = Rural Male, R_F = Rural Female U_M = Urban Male, U_F = Urban Female.

Sources : National Sample Survey Organization Report No. 515, Employment and Unemployment Situation in India, 2004-05.

Pre-reform changes in Workforce Participation Rate discussed here in terms of the usual (principal+subsidiary) status data is defined as the number of workers per thousand persons. Table 1 presents the data on workforce participation rates according to usual status during 1972-73 to 2004-05. It is evident from the data that the all India level, the workforce participation registered a marginal decline before the reform period.

The crude WPRs estimated on the basis of the quinquennial surveys, namely, 27th round (1972-73), 32nd round (1977-78), 38th round (1977-78), 43rd round (1987-88), 50th round (1993-94), 55th (1999-2000) and 61st (2004-05) separately by rural-urban residence and gender. (The rates are called "crude" because the denominator refers to the population of all ages together, rather than only the population of working ages).

The WPRs show a reasonable stability in the rates for rural males around 53-55 per cent and for urban males around 49-54 per cent. The rates for females have tended to fluctuate between 31 to 33 per cent in rural areas and between 13 and 16 per cent in urban India. After the decline in WPR during 1999-2000 the recovery seems to have taken place during the period 2004-05. After the period of jobless growth an increase in WPR could be observed across all four segments. As far as the WPR in rural sector is concerned it increased from 53.1 per cent to 54.6 per cent during the period from 1999-2000 to 2004-05 in case of males while in case of females the increase was from 29.9 per cent to 32.7 per cent over the same period. Similarly, the male WPR in the urban areas increased from 51.8 per cent to 54.9 per cent and for females from 13.9 per cent in 1999-00 to 16.6 per cent in 2004-05. Thus both male and female participation rates have registered an increase.

Since 1972-73 the highest recorded rural female WPR was in the period 1983 at 33.0 per cent, then start declined, the year 1987-88 was a year of severe drought in the economy. Since that peak the rural female WPR had continuously declined. 43rd round of NSSO is an abnormal year in terms of a bad monsoon and also a bad year and it should depress growth rate of employment.

SECTOR-WISE DISTRIBUTION OF WORKERS

The usual status workers are classified on the basis of broad industrial category namely: primary, secondary, tertiary and non-farm sectors. Table 2 has given sectoral distribution of rural and urban workers. In rural India the proportion of male workers engaged in the primary sector has been steadily declining from 83.2 per cent in 1972-73 to 78.1 per cent in 1983 to 74.8 per cent in 1993-94 to 71.4 per cent in 1999-00 and 66.5 per cent in 2004-05. The proportion of their employment in secondary, tertiary and total NFS has witnessed a steady increase. Their excessive dependence on agriculture as a source of livelihood has steadily been melting down and their employment clearly witnessed a modest degree of diversification. The rural female workers dependence on the primary sector employment too witnessed a steady decline rural females share dropped from 89.7 per cent in 1972-73 to 87.8 per cent in 1983. But it was much worse in that during post-1987 (85.1 per cent) years, their proportion in primary sector workforce actually increased instead of declining in 1993-94 (86.6 per cent) from 85.3 per cent in 1999-00 and further to 83.3 per cent in 2004-05. Rural female workers remains heavily tagged with agriculture. In secondary sector rural male employment witnessed a steady increase from a mere 7.8 per cent in 1972-73 to 9.4 per cent in 1983 to 10.5 per cent in 1993-94 to 12.6 per cent in 1999-00 and to 15.6 per cent in 2004-05. Rural female employment witnessed a steady increase from a mere 6.0 per cent in 1972-73 to 7.1 per cent in 1983 to 7.9 per cent in 1993-94 to 9.0 per cent in 1999-00 to 10.2 per cent in 2004-05. In tertiary sector rural male workers share increased from 9.0 per cent in 1972-73 to 12.5 per cent in 1983 to 14.7 per cent in 1993-94 to 16.0 per cent in 1999-00 and increased further to 17.9 per cent in 2004-05. The share of rural females in tertiary sector did not show any noticeable increase and remained around 5-6 per cent. Rural male NFS employment has expanded from as low 16.8 per cent in 1972-73 to 21.9 per cent in 1983 to 25.2 per cent in 1993-94 to 28.6 per cent in 1999-00 to 33.5 per cent in 2004-05. Rural female NFS employment increased from 13.4 per cent in 1993-94 to 14.7 per cent and 16.7 per cent in 2004-05. It is clear that the rural

TABLE 2

Sectoral Distribution of Usual Status (US+SS) Workers in India by Sex and Residence : NSS Data : 1972-73/2004-05

(percent of pop)

NSS Round	*Rural Male*				*Rural Female*			
	PS	*SS*	*TS*	*NFS*	*PS*	*SS*	*TS*	*NFS*
(1)	*(2)*	*(3)*	*(4)*	*(5)*	*(6)*	*(7)*	*(8)*	*(9)*
27 (1972-73)	83.2	7.8	9.0	16.8	89.7	6.0	4.3	10.3
32(1977-78)	81.1	8.3	10.6	18.9	88.3	6.6	5.1	11.7
38(1983)	78.1	9.4	12.5	21.9	87.8	7.1	5.1	12.2
43(1987 -88)	75.2	11.4	13.4	24.8	85.1	9.6	5.3	14.9
50(1993-94)	74.8	10.5	14.7	25.2	86.6	7.9	5.5	13.4
55(1 999-00)	71.4	12.6	16.0	28.6	85.3	9.0	5.7	14.7
61 (2004-05)	66.5	15.6	17.9	33.5	83.3	10.2	6.6	16.7
	Urban Male				*Urban Female*			
27(1972-73)	11.7	32.1	56.2	88.3	30.0	30.3	37.7	68.0
32(1977-78)	11.5	32.8	55.7	88.5	32.4	32.1	35.5	67.6
38(1983)	11.4	32.9	55.7	88.6	31.7	30.1	38.2	68.3
43(1 987-88)	10.4	32.7	56.9	89.6	30.2	30.9	38.9	69.8
50(1993-94)	10.3	31.7	58.0	89.7	25.2	28.5	46.3	74.8
55(1993-94)	6.6	32.8	60.6	93.4	17.7	29.4	52.9	82.3
61 (2004-05)	6.1	34.5	59.4	93.9	18.1	32.4	49.5	81.9

Note : PS—Primary Sector, SS—Secondary Sector, TS—Tertiary Sector and NFS—Non-farm Sector

Source : National Sample Survey Organization Report No. 515, Employment and Unemployment Situation in India, 2004-05.

worker's shift to non-agricultural sectors were relatively sharper during the post-reform periods.

In urban India male employment in primary sector did not increase, but registered a fall from 11.7 per cent in 1972-73 to 11.4 per cent in 1983 to 10.3 per cent in 1993-94 to 6.6 per cent in 1999-00 to 6.1 per cent in 2004-05. Urban female share also declined from 32.0 percent in 1972-73 to 31.7 per cent in 1983 to 25.2 per cent in 1993-94 to 17.7 per cent in 1999-00 then increased to 18.1 per cent in 2004-05. In the secondary sector male participation increased from 32.1 per cent. in 1972-73 to 32.9 per cent in 1983 but then it declined to 31.7 per cent in 1993-94 increased to 32.8 per cent in 1999-00 to 34.5 per cent in 2004-05. Female participation also declined from 30.3 per cent in 1972-73 to 30.1 per cent in 1983 to 28.5 per cent in 1993-94 increased from 29.4 per cent in 1999-00 to 32.4 per cent in 2004-05. Tertiary sector share declined from 56.2 per cent in 1972-73 to 55.7 per cent in 1983 then it increased to 58.0 per cent in 1993-94 share increased from 60.6 per cent in 1999-00 then it declined to 59.4 per cent in 2004-05, female share increased from 37.7 per cent to 1972-73 to 38.2 per cent in 1983 to 46.3 per cent in 1993-94 to 52.9 per cent in 1999-00 and decline to 49.5 per cent in 2004-05.. NFS share of male increased from 88.3 per cent to 1972-73 to 88.6 per cent in 1983 to 89.7 per cent in 1993-94 to 93.4 per cent in 1999-00 and to 93.9 per cent in 2004-05, female increased 68.0 per cent in 1972-73 to 68.3 per cent in 1983 to 74.8 per cent in 1993-94 to 82.3 per cent in 1999-00 and then declined to 81.9 per cent in 2004-05.

There was a distinct shift of rural work force (both male and female) in favour of secondary and tertiary sector after 1987. This could have been possible because of the beginning of liberalization of Indian economy since mid eighties which resulted in the robust growth of these sectors. Such shift was not evident in case of rural female workers which could be attributed to the social and cultural value system the rural society adheres to. There educational status could also be the cause for such state of affairs as the prevalence of bias against women in the matter of education could be found more in rural society than the urban one. When we compare the distribution of male and female workers, it is noticed that the percentage of female workers are higher in primary sectors whereas the percentage of male workers are higher in both the secondary and tertiary sector. It is since the 1970s that the rural economy began to diversify and as a result employment opportunities opened up in the activities other than agriculture. Therefore the absorption of workforce in non-agriculture in noticeable magnitude could be observed in this period. Rural workers shift to non-agricultural sector during economic reforms periods. There are some surprises that there has been a significant decline in the share of primary sector in rural areas. Instead the more noteworthy shift for rural males has been to tertiary sector and secondary sector.

NATURE OF EMPLOYMENT

In Table 3 employed persons are categorized into 3 broad groups according to their status of employment. These broad groups are : (1) Self-Employed, (2) Regular Employees, (3) Casual Labour, table highlights the distribution of usually employed by these broad groups of employment in India during the period from 1972 to 1993-94 in rural and urban areas.

In rural India, the incidence of self employment, as is indicated from the findings of various rounds of NSS from 27th to 50th round, had consistently declined both for male and female workers. For males it declined from around 65.9 per cent in 1972-73 to 60.5 per cent in 1983 to 57.9 per cent in 1993-94 to 55.0 per cent in 1999-00 and then increased to 58.1 per cent in 2004-05 and for rural females it dropped from 64.5 per cent in 1972-73 to 61.9 per cent in 1983 to 58.5 per cent in 1993-94 to 57.3 per cent in 1999-00 and increased to 63.7 per cent in 2004-05. In urban India it has been increasing from 39.2 per cent in 1972-73 to 40.9 per cent in 1983 to 41.7 per cent in 1993-94 decline to 41.5 per cent in 1999-00 and increased to 44.8 per cent in 2004-05 for male workers for male workers and urban females it faced a sizeable decline from 48.4 per cent in 1972-73 to 45.8 per cent in 1983 to 45.4 per cent in 1993-94 to 45.3 per cent in 1999-00 and increased to 47.7 per cent in 2004-05 respectively.

Regular salaried jobs have declined in the period under reference both for rural males and rural females. The relative

TABLE 3

Composition of Usual Status (PS+SS) Workers by Sex and Rural-Urban Residence: Data 1972/1993-94 : All India

(*Percent in Pop.*)

Worker's Residence	*Workers Sex*	*NSS Round*	*Mode of employment*			
			SE	*RWL*	*CL*	*Index of Casualisation*
(1)	(2)	(3)	(4)	(5)	(6)	(7)
Rural	Male	27(1972-73)	65.9	12.1	22.0	182
		32(1977-78)	62.8	10.6	26.6	251
		38(1983)	60.5	10.3	29.2	283
		43(1987-88)	58.6	10.0	31.4	314
		50(1993-94)	57.9	8.3	33.8	407
		55(2004-05)	55.0	8.8	36.2	411
		61(2004-05)	58.1	9.0	32.9	366
Rural	Female	27 (1972-73)	64.5	4.1	31.4	766
		32(1977-78)	62.1	2.8	35.1	1254
		38(1983)	61.9	2.8	35.3	1261
		43(1987-88)	60.8	3.7	35.5	959
		50(1993-94)	58.5	2.8	38,7	1382
		55(1999-00)	57.3	3.1	39.6	1277
		61(2004-05)	63.7	3.7	32.6	881

Urban	Male	27 (1972-73)	39.2	50,7	10.1	20
		32(1977-78)	40.4	46.4	13.2	28
		38(1983)	40.9	43.7	15.4	35
		43(1987-88)	41.7	43.7	14.6	33
		50(1993-94)	41.7	42.0	16.3	39
		55(1999-00)	41.5	41.7	16.8	40
		61(2004-05)	44.8	40.6	14.6	36
Urban	Female	27 (1972-73)	48.4	27.9	23.7	85
		32(1977-78)	49.5	24.9	25.6	103
		38(1983)	45.8	25.8	28.4	110
		43(1987-88)	47.1	27.5	25.4	92
		50(1993-94)	45.4	28.6	26.0	91
		55(1999-00)	45.3	33.3	21.4	64
		61 (2004-05)	47.7	35.6	16.7	47

Notes : 1. SE—Self-Employed, RWS—Regular wage/Salaried, CL—Casual Labour

2. Index of casualisation shows the number of casual wage earners for every one-hundred of regular salaried jobs; 4= (3/2) 100.

Source : National Sample Survey Organization Report No. 515, Employment and Unemployment Situation in India, 2004-05.

decline rural males was greater as can be gauged from the fact that it declined from 12.1 per cent in 1972-73 to 10.3 per cent in 1983 and further to 8.3 per cent in 1993-94 further increased to 8.8 per cent in 1999-00 and to 9.0 per cent in 2004-05.. Rural females declined from 4.1 per cent in 1972-73 to 2.8 per cent in 1993-94 increased to 3.1 per cent in 1999-00 and to 3.7 per cent in 2004-05. Urban females started increasing sluggishly from 27.9 per cent 1972-73 to 28.6 per cent in 1993-94 to 33.3 per cent in 1999-00 to 35.6 per cent in 2004-05. But urban males declined from 50.7 per cent in 1972-73 to 43.7 per cent in 1983 to 42.0 per cent in 1993-94 to 41.7 per cent in 1999-00 to 40.6 per cent in 2004-05.

Employment under casual labour basis has increased for all the 4 categories of workers. Rural male increased from 22.0 per cent in 1972-73 to 29.2 per cent in 1983 to 33.8 per cent in 1993-94 to 36.2 per cent in 1999-00 and then declined to 32.9 per cent in 2004-05, rural female increased from 31.4 per cent in 1972-73 to 35.3 per cent in 1983 to 38.7 per cent in 1993-94 to 39.6 per cent in 1999-00 and declined to 32.6 per cent in 2004-05. Urban male increased from 10.1 per cent in 1972-73 to 15.4 per cent in 1983 to 16.3 percent in 1993-94 to 16.8 percent in 1999-00 and declined to 14.6 per cent in 2004-05 and urban female increased from 23.7 per cent in 1972-73 to 26.0 per cent in 1993-94 declined from 21.4 per cent in 1999-00 to 16.7 per cent in 2004-05.

As far as casualisation is concerned it could found more in case of male while for females trend did not exhibit a uniform pattern. Increasing casualisation of wage labour is evident from significant increase in casually employed males compared to their female counterparts. The extremely high incidence of casualisation for rural female workers and its rise over time is noticeable through the rough index of casualisation given in table. The low share in regular salaried jobs for rural workers (in 1987-88 10.0 per cent to 8.3 per cent in 1993-94 against 31.4 per cent and 33.8 per cent under casual labour for rural males and 3.7 per cent only in 1987-88 to 2.8 per cent in 1993-94 against 35.5 per cent and 38.7 per cent and under casual labour for rural females) tells the story of their relative disadvantage. The marked rural-urban differences in terms of the proportion of workers engaged as casual wage earners at once confirm numerous disadvantage (e.g. low wage rates, irregularity and uncertainty in employment, uncongenial work conditions) of rural workers, most visibly the females among them. But index of casualisation declined in both the rural and urban areas from 1999-2000 to 2004-05.

In the round 61st of NSS the increase in self-employment for both males as well as females could be observed in rural areas. It should be noted that the share of casual employment has declined for both females and males. The share of regular employment has increased for both males and females. Recent NSS (61st round) shows distinct changes from the earlier pattern which was observe in 55th round (1999-2000). In urban areas a somewhat different trend has been noticed for both males and females. The share of self employment has increased for both males and females. The regular employment share declined for males but increased for females. The share of casual employment has decreased relatively faster for females compared to that of males in urban areas.

CONCLUSION

It shows a reversal of the declining trend in employment in the period 1993-94 to 1999-2000 and 1999-2000 to 2004-05. There was a turnaround in employment growth in rural India after a phase of jobless growth during the 1990s. The employment scenario has undergone a change. As per the 61st round of NSS employment in the period 1999-2000 to 2004-05 has increased. Along with the sharp increase in employment the work force has also increased dramatically. Increased participation of female population and aged population in the labour market.

References

Abraham, V., (2009), "*Employment Growth in Rural India: Distress-Driven?*" *Economic and Political Weekly*, Vol. 44, No. 16, April, 18-24.

Bhaumik, S.K. (2002), "*Employment Diversification in Rural India : State Level Analysis*", *Indian Journal of Labour Economics*, Vol. 45, No. 4, pp. 719-44.

Datt, R. and Sundharam, K.P.M. (1993), *"Indian Economy"*, S. Chand and Company, New Delhi.

Datt, R. (2003), *"Economic Reforms, Labour and Employment"*, Deep and Deep Publications Pvt. Ltd., New Delhi.

Desai, S.S.M. (1987), *"Economic History of India"*, Himalaya Publishing House, Bombay.

Dev, S.M. (1990), *"Non Agriculture Employment in Rural India: Evidence at a Disaggregate Level"*, Economic and Political Weekly, Vol. XXV, No. 28, July 14, pp. 1526-1536.

Dhindas, K.S. and Sharma, A. (ed.) (2006), *"Economic Reform and Development"*, Concept Publishing Company, New Delhi.

Ghuman, R.S., (2005), *"Rural Non-farm Employment Scenario", Economic and Political Weekly*, Vol. 40, No. 41, Oct. 8-14.

Misra, S.K. and Puri, V.K., (1994), *"Indian Economy: Its Development Experience"*, Himalayas Pub. House, New Delhi.

Sundharam, KP.M. (1992) *"Introduction to Indian Economy"*, New Delhi, S. Chand and Co.

Dnni, I. (1998), *"Non-Agricultural Employment and Poverty in Rural India: A Review of Evidence"*, Economic and Political Weekly, Vol. XXXIll, No. 13, March 28, April 3, pp. A.36-A.44.

CHAPTER

15

Outsourcing and Employment Prospects in India

SHEHROZ A. RIZVI AND JEEGYASA

INTRODUCTION

Now-a-days Business Process Outsourcing (BPO) is the burning issue in the e-business World, and is better known as "outsourcing". BPO companies having full advantages of the realities of globalization by exporting certain business process to outside providers, which can be provided cheaper, faster and better by outsourcing service. To grow up in internal communication, technology, web increased focus on core competencies, cost reduction has stimulated the growth of the IT enabled services at globally most optional location, and it provides contract to customers everywhere in the world.

Outsourcing began in early eighties when the organizations started to delegate their non-core functions to external organizations, which was providing a particular services, functions or management of the outsourced function.

Actually, off-shoring is described to provide services to the other countries.

There are two approaches to specify on outsourcing contract First: we can specify the contract in terms of output, for example, number of calls answered by a call centre and second we can specify an activity to cover a service level, i.e. the provision of call centers that answer call within a stipulated period. (Vagadia, 2007)

According to Gartner Dataquest, "business process outsourcing" can be defined as the delegation of one or more IT intensive business processes to an external providers, who in turn owns, administrates and manages selected processes based upon defined measurable performance metrics".

As per the stated definition, the process of outsourcing generally encompasses five stages—

- Strategic thinking, to develop the Organization's Philosophy about the role of Outsourcing in the activities.
- Evaluation and selection to decide on the appropriate outsourcing project and potential locations for the work to be done and service providers to do it.
- Contract development to work out the legal processing and service level agreement (SLA) terms.
- Outsourcing management or governance to refine the ongoing working relationship between the client and outsourcing service providers.
- It starts its work at any place where resources are fully available and they can provide efficiency at relatively cheaper price.

As per the annual report of Planning Commission, 2007-08, the year 2006-07 witnessed a revalidate of the Indian Information Technology business process outsourcing (IT-BPO) growth story driven by maturing appreciation of India's role and growing importance in global service trade. (Annual Report 2007, 2008)

There are many advantages in outsourcing, so most of the organizations choose it. When organizations outsource to countries like India they benefit from lower cost and higher quality services. Outsourcing can help the organizations to make better use of their resources, time, and infrastructure.

R.B.I. report 2004-05, the robust expansion in services output since 2005-06 has also been driven by services export. While the direct contribution of global factors to the services sector output in India came from set-up in external demand for Indian services, the direct contribution came through increasing globalization and the resultant pressure on cost efficiency, which expanded the scopes for outsourcing of business process activities and knowledge. (R.B.I report, 2004-05)

Typical objectives desired from outsourcing include:

- Improving strategic position through increased focus on core function.
- Improving competitiveness through operational performance improvement and access to cheaper and more specialists.
- A reduction is cost through new technology or innovative practices and improves the jobs.

SCOPE OF OUTSOURCING

Scope of Outsourcing can be divided into two services horizontal services and vertical services. (See diagram on the next page)

Horizontal Services

The horizontal services include a vide range of services like H.R. finance and accounting, financial research higher engineering. CRM, which are common to industries. Outsourcing is more common in these processes as they are typically back offices process and do not have any competitive advantages.

Vertical Services

The vertical services are specific claim processing is specific to the insurance sector. Industries that have taken the lead in outsourcing and dominate vertical focused services and financial service, insurance service, health care, etc.

Scope of Outsourcing

Horizontal Service

(A) Financing and Accounting
- (i) transaction management
- (ii) billing
- (iii) tax management
- (iv) financial analysis
- (v) risk
- (vi) financial report management

(B) Customer Service
- (i) inbound calls
- (ii) outbound calls
- (iii) email support
- (iv) market service

(C) Transaction processing
- (i) claim processing
- (ii) credit card processing
- (iii) loan processing
- (iv) cheque processing
- (v) collection

(D) HR
- (i) Payroll
- (ii) benefit administration
- (iii) recruitment
- (iv) training
- (v) E-learning
- (vi) record management

(E) Content Development
- (i) Animation
- (ii) Gaphics

(F) Financial research
- (i) data processing
- (ii) decision support
- (iii) data mining
- (iv) asset management
- (v) market research

(G) Higher Engineering
- (i) Engineering
- (ii) R and D

Vertical Service
- (i) health care
- (ii) financial Service
- (iii) insurance Service
- (iv) airlines services

Health Care Services

It provides health care organization opportunities to improve quality and reduce cost while presenting challenges to the security and privacy of patient data. Health care provider sector is opening its mind to possibilities of business transformation through B.P.O.

Insurance Services

Insurance companies have outsourced process like application processing underwriting claims, adjudication and customer care, etc.

Airlines Services

Airlines have been forerunner in the field of outsourcing customer care, data service, loyalty programs, revenue accounting, and cargo support and revenue recovery process.

There is a rapid transformation during for last few years in these services like governance, retail service, pharmaceutical research service.

E-governance Service

It means an electronic service, which delivers the government's service and information to the public. In this service, outsourcing held the government generally lacking in knowledge of current and emerging. It technology to make the height choice of technology

Retail Service

It implies geographical border and time zones means little for retail and manufacturing enterprises competing for present taste of the people and market share in the global economy.

PHARMACEUTICAL RESEARCH SERVICE

Outsourcing provides the small pharmaceutical company the ability to take new drugs through clinical trials by avoiding the investment in infrastructure, which would normally be necessary. (Patel, 2005)

Global Outsourcing

Global outsourcing is purely enabling business around the world without any hindrance or barriers. As enterprises, think to globalize their outsourcing models. Outsourcing is not a longer process, it is just a short-term quick fix, which achieves cost reduction, and it is a solution to achieve strategic business objectives for the outsourcing companies. They provide job in the area like "chief globalization officer" and "strategic service manger".

TABLE I

World ITES BPO—BPO but Segments, 2001-06

(Figures in US $ million)

	2001	*2006*	*2001-06 (AGR)*
Human resources	7,373	25,555	28.2
Logistics	140,700	308,657	17
Purchasing	5,288	12,185	18.2
Engineering/R and D	69,798	123,882	12.2
Marketing	76,666	108,340	7.2
Sales	107,412	165,736	9.1
FOM	120,635	172,329	7.4
Administration	36,644	53,396	7.8
Legal	111,273	163,962	8.1
Finance/Accounting	36,356	64,872	12.3
Total	712,145	1,198,908	11

Source : IDC.

World-wide spending on (ITES/BPO) service approximately US $ 712 million in 2001. IDC projecting that by 2006, the potential ITES-BPO market may increase to US $1.2 trillion, with an overall compounded annual growth rate CAGR of 11% while traditionally the driver for ITES-BPO activity has been cost reduction companies an increasing vowing these services as strategic and essential elements for organic growth.

TABLE 2

World-wide ITES-BPO Spending by Region 2001-06 and 2002-06

(Figure in US $ million)

Region	*2001*	*2002*	*2006*	*2001-06 (AGR %)*	*2002-06 (AGR%)*
America	4,50,947	4,84,732	6,47,427	9.9	7.5
Europe, Middle East and Africa	1,54,423	1,71,303	2,37,390	11.5	8.5
Asia Pacific	1,06,775	1,17,622	1,94,228	14.7	13.0
World-wide	7,12,145	7,73,657	1,0,79,054	11.0	8.6

Source : International Data Corporation @ NASSCOM.

Table 2 shows that in 2002-06, the CAGR of America is 7.5% Europe, Middle-east and Africa is 8.5% and Asia Pacific is 13.8%. In comparison to America and Europe Middle-East and Africa, Asia Pacific is higher.

OUTSOURCING IN INDIA

Outsourcing Provided in India is outsourcing to India (021)" located in Bangalore, the silicon valley of India. At outsourcing to India, we provided an entire range of technology driver service that can give our organization a competitive advantage. Outsourcings to India have also saved our time, effort, work force and at the same time give as access to skilled and professional service.

The booming Information technology IT segment comprising ITES (Information Technology Enable Service/BPO) are the core sectors that have derived the country into the epicenter of change.

The liberalization of the Indian telecom sector in 1994 gave unexpected boost to the ITES-BPO industry. In no time, India has turned into a hot aim for global offshore outsourcing companies. The expansion in the sector can be attributed to leading. IT gains, captive player can third party service providers, who dominate the Indian ITES-BPO.

India is the most favored IT/BPO destination of the world and the big MNCs interested in outsourcing their operation to BPO in India. India has large skilled human resource. India has inherent strength, which have achieved it a major success as an outsourcing destination because India has large number of graduates in the world.

Global organization choosing outsourcing in India for some reasons, such as:

- Cost effecting services
- Increased efficiency
- Increased productivity
- Shared risk
- Reduced operating cost
- Increased quality
- Better service
- More time to focus on core competencies

Top Drivers for Outsourcing in India

- Two million graduates each year (Existing pool: over 25 million)
- 120,000 Engineers every year vs. 63000 in US.
- Cost reduction up to 50%.
- English speaking and IT knowledge workforce.
- Government support for IT and BPO industry.
- Improving telecom infrastructure.
- Lower infrastructure costs.
- Favorable time lag: 12 hours with US and 5 hours with Europe.
- Overnight turnarounds possible
- Resource with experience of financial and legal systems similar to the west.
- Strong domestic IT services industry to support IT led BPO

The Table 3 shows the increment in India ITES-BPO scenario, due to increase in export and domestic market revenue. The India ITES-BPO in 2004 was 3.4% US $ billion where export was 3.1 and 0.3 which further increased in 2005

TABLE 3
The Indian ITES-BPO Scenario

	2004	*2005*	*2006*	*2007*	*2008*	*Growth % 2008-09*
ITES-BPO US	3.4	5.2	7.2	9.5	12.5	32
Export	3.1	4.6	6.3	8.4	10.9	30
Domestic	0.3	0.6	0.9	1.1	1.6	41

Source : IDC.

to 5.2 from 3.4 and in 2008 is become 12.5, when export and domestic market revenue was 10.9 and the growth percent increased in 2007-08 was 30%, 41% and ITES-BPO was 32%.

A research conducted by Gartner in the year 2008 that India is the undisputed leader in the outsourcing space. India's most prized resource is its rapidly available technical work-force. India has the second largest English speaking scientific professional in the world; secondly, only the U.S. is estimated that India has over 4 million technical workers, over 1832 educational institutions and polytechnic, which train more than 67,785 Computer Software Professionals every year. The enormous base of skilled workers is major draw for global customers.

India is the second largest firm for example Infosys technologies could potentially overtake information technology (IT) services in overall Profit. Infosys BPO currently contributes to around 5% to the firm's total revenue. As stated by Chief Executive S. Gopal Krishna, "If we take BPO alone, it can be ever bigger than IT service because the address sometime like 15-20% spend for over clients organizations, whereas IT usually address 2-5% of expenditure for our clients."

In India, the BPO sector has given another wave in the global outsourcing like KPO-LPO and PPO. KPO-Knowledge Processing Outsourcing, can be defined as pursuing goal by applying the province knowledge, experienced and skill. LPO (Legal Processing Outsourcing) is the global outsourcing. The work is done by to law graduates and experienced lawyers.

The IT Sector can be Classified into four Broad Categories

- *IT Sector*: It can further be categorized into information service. In outsourcing package, software support and installation system, integration processing services, hardware support, installation and IT training education.
- *Engineering Services*: It include mechanical design, electronic system design (including chip/board and embodied software design), design validation testing industrialization and prototyping.
- *IT Enabled Services*: These are services that use telecom network or the internal for example, remote maintenance, back office operation, data processing call centres, business process outsourcing.
- *E-Business*: (Electronic businesses) is carrying out business on the internet including buying-selling serving customer and collaborating with business on the internet.

Top BPO Companies in India

- TCS BPO
- Infosys BPO Ltd.
- Wipro BPO Solutions
- Cognizet
- HCL BPO
- HP India
- Mphasis BPO

For the younger generation, the call centers are a welcome habit. The Industry has bamboo such magic around the entire nation that these days a city without a call centre would be unreachable. Call centres contribute a fair share to the revenue of the Indian BPO industry about 70% of the BPO industry's profit come from call centres, 20% from high-volume, low value data work and the remaining 10% from higher value information work.

Outsourcing success depends on three factors: executive levels support in the client organization for the outsourcing mission; ample communication to affected employees; and the client's ability to manage its service providers. The outsourcing professionals in charge of the work on both the client and provider sides need a combination of skills in such areas as negotiations, communication, project management, the ability to understand the farm and conditions of the contract and service level agreement (SLA's).

CONCLUSION

The business process outsourcing is becoming a cause of great concern for some big economies specially U.S.A. where government is forced to take stern action against outsourcing. However, at the same time outsourcing is being proved good for countries like India, China and Philippines to name few. China has built a strong base to provide the required service to outsourcer through which China is able to provide jobs to its youth brigade in a quite good number. Many outsourcers for getting their work done better by the qualified persons are also approaching India. India has plenty of skilled workers like engineers and others with the good command over English language and knowledge of computer education. In fact, India is considered a very good destination for outsourcers to get their work efficiently done. India also gains a lot from it as lots of job opportunities are created for Indian youth. If India concentrates on improving its services it will be advantage one for the country to attract more outsourcers. This act is certainly going to have positive impact on Indian job market.

REFERENCES

Patel, Alpesh (2005), Outsourcing Success, Palgrave MacMillan.

Bharat Vagadia (2007), Outsourcings of India : A Legal Handbook, (amazons.com), Springer.

Blokdijk Gerard (2008), Outsourcing 100 success secret: (lulu.com)

IDC report : (2006), NASSCOM.

Reserve Bank of India report : (2004-05), Reserve Bank.

Annual report of Planning Commission : (2007-08), Government of India.

www.nasscom.com

www.garter.com

http: / /enwikipedia.org

http: / /ezinearticals.com

CHAPTER

16

Labour Force Growth and Employment Requirements in Non-Farm Sector of Rural India

K.G. Suresh Kumar

INTRODUCTION

Rural non-farm economy, in recent times, is considered as an effectual strategy for decentralization of economic activities to rural India. The Economic Census of India estimates that around 41.89 million rural people are employed in non-agricultural establishments which registered a growth rate of 4.56% during 1998-2005. However, the sector has been contending with a number of factors like inadequate rural infrastructure, particularly roads, electricity and communication facilities, lack of sufficient skilled labour and adequate access to credit, information and training facilities, etc. The present study investigates the strengths and weaknesses of the rural-non-farm-sector of India analyzing the structure and growth of rural-non-farm-sector and its trends towards employment and income generation to arrive at certain inferences like formulation of possible approaches with a view to promote rural-non-farm-sector self-sustaining in the changing competitive environment.

Farm activity means agricultural activity and non-farm activity is used synonymously with non-agricultural activity. There are two alternative approaches to define rural-non-farm activities (Saith, 1992). The first is the locative approach in which the primary criterion is that a Rural Non-farm activity is performed in a location which falls within a designated rural area. The second is based on the linkage approach where an industrial enterprise generates significant development linkages with the rural areas. For purposes of this study we are using the first. Rural-Non-Farm-Sector includes all economic activities viz., household and non-household manufacturing, handicrafts, processing, repairs, construction, mining and quarrying, transport, trade, communication, community and personal services, etc. in rural areas. Rural-Non-Farm-Activities thus, play an important role to provide supplementary employment to small and marginal farm households, reduce income inequalities and rural-urban migration.

Though, agricultural sector has played a very significant role for generation of rural employment in the Asia and Pacific region, its contribution to the overall economy has greatly reduced in the recent past (Asian Productivity Organization, 2004). Therefore, development of various non-farm-activities can effectively be exploited as a potent stimulator for further economic growth offering rural communities better employment prospects on a sustainable basis.

IMPORTANCE OF EMPLOYMENT IN RURAL NON-FARM SECTOR

The non-farm sector, particularly in rural areas is being accorded wide recognition in recent years for the following reasons:

- Employment growth in the farm sector has not been in consonance with employment growth in general.

- A planned strategy of rural non-farm development may prevent many rural people from migrating to urban industrial and commercial centers.
- When the economic base of the rural economy extends beyond agriculture, rural-urban economic gaps are bound to get narrower along with salutary effects in many other aspects associated with the life and aspirations of the people.
- Rural industries are generally less capital-intensive and more labour absorbing.
- Rural industrialization has significant spin-offs for agricultural development as well.
- Rural income distribution is much less unequal in areas where a wide network of non-farm avenues of employment exists; the lower strata of rural societies participate much more intensely in non-farm activities, though their involvement is much less remunerative as compared with that of the upper strata.

LABOUR FORCE GROWTH AND EMPLOYMENT REQUIREMENTS

To provide employment for additional labour force which is estimated to grow at the rate of 2.51% per annum during the Tenth Plan period (2002-07), besides reducing the backlog of unemployment accumulated from the past, is a daunting challenge for India. Despite an expected reduction in the growth rate of population to 1.63% per annum by 2002-07, the labour force growth reached 2.51% per annum. This is attributed to change in the age structure of the population and an increase in the population in the most active working age group of 15-59 (Tables 1 and 2). Growth or decline in the labour force participation rates (LFPRs) depends on certain factors. With the increasing thrust on education, LFPRs in the age group 15-19 years will decline. On the other hand, with improved health and longevity, LFPRs in the older age groups, particularly 50+ years, will increase by 7.9-8.9% during the Eleventh Plan period (Table 1). The labour force projected to increase by 40.02 million in special group and 55.82 million in working age group (15+) during the period of 2007-12 implies the need for an increase in the pace of creation of additional work opportunities commensurate with the growth of labour force (Table 2).

TABLE I

Age Structure of Population

Age group	*2001*	*2006*	*2011*	*2016*
0-14	35.6	32.5	29.7	27.1
15-59	58.2	60.4	62.5	64.0
60+	6.3	7.0	7.9	8.9
All age groups	100.0	100.0	100.0	100.0
Population	1,027.0	1,113.7	1,194.4	1,267.5

Note : Age distribution in per cent, population in million.
Source : Planning Commission, Govt. of India, Tenth Five Year Plan, 2002-07.

TABLE 2

Growth in Population and Labour Force Projection by Age Group

Population/labour force	*2002-07*	*2007-12*	*2012-17*
Population (all age groups)	1.63	1.41	1.20
Population (15-59 years)	2.41	2.08	1.70
Labour force (15-59)	2.42	2.15	1.78
Population (15+)	2.57	2.26	1.93
Labour force (15+)	2.51	2.25	1.92

Note : (1) Data for respective years are per cent per annum (2) Labour force projections here are on the basis of labour force participation rate for each quinquennial age group remaining unchanged, i.e. the changes in labour force growth in relation to population are due to changes in the age composition of the population.
Source : Planning Commission, Govt. of India, Tenth five year plan, 2002-07.

Unemployment is estimated at 21.15 million, 5.11% of the total population (Table 3). To achieve full employment by 2011-12, it is estimated that employment should grow at 2.7% per annum based on the proposed policy and programmes in the Tenth Plan. This would require GDP to grow at 8% per annum. It is observed that the proportion of person-days of the usually employed, utilized for work, is lower for females as compared to the males throughout the period 1987-88 to 1999-2000. During 1999-2000, this proportion was estimated at about 68 per cent for females as against 90 per cent for males in rural India. The distribution obtained from the 1999-2000 survey is presented in Table 4. Distribution of male and female work force in non-farm activity in rural areas during 1983-2005 has been depicted in Figure 1.

TABLE 3

Increase in Labour Force and Working Age Population

Basis of scenario	*2002-07*	*2007-12*
Increase in labour force (Specific group)	35.29	40.02
Increase in working age population (15+)	55.25	55.82

Note : Data for respective five-year blocks in million.

Source : Report of Planning Commission Special Group on creation of 10 million employment opportunities per year since 2002.

TABLE 4

Labour Force, Employment and Unemployment

Parameter	*1999-2000*	*2001-02*	*2006-07*	*2011-12*	*% per annum*
Labour force	363.33	378.21	413.50	453.52	1.80
Employed	336.75	343.36	392.35	451.53	2.70
Unemployed	26.58	34.85	21.15	1.99	-9.50
Unemployed rate (%)	7.32	9.21	5.11	0.44	—

Note : (1) Data for respective years in million, (2) Special group estimates on CDS basis.

Source : Planning Commission, Govt. of India, Tenth five year plan, 2002-07

TABLE 5

Distribution of Male/Female per 1000 Person-days Usually Employed in Rural India

Current daily status	*Male*			*Female*		
	1999-2000	*1993-94*	*1987-88*	*1999-2000*	*1993-94*	*1987-88*
Employed	897	909	926	676	663	638
Unemployed	53	40	27	41	30	26
Not in labour force	51	51	47	283	306	336
All	1000	1000	1000	1000	1000	1000

Source : NSSO, India.

FIG. 1

Distribution of Rural Workforce in Non-farm-activities

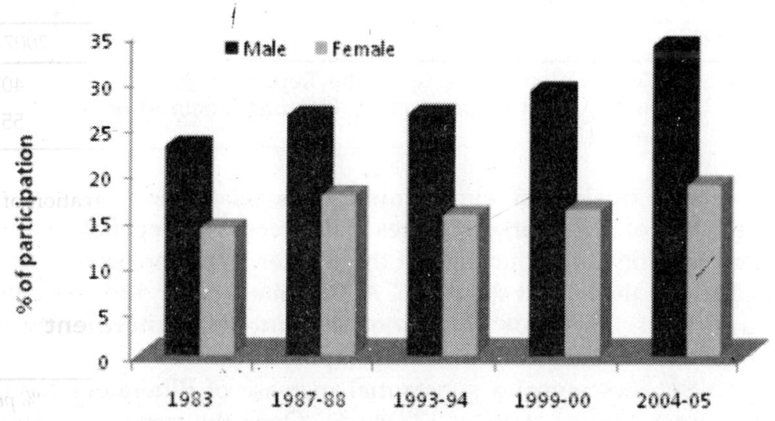

Source : NSS Report No. 522 (62/10/01), Employment and Unemployment situation in India, July 2005 to June 2006.

QUALITY OF EMPLOYMENT

Nearly 51.3% of the workforce is either illiterate or educated below the primary level (Table 6). Even in industries where skill up-gradation for raising productivity requires a reasonable level of educational standard, 39.6 and 74.0% of the workforce consisting of male and female respectively was illiterate. The pattern of development of the Indian economy requires skill and education levels not immediately available thereby leading to a mismatch between the demand and supply of labour services—an increasing level of unemployment in one segment of the labour market coupled with a labour shortage in the other.

TABLE 6

Distribution of Workers in the Rural Area by the Level of Education (%), 1999-2000

Rural area	*Level of general education*				*Share in workforce*
	Illiterate	*Up to Primary*	*Secondary and above*	*Total*	
Male	39.6	27.3	33.1	100	49.74
Female	74.0	15.5	10.5	100	25.77
Person	51.3	23.3	25.4	100	75.51

Source : This section draws upon the Report of the Task Force on Employment Opportunities, Planning Commission, Govt. of India (2001).

The quality of employment also relates to wages and security of the worker. Wages paid/received depend on the productivity and education of the worker. The more skilled and educated the higher the wage. A full-time worker, illiterate and unskilled, may not earn adequate income. There is high incidence of this kind of underemployment.

Surveys reveal a substantial increase of illiterates among the unemployed persons (Table 6). Over the period of 1993-2005, the proportion of those with educational level up to primary among the unemployed increased from 1.9-3.0% and 2.6-3.1% in case of male and female respectively while unemployed decreased from 8.3 to 6.5% among the secondary school or higher educational level and increased from 9.8 to 18.2% in case of female. It shows not only a lack of sufficient efforts and resources to educate the workforce, but also a mismatch between the kind of job opportunities that are needed and those that are available in the job market. The situation also indicates the need for more jobs requiring skilled labour rather than the simple low productive manual labour.

TABLE 7

Education Profile of the Unemployed in India (%)

Year	*Illiterate*	*Literate*		*Total*
		Up to Primary	*Secondary and above*	
Rural Male				
1993-94	1.8	1.9	8.3	3.1
1999-00	3.0	3.0	7.3	3.9
2004-05	2.7	3.0	6.5	3.8
Rural Female				
1993-94	2.2	2.6	9.8	2.9
1999-00	2.7	2.6	16.9	3.7
2004-05	2.5	3.1	18.2	4.2

Source : NSSO Report No. 515 (61/10/1).

STRENGTHS AND WEAKNESSES OF NON-FARM SECTOR

Non-farm activities either keep the poor falling into deeper poverty or are advantageous in lifting the poor above the poverty line. Keeping this in view, it becomes imperative to identify the strengths and weaknesses of the non-farm sector in India to focus on, in order to alleviate poverty. The strengths and weaknesses of rural non-farm sector in India as highlighted by Mukherjee and Zhang (2005) have been discussed below.

Strengths

Institutional Basis for Rural Non-farm Sector

In India, the institutions underlying the development of the rural non-farm sector are very strong. These include secure property rights; a well-developed financial system with preferential access to credit for the sector; supporting institutions such as the KVIC, State Khadi Board, NHHDC, Small Industries Development Bank of India (SIDBI), State industrial corporations; policies and programs promoting linkages with agriculture, especially agro-industries; domestic marketing channels for rural non-farm production; as well as government support in export promotion. The institutional mechanisms for a rapid growth of the rural non-farm sector are already in place.

Decentralization Process

Over the last two decades the State governments in India have been able to exercise far more independence in decision-making than in the pre-1980 period. Regional parties are an integral part in coalition governments at the Center. In turn, they have negotiated economic autonomy in the formation of state specific policies for development. Moreover, with the opening up of the economy in 1991, foreign direct investment (FDI) has come to play an important role in the overall policy environment. State governments are in competition with one another to attract higher FDI levels both in manufacturing and infrastructure. In some ways, it mirrors the path followed by China, although the volume of FDI coming to India is less than 10 percent of what is flowing into China. On the positive side, however, this creates an opportunity for higher levels of investment in the future.

Weaknesses

Infrastructure

The most significant bottleneck in generating higher levels of rural non-farm activity in India is the quantity, quality and reliability of infrastructure. For example, the World Bank Investment Climate Survey for India indicates that power outages were one of the most serious obstacles to the development of the non-farm sector (Economist, 2005; World Bank, 2005). Although corrective steps are now being taken, increased infrastructure remains the most important priority for the future. To achieve a sustained growth rate of 8-9 percent, the investment rate has to be stepped up from the current level of 24 percent to nearly 35 percent over the next decade, with investment directed at the rural sector (Planning Commission, 2000).

Regulatory Restrictions on Small-scale Sector

Regulation of the small-scale sector constitutes an important aspect of non-farm development policy in India. In the initial stages, capital investment restrictions were imposed to protect the small-scale sector, especially in rural areas, from predation by large industry. Reservation of products for the sector was initiated to create a domestic market and quantitative restrictions imposed to protect them from competition from imports. At the end of the 1990s, however, these very policies have become detrimental to the dynamism of the small-scale sector, especially in the rural areas. Capital investment limits have discouraged economies of scale, and concessions offered to small industry have created adverse incentives against re-investment. Several official reports have recommended a substantial increase in the capital investment limit (from the present level of around $200,000) to make better use of technology and improve productivity (Planning Commission, 2000). However, no such policy announcement has been made as yet.

Reservation of products for the small-scale sector has gradually reduced in significance, although this has created rents within the system. The decision of the government to put all the reserved items in the open general license category from April 2005 would mean free import of such items at the prevailing tariff rate. With the latter slated to come down over time to around 20 percent as per the WTO norms, this will effectively signal the end of protection for the small-scale industry.

Quality of Manpower

High levels of illiteracy in rural India have hampered the growth of the rural non-farm sector. Education has both intrinsic and instrumental value. Apart from having a positive correlation with wages, a minimum basic standard of education is necessary to apply for credit, to be aware of one's rights and responsibilities and to deal with instances of corruption and malpractice. Often, a lack of education is intrinsic to poverty, which seems to have been the case in India until recently.

In the rural areas, lack of education leads to labor being stagnant in agriculture, or moving to casual work occupations in the non-farm sector, and not to salaried employment with higher wages and benefits. Together with lack of technical skills, there is little incentive for rural firms to invest in technology, leading to low levels of labor productivity in the rural manufacturing sector compared to urban manufacturing (Chadha, 2003). The same is true of the service sector as well, which has the potential for expansion given the already strong base in the urban economy. Higher investment to improve both the quality and the access to education (primary, secondary and above) needs to be a priority for policymakers.

Forward and Backward Linkages

Absence of appropriate forward and backward integration greatly affects performance of non-farm activities in rural areas. Forward linkages of the rural non-farm sector serve as inputs to other sectors. Also, in backward linkages the rural non-farm sector demands the outputs of other sectors. Empirical studies indicate that forward linkages from rural non-farm activities to agriculture (rurally produced agricultural inputs) are particularly important where traditional agricultural technologies are utilized, while in case of backward linkages between rural non-farm activities and agriculture, especially the linkages between rural agricultural processing and the agriculture sector and between rural transport and rural marketing activities are quite significant for rural economic development. However, gaps in the integration of the production linkages brought about by poor infrastructure, low accessibility of market, support service weaknesses and intervention of middle men have constrained the development of non-farm enterprises in India.

CONCLUSION

The rural non-farm sector is increasingly playing an

important role in the development of rural areas in Asia and the Pacific region. Specifically, as agriculture in the region declines in importance in terms of its contribution to the economy, the rural non-farm sector will need to become more and more a major provider of employment and income to many rural folks. It should be noted, however, that rural non-farm enterprises are not a substitute for employment in agriculture but rather as a supplementary measure. Agricultural development is still important and should be pursued as a necessary precondition. The promotion of rural non-farm enterprises also should be undertaken within the broader context of rural development.

Efforts are needed to identify appropriate and effective institutional vehicles for development of non-farm sector policy and interventions for creating employment opportunities. Many strategies and programs to promote rural non-farm enterprises have been formulated in various countries. China's labour-intensive township and village enterprises, for example, often described as the "engine of growth" behind that country's remarkable growth during the past decades represents the vanguard in China's new capitalism. The township and village enterprises are hybrid institutions generally unusual alliances between township and village enterprises entrepreneurs and local government officials (acting in the capacity of "owners" of township and village enterprises). In this regard, the role of government is crucial, especially in the provision of necessary infrastructure and other support services in the country. It is also vital to improve the marketing links between the village entrepreneurs and the larger business firms located in the towns/cities. Such strategic alliances or partnerships can contribute to the sustainability of small village and tiny enterprises in the rural areas. Other important considerations that need to be focused on may include human resource development, financial/credit facilities, research and development and women's participation with a view to making the activities self-sustaining in the changing competitive environment.

References

Asian Productivity Organization (2004), Non-farm Employment Opportunities in Rural Areas in Asia, Tokyo, p. 13.

Chadha, G.K. (2003), Rural Non-farm Sector in the Indian Economy: Growth, Challenges and Future Direction, *(Mimeo)*, International Food Policy Research Institute, Washington D.C.

Economist (2005), India's Electricity Reforms: A Power Shortage May Thwart India's Rush to Modernity, *Economist*, September 22: 83-84.

Mukherjee, A. and Zhang, X. (2005), Rural Non-farm Development in China and India: The Role of Policies and Institutions, Development Strategy and Governance Division, International Food Policy Research Institute, pp. 33-35.

Planning Commission (2000), Report of the Task Force on Employment Opportunities, Government of India: New Delhi.

Saith, A. (1992), The Rural Non-farm Economy: Processes and Policies, International Labour Force, Geneva (mimeo), pp. 12-16.

Saxena, M. (2004), India (1) In: Non-farm Employment Opportunities in Rural Areas in Asia, Tokyo, p. 90.

World Bank (2005), Investment Climate Survey Online http://iresearch.worldbank.org/ics/jsp/index.jsp

CHAPTER

17

Rural Non-Farm Employment and Women in India

ARIFA SALEEM AND SEHAR SHABZAD

INTRODUCTION

As mentioned above, the economy consist of—farm and non-farm sectors. Farm sector mainly consist of farm activities which means agricultural activities, whereas, non-farm sector consist of non-farm activities which is used synonymously with non-agricultural activities. There are two alternative approaches to define rural non-farm activities. The first approach is the Locative Approach in which the primary criterion is that a rural non-farm activity is performed in a location which falls within a designated rural area. The second approach is based on the Linkage Approach where an industrial enterprise generates significant development linkages with rural areas. For rural non-farm sector includes all economic activities like—household and non-household manufacturing, handicrafts, processing, construction, mining, quarrying, transport, trade, communication, community and personal services, etc. in rural areas.

IMPORTANCE OF RURAL NON-FARM SECTOR

The non-farm sector, particularly in rural areas is being accorded with recognition in recent years for the following reasons:

- Employment growth in the farm sector has not been in consonance with employment growth in general.
- A planned strategy of rural non-farm development may prevent many rural people from migrating to urban industrial and commercial centers.
- When the economic base of the rural economy extended beyond agriculture, rural-urban economic gaps are bound to get narrower along with salutary effects in many other aspects associated with the life and aspiration of the people.
- Rural industries are generally less capital intensive and more labour-absorbing.
- Rural industrialization has significant spin-offs for agricultural development as well.
- Rural income distribution is much less unequal in areas where a wide network of non-farm avenues of employment exists; the lower strata of rural socièties participate much more intensely in non-farm activities through their involvement is much less remunerative as compared with that of the upper strata.

Thus, rural non-farm economy, in recent times is considered as an effectual strategy for decentralization of economic activities to rural India. Over the years, it was regarded as a potential source of employment generation in the economy

GROWTH OF NON-FARM EMPLOYMENT IN INDIA

The key to India's development lies in the development of its rural areas. The agricultural sector occupies a pivotal place in the national economy both in terms of its contribution to the GDP and employment generation. However, segmenting rural employment growth into the farm and non-farm employment growth had been significantly higher than farm sector employment growth. According to the economic census, the share of employment in the farm sector was as high as around 75 percent during 1971. By 1991, it had declined to around 65.5 percent. Consequently, the corresponding share of non-farm employment had registered an increase from around 25 percent to 35 percent between 1970 and 1991. It was estimated that around 41.89 million persons worked in rural non-farm sector which constitute 46.55 percent of the total employment in non-agriculture sectors including both rural and urban areas.

Fig. 1

Share of Rural Non-farm Workers during the Period 1977-78 to 2004-05 (in percentage)

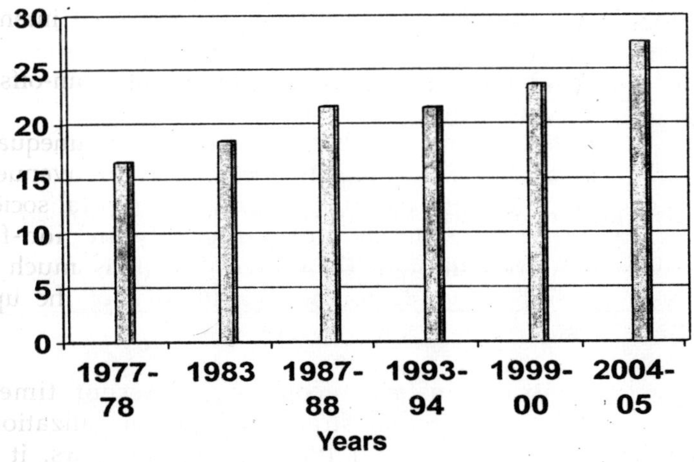

Source : Various Relevant NSSOI Rounds.

Figure 1 establishes that the proportion of non-farm employment in total rural employment has risen from 16.6 percent in 1977-78 to 18.5 percent in 1983, 21.7 percent in 1987-88, 23.7 percent in 1999-2000 and finally to 27.6 percent in 2004-05. It must be noted that for a brief period of stagnation between 1987-88 and 1993-94, the figure rose consistently to record a total increase of 11 percent points between 1997-78 and 2004-05. The increase during the approximately 10 years long sub-periods between 1977-78 and 1987-88 and 1993-94 and 2004-95 amounted to 5.1 percent and 6 percent points respectively.

WOMEN IN RURAL NON-FARM SECTOR IN INDIA

Women are considered to be the important part of the economy as well as for non-farm sector. It plays important and crucial role in the process of development by contributing monetary and non-monetary inputs towards national income. The development of women is considered as the barometer of nation's development. According to census 2001, around 25.6 percent of female population is workers. Majority are employed in rural areas. The contribution of female workers in non-farm sector was just below 4 percent in 1971 while it was increased to 7 percent in 1991. However, over the years, the concentration of women workforce has been increasing at much faster rate than male in non-farm employment. This can be analyzed from the Table 1.

TABLE I

Distribution of Rural Workforce in Non-farm Activities

Year	*Person*	*Males*	*Females*
1972-73	N.A	16.7	10.3
1977-78	16.6	19.3	11.8
1983	18.5	22.2	12.5
1987-88	21.7	25.4	15.3
1993-94	21.6	26.0	13.8
1999-2000	23.7	28.6	14.6
2004-05	27.6	33.5	16.7

Table 1 reveals a much higher rate of increase in non-farm employment among male workers than their female during 1977-78 and 2004-05. The percent of male rose from 16.7

percent in 1972-73 to 33 percent in 2004-05 whereas that of female registered a rise from 10.3 percent to 16.7 percent. The increase in both categories was noticeably sharper during period 1999-2000 to 2004-05 than in earlier periods. Quite significantly, the male workers increased by 4.9 percent points while the female by only 2 percent points.

The contribution of women in non-farm sector or in employment can also be examined with the help of "Labour Force Participation Rate" and "Work Force Participation Rate". It is given in Table 2.

TABLE 2

Labour Force Participation Rate and Workforce Participation Rate in Rural India (UPS)

Year	*Labour force participation rate*		*Work force participation rate*	
	Male	*Female*	*Male*	*Female*
1983	540	252	528	248
1987-88	532	254	517	245
1993-94	549	237	538	234
1999-2000	533	235	522	231
2004-05	546	249	535	242

Source : NSS Report of Employment and Unemployment Situation in India.

From the Table 2, it was observed that after the decline in labour force participation and work force participation rates during 1993-94 to 1999-2000, both these indicators recovered during the period 1999-2000 and 2004-05 for both male and female. The male labour force participation rate in the rural sector increased from 533 per thousand to 546 while that of female increased from 235 per thousand to 249. Similarly, the work force participation rate of male is increased from 522 per thousand to 535 and for female it increased from 231 to 242 per thousand

Since 1983, the highest recorded female labour force participation rate was in the period 1987-88 at 2.54 per

TABLE 3

Industrial Composition of Rural Workers (UPS)

	Year	0	1	2 and 3	4	5	6	7	8	RNFS
(1)	(2)	(3)	(4)	(5)	(6)	(7)	(8)	(9)	(10)	(11)
Rural										
Male	1983	77.2	0.6	7.1	0.2	2.3	4.4	1.7	6.2	22.8
	1987-88	73.9	0.7	7.6	0.3	3.7	5.2	2.1	6.4	26.1
	1993-94	73.7	0.7	7.0	0.3	3.3	5.5	2.2	7.1	26.3
	1999-2000	71.2	0.6	7.3	0.2	4.5	6.8	3.2	6.1	28.8
	2004-05	66.2	0.6	8.0	0.2	6.9	8.3	3.9	5.9	33.8
Rural										
Female	1983	86.2	0.4	6.5	0	0.9	2.2	0.1	3.4	13.8
	1987-88	82.5	0.5	7.5	0	3.2	2.4	0.1	3.7	17.5
	1993-94	84.7	0.5	7.5	0	1.1	2.2	0.1	4.0	15.3
	1999-2000	84.1	0.4	7.7	0	1.2	2.3	0.1	4.3	15.9
	2004-05	81.4	0.4	8.7	0	1.7	2.8	0.2	4.6	18.6

Note : Agriculture (0), Mining and quarrying (1), Manufacturing (2 and 3), Electricity and water (4), Construction (5), Trade, Hotel and Restaurant (6), Transport, Storage and Communication (7), Other Services (8), Rural non-farm sector (RNFS).

Source : NSS Report No. 515, Employment and Unemployment Situation in India.

thousand. It is common knowledge that 1987-88 was a year of severe drought in the economy, when the female labour force participation rate peaked. Since that peak the female labour force participation rate has continuously declined to reach the lowest in 1999-2000 at 235 per thousand. Thereafter, it suddenly shot up to 249 per thousand in 2004-05. Similarly, female work force participation rate which has secular decline in 1999-2000 suddenly increased in 2004-05 to 242 per thousand.

The non-farm sector consists of various industries. To analyze the employment of women in non-farm sector we have to analyze the women employment in these industries. This can be analyzing from Table 3.

Analysis of the industrial composition of rural workers shows that the inertia among the rural male workers against inter-sectoral mobility seems to be gradually reducing. The total share of employment in the agriculture sector had decline from 77 percent to 66 percent from 1983 to 2004-05. The largest decline was in the period of 1999-2000 to 2004-05. Correspondingly, the rural non-farm sector employment share among male increased from 23 percent in 1983 to 34 percent in 2004-05. This increased in rural non-farm sector employment was spread within the manufacturing sector, construction, trade, hotel and restaurant and transport, storage and communication.

Similarly, the dependence of female on agriculture sector decline from 86.2 percent to 81.4 percent from 1983 to 2004-05, that is, just by 5 percent points. From table we observed that, there are hardly any shifts to rural non-farm sector in the case of female employment. The share of female employment in rural non-farm sector has increased from 13.8 percent to 18.6 percent during the entire period. This shift was spread within the manufacturing sector, construction and other services.

WEAKNESSES OF RURAL NON-FARM SECTOR IN INDIA

Infrastructure

The most significant bottleneck in generating higher level of rural non-farm activity in India is the quality, quantity and reliability of infrastructure like poor condition of roads, non-availability of electricity, etc. which adversely affect the production and productivity of the sector.

Regulatory Restrictions on Small Scale Sector

Regulation of the small scale sector constitutes an important aspect of non-farm development policy in India. In the initial stages, capital investment restrictions were imposed to protect it. Reservation of products for the sector was initiated to create a domestic market and quantitative restrictions are imposed to protect them from competition. All these restrictions adversely affect the non-farm economy in India

Quality of Manpower

Higher levels of illiteracy in rural India have hampered the growth of the rural non-farm sector. It lead to restrict the labour in agriculture, a basic standard of education is necessary to apply for credit, be aware of one's rights and responsibilities and to fight against corruption and malpractices in India.

Lack of Technical Skill

There is little incentives for rural non-farm sector to invest in technology, leading to low level of labour productivity in the sector as compared to other sectors like agriculture, urban manufacturing, etc. the same is also true for services sector also.

Forward and Backward Linkages

Absence of appropriate forward and backward integration greatly affects performance of non-farm activities in rural areas. Forward linkages of the rural non-farm sector serve as inputs to other sectors. Also, in backward linkages the rural non-farm sector demands the outputs of other sectors.

Lack of Credit Facilities

In spite of new source of credit, such as credit programs and cooperatives after sponsored by non-government organisations (NGOs), the non-farm sector is still experiencing credit constraint which will adversely affect the growth and development of the sector.

Lack of Women Work Participation in Rural Non-farm Sector in India

The important constraint on rural non-farm sector is the ineffective participation of women in the sector due to various social, cultural and religious factors.

Lack of Marketing of Rural Non-farm Products

The important constraint of rural non-farm employment is the lack of marketing of its products in the economy. Since the products were not matched with the existing competition in the market by its quality, thus lack in marketing, etc.

ROLE OF GOVERNMENT IN NON-FARM SECTOR

An important objective of development planning in India, has been to provide for increasing employment opportunities in the economy. To fulfill this objective government announced various schemes, policies and programmes time to time. Government also announced various schemes and polices for the rural employment growth and development. Some of them are given below.

Rural Landless Employment Guarantee Programme (RLEGP)

The rural landless employment guarantee programme was started in the rural areas on 15th August, 1983, its objective of creating employment, constructing the productive projects and improving the rural life. The total expenditure of this programme is financed by the Central Government.

Prime Minister's Rozgar Yojana (PMRY)

The scheme was launched on October 2, 1993 and initially was in operation in urban areas. From April 1994, the scheme is being implemented throughout the country. Self-Help Groups (SHGs) are considered eligible for financing under PMRY effective from December 8, 2003.

Note : The newly created Ministry of Micro, Small and Medium Enterprises (MSME) has recently submitted a proposal to the Planning Commission for the merges of PMRJ and REGP into the single employment generating scheme for the youth. MSME ministry has sought this amalgamation for the implementation of employment generation schemes at the grass-root level.

Jawahar Gram Samridhi Yojana (JGSY)

It was launched on April 1, 1990 to ensure the development of rural infrastructure at the village level. The objective of JGSY was generation of wage employment for the unemployed poor in the rural areas.

Swarna Jayanti Shahari Rozgar Yojana (SJSRY)

In the Golden Jubilee of Independence a new scheme named Swarna Jayanti Shahari Rozgar Yojana (SJSRY) introduced which became operational since December, 1, 1997. It provides gainful employment opportunities in the economy.

Sampoorna Grameen Rozgar Yojana (SGRY)

The ongoing schemes—the Employment Assurance Scheme (EAS) and the Jawahar Gram Samridhi Yojana (JGSY) was merged into the Sampoorna Grameen Rozgar Yojana (SGRY) on September 25, 2005, the objective of the programme is to provide additional wage employment in the rural areas and also it provide food security, along-side creation of durable community, social and economic assets in the areas.

National Rural Employment Guarantee Programme (NREGP)

After the notification of the National Rural Employment Guarantee Act on September 7, 2005, a new scheme named "National Rural Employment Guarantee Programme (NREGP)" has been launched on February 2, 2006. The on-going programmes of Sampoorna Grameen Rozgar Yojana (SGRY) and National Food for Work Programme (NREGP) are being merged within the NREGP in the 200 districts in the initial stages. Later on, the government announced to cover all districts of the country under NREGP from the year 2008-09.

CONCLUSION

The rural non-farm sector is increasingly playing an important role in the development of rural areas. Specifically, as agriculture in the region declines its importance in terms of its contribution to the economy, the rural non-farm sector will need to become more and more a major provider of employment and income to many rural folks. It should be noted; however, that rural non-farm sector is not a substitute for the employment in agriculture but rather as a supplementary measure.

Efforts are needed to identify appropriate and effective institutes and vehicles for the development of non-farm sector policy and interventions for creating employment opportunities. There is need to focused on the human resource development, financial/credit facilities, research and development and women participation with a view to making the activities self-sustaining in the changing competitive environment.

REFERENCES

Basant, R. and B.L. Kumar (1989): Rural Non-Agricultural Activities in India: A Review of Available Evidence, *Social Scientist*, Vol. 17, Nos. 1-2.

Basu, D.N. and S.P. Kashyap (1992), Rural Non-Agricultural Employment in India—Role of Development Process and Rural-Urban Employment Linkages, *Economic and Political Weekly*, Vol. 27, Nos. 51/52.

Basant, R., B.L. Kumar and R. Parthasarthy (1998), (Edited). Non-Agricultural employment in Rural India: The Case of Gujarat, Rawat Publications, Jaipur, India.

Centre for Monitoring Indian Economy, CMIE (2002), Profiles of districts, October Issue, Centre for Monitoring Indian Economy (CMIE), Mumbai.

Central Statistical Organization (CSO), Annual Series of Income—various Issues, New Delhi.

Dev, S.M. (1990): Non-agricultural Employment in Rural India—Evidences at a Disaggregate Level, *Economic and Political Weekly*, Vol. 25, No. 28.

Disaggregate Level, *Economic and Political Weekly*, Vol. 25, No. 28.

———, (2001), Economic Census, (1998), All India Report, New Delhi.

Fisher, T. and V. Mahajan (1998), The Forgotten Sector: Non-farm Employment and Enterprises in rural India, London: Intermediate Technology Publications.

Four Sub Rounds of the NSS 27th round Survey on Employment Unemployment, 1972-73, *Sarvekshna*, Vol. 1, No. 2.

Hazell, P. and S. Haggblade (1991), rural Growth Linkages in India, *Indian Journal of Agricultural Economics*, Vol. 46, No. 4.

National Sample Survey Organization (NSSO) (1977), employment Unemployment Situation at a Glance: A Brief Note Based on Data of All the National Sample Survey Organization (NSSO) (1981), Survey Results of the Second Quinquennial Survey on Employment and Unemployment: All India NSS 32nd Round, *Sarvekshna*, Vol. 5, Nos. 1 and 2.

National Sample Survey Organization (NSSO) (1987), Report on the third Quinquennial Survey on Employment and Unemployment Survey Results—All India, Department of Statistics, Government of India.

National Sample survey Organization (NSSO) (1990), Results of the Fourth Quinquennial Survey on Employment and Unemployment, NSS 43rd Round, *Sarvekshna*.

National Sample Survey Organization (NSSO) (1997), Key Results on Employment and Underemployment, Department of Statistics, Government of India.

National Sample Survey Organization (NSSO) (2001), Employment and Unemployment Situation in India, 1999-2000, NSS 55th Round (July 1999-June 2000), NSS Report No. 458, May.

National Sample Survey Organization (NSSO) (2002), Unorganised Manufacturing in India, 2000-01, NSS 56th Round, NSS Report N.o. 479.

National Sample Survey Organization (NSSO) (2003), Unorganized Service Sector in India, 2001-02, NS National Sample Survey Organization (NSSO) (206), Employment and Unemployment Situation in India, 1999-2000, NSS 61st round, NSS Report No. 515, September.

Vyas, V.S. and P. Bhargava (1995), Public Intervention for Poverty Alleviation: An Overview, *Economic and Political Weekly*, Vol. 30, No. 41/42.

Unni, J. (1998), Non-Agricultural Employment and Poverty in Rural India—A Review of Evidence, *Economic and Political Weekly*, Vol. 33, No. 13.

———, (1991), Regional Variations in Rural Non-Agricultural Employment—An Exploratory Analysis, *Economic and Political Weekly*, Vol. 26, No. 3.

CHAPTER

18

Rural Non-Farm Employment in Bihar and Jharkhand : Trend and Issues

CHANDRA PRAKASH AZAD AND BHAWANA JHA

INTRODUCTION

Debates about rural development attach increasing importance to the rural non-farm sector. Traditionally, rural households in developing countries have been viewed as though they were exclusively engaged in agriculture. There is mounting evidence, however, that rural households can have highly varied (and often multiple) sources of income. Rural households can, and do participate in a wide range of non-agricultural activities, such as wage employment and self-employment in commerce, manufacturing and services, alongside the traditional rural activities of farming and agricultural labour. Such non-farm incomes can contribute significantly to total incomes of farming households in developing countries.

The Indian economy grew at an impressive rate in the last decade and demographic pressure also slowed. Yet, the incidence of unemployment (CDS) towards the end of the 1990s was more than seven percent. The situation is especially disconcerting in the rural sector. Employment in rural sector, which is associated mostly with agriculture, has stagnated during the 90s (Jha 2006). Considering the increased pressure on land there exists limited scope for increasing employment in agriculture so that employment in the non-farm sector becomes an important option

An analysis of the employment situation in India during the 1990s shows that there has been reasonably high growth in non-farm employment (NFE) in rural areas since 1993. Unfortunately, this has not been associated with an increase in total employment, with the growth rate registering a dramatic decline compared to the preceding decade of 1983-93. One also observes a decline in the percentage of subsidiary employment as also self-employment in rural areas, the jobs that arguably helped households in finding a survival strategy. The growth of casual employment during 1993-2000, also slowed down, compared to the preceding decade. More importantly, the growth rate of NFE during 1993-2000, although higher than that of the population and labour force, is much below the figure of the previous decade, 1983-93. Further, the lack of demand for food grains, on the face of a decline in off-take from PDS and a dwindling intake of major nutrients in per capita terms, is disturbing. All these do not fit in well with the proposition that rural poverty has declined substantially during the later half of the 1990s, as suggested by the data on consumption expenditure. The rate of growth of NFE in rural areas was below that of urban areas during the 1980s and the gap has widened during 1993-2000. Furthermore, the growth rate has declined significantly in the late 1990s compared to the previous period, as noted above. Similarly, the decline in the growth rate of NFE for females is conspicuous both for rural and urban areas, compared to the male counterpart. As a consequence, the share of female NFE in total (female) employment in rural areas in 1999-2000 works out as less than that of 1987–8. The slowing down of the process of sectoral diversification can thus be seen to have adversely affected the more vulnerable sections of the population, such as women and the rural population, much more than the others. Indeed, the growth in NFE has taken place largely within the urban informal sector. Here, too, males have the major share in incremental employment. The relationship between the incidence of NFE in rural areas with levels and nature of employment, unemployment and poverty *at state level* suggests that a high share of NFE does not necessarily imply healthy economic development. The former is associated positively neither with per capita income nor with the percentage of non-poor population. Further, rural NFE reflects no significant relationship with levels of urbanisation. Its correlation with the growth of the urban population in the 1980s also works out as insignificant, which in the 1990s proves negative as well as significant. One can, therefore, argue that people engaged in traditional occupations, such as artisans, craftsmen, carpenters, goldsmiths, blacksmiths, etc are hit badly in the relatively urbanised regions.

Spatial variation in development and socio-economic indicators around urban centres presents adisturbing situation. There is a sharp decline in the levels of per capita income in rural areas within a distance of 0–15 km, despite having a high percentage of NFE. One may argue that the economic activities pushed out from the city to the rural hinterland have significantly lower earning/productivity than their counterparts in urban areas. Further, the agricultural wage rates both for male and female labourers decline very sharply in the immediate hinterland, along with the size of the landholdings. Indicators of social development, such as enrolment at schools, per capita expenditure on education, etc. also show a sharp decline, while infant mortality (IMR) shows a rise. All these seriously question the hypothesis regarding the rural-urban continuum and healthy interdependencies between urban centres and their hinterlands. It appears that the peripheries of cities and towns are degenerating as low productive activities are pushed out and low skilled rural migrants are absorbed. The decline in the growth of NFE during the 1990s need not itself be an alarm signal, as people have generally been seeking shelter in activities outside farming, as a residual sector. It would be erroneous to encourage the growth of NFE in rural areas or be complacent about it, unless the productivity of the workforce engaged in this sector can be increased. Industries and some of the service activities that have high employment potential and are linked with modern sectors should be encouraged to bring about sectoral diversification, as these can enhance levels of productivity. Unfortunately, the capacity of the government to generate such employment directly through anti-poverty and other programmes is limited. It is, therefore, recommended that these programmes should primarily be focused on the creation of an economic infrastructure, the provision of basic amenities and the strengthening of rural-urbán (RU) linkages. The responsibility of job creation can be left to the market, the state setting up a framework for legislating and monitoring wages and working conditions in the private sector.

2. EMPLOYMENT PROFILE OF BIHAR

The employment profile of Bihar presents a picture of the main features of non-agricultural employment in the state. The distribution of workers according to sex under different sectors of employment in Bihar is presented. (Table 1)

Table 1 reveals that the share of workers in agricultural sector and non-agricultural sector (from 1961 to 2001 census year) is almost constant throughout the period. The same trend was observed for non-agricultural employment during this period. It may be concluded that there is no significant increase or decrease in employment in both sectors in the state.

I. Division of Workers

Division of workers are one of the basic tools to discuss the non-farm employment. Information relating to of total workers in the state has been given in Table 2

The Table 2 shows a continuous decline in the share of workers who were engaged in mining, quarrying, livestock, forestry, etc. over the entire period of four decades. But in case of construction, trade and commerce and in other services

TABLE I
Distribution of Workers According to Sex under Agricultural and Non-Agricultural Sectors in the State of Bihar

(*In %*)

Census Year	*Agricultural Sector*			*Non-Agricultural Sector*			*All Sectors*	
	TW	*M*	*F*	*TW*	*M*	*F*	*M*	*F*
(1)	(2)	(3)	(4)	(5)	(6)	(7)		
1961	80.23	64.28	35.72	19.77	79.81	20.19	67.35	32.65
1971	84.69	84.56	15.44	15.31	94.16	5.84	86.03	13.97
1981	81.18	79.89	20.11	18.82	95.18	4.82	84.56	15.42
1991	81.10	76.69	23.31	18.90	84.75	15.25	83.98	16.02
2001	81.22	70.19	29.81	18.78	87.02	12.98	73.35	26.25

Source : Various Issues and Census of India of Bihar through figure Bihar ANK DO MEIN.

TABLE 2

Division of Total Workers According to Census (1961-2001) in Bihar

(*In %*)

Division of Workers	*1961*			*1971*			*1981*			*1991*			*2001*
	P	*M*	*F*	*P*	*M*	*F*	*P*	*M*	*F*	*P*	*M*	*F*	*P*
(1)	(2)	(3)	(4)	(5)	(6)	(7)	(8)	(9)	(10)	(11)	(12)	(13)	(14)
In mining, Quarrying, Livestock, Forestry, etc.	3.39	83.03	16.97	2.44	91.84	8.16	2.42	86.79	13.21	1.65	86.28	13.72	1.43
In Household industry	5.49	57.09	42.91	2.47	86.22	13.78	1.95	79.93	20.07	1.74	76.29	23.71	1.66
In Manufacturing other than Household Industry	2.21	90.49	9.51	2.60	95.49	4.51	2.38	93.18	6.82	2.26	93.26	6.64	2.13
In construction	0.56	91.68	8.32	0.56	94.10	5.90	0.75	92.03	7.97	0.63	91.62	8.38	0.58
In Trade and Commerce	2.72	85.76	14.24	3.26	96.21	3.79	4 18	89.56	10.42	4.0	90.46	9.54	3.86
In Transport, Storage and Communications	1.19	98.79	1.21	1.17	97.73	2.27	1.77	96.27	3.73	1.28	96.53	3.47	1.36
In Other Services like Repair, etc.	7.57	98.94	21.06	4.89	91.53	8.47	5.49	83.32	16.68	7.12	84.09	15.91	6.32

Source : Census of India, Various Issues (1961-2001), here, P stands for person, M male and F female.

sectors it has increased marginally. The table also indicates that the composition of the workforce is changing gradually.

3. OBJECTIVES

The study covers the following objectives:

I. To present the trend and issue of non-farm employment in Bihar and Jharkhand
II. To analyse the incidence and growth patterns of rural non-farm employment in India as well as in the state of Bihar and Jharkhand
III. To identify the obstacle in the growth of no-farm employment.
IV. To find out the factors for the generation of non-farm employment.
V. To examine the share of male and female workers in rural non-farm employment in study area.
VI. To suggest some measures for growth in rural non-farm employment.

4. METHODOLOGY

The study has been undertaken on the basis of both secondary and primary data. Secondary data were collected from state headquarters, district headquarters and respective sampled towns. Censuses of India have also been an important source data. Major sources of our data for Bihar have been through figures (Government of Bihar), Statistical Report (Government of Bihar), Bihar Ek Jhalak (Government of Bihar), Economic Census (Government of India), Center for Monitoring Indian Economy (CMIE), etc.

5. REVIEW OF LITERATURE

Studies on Rural Non-farm employment have been an important topic of discussion for our academicians, planners and politicians. In the following pages a brief review of studies undertaken during the recent past, has been presented. Though literature on the exact topic is scanty, however, an attempt has been made to review whatever relevant literature was available.

Bhalla and Hazell (2003), in this study on "Scenarios regarding employment and income growth in agriculture and non-agriculture in rural and urban areas" observed that during the period, 1972-73 to 1999-2000, the trend of non-agricultural sector in both rural and urban areas had recorded higher growth rate in employment than agriculture. They also found that in rural areas during the period, the growth rate of employment in the non-agricultural sector was found higher than that of agriculture. The analysis also revealed that the manufacturing, trade, transport and service sectors were emerging as important sources of rural employment.

Mellor (1976) in his study "Important of Farm and Non-farm Sectors in Indian Economy" found that most of the farmers possessed small land holdings and much non-farm activities were located in small town to class II town. Therefore, these farmers migrated from village to town in almost all parts of the country.

Hazell and Huggblade (1991) examined the labour market interactions and the composition of non-farm activities to find that labour market interactions shifted in the composition of non-farm activity. Along with very labour-intensive, low-return activities were noticed more skilled, higher investment, higher return activities.

Bhalla *et. al.* (2003) in their study "Growth of Non-farm Employment Upto 2050" found that about one-third of the workforces would still be engaged in relatively low productivity agricultural works. They suggested the need to develop a strategy to eliminate poverty in rural areas within a generation through accelerating the growth of agriculture for directly providing productive employment in agriculture and non-agriculture through diversification of agriculture.

Rao (2000) in his study "Growth of Non-farm Employment in Developing Countries" observed that in the earlier stage of development, towns provided non-farm employment in urban areas.

Okafar (1983) in his "Involvement of Adult and Children in Non-farm Activities" revealed that two factors established a positive relationship due to population pressure. He further observed that employment diversification has received scanty attention along with adults and children's participation being a common affair.

Viverberg (1990) found in the study "Role of Non-farm Employment" that wage employees are indeed privileged to work in non-farm activities and add new dimension to the non-farm sector analysis. He also found that rural households with non-farm enterprises trend to be better than those without it.

Dev (1990) examined the level of NSS (1977-78) inter-regional variation in the share of rural non-agricultural employment. He found that unemployment rate was positively related to non-agricultural employment having, differed from the residual sector.

Prakash and Srivastava (2000) found in their study entitled "Scenario of Male and Female Non-Agricultural Employment in Rural Areas" that the non-agricultural employment among rural males has expanded at an annual rate of 4.7 per cent, while agricultural employment has expanded at an average annual rate of 1 per cent only. The corresponding rates of growth of rural male and female employment were 4.6 and 1.5 per cent respectively.

Visaria and Basant (1994), Mahendra Dev (1994) Fisher, *et. al*. (1997), examined the Supply of basic physical infrastructure, such as roads, electricity, water and social infrastructure, education etc. They found that the states with higher literacy rate were enjoying higher percentage of RNFS employment. The study also reveals households participating in non-farm sector due to push factors. The difference in these correlates across various regions gave rise to variation in RNFS employment in these regions. They again observed that within the RNFS, there was a wide variety among the sub-sectors as far as employment and its rate of growth and wages were concerned. Some sectors, like service and manufacturing, were fast growing and engaged a major chunk of RNFE workers, other like: construction, transports communications were found to have generally engaged relatively less number of workers in rural areas.

6. STUDY AREA PROFILE

Bihar has a total geographical area of 93.60 lakh hectares on which it houses a population of 82.9 million, thereby generating a population density of 880 persons per sq. km. (census 2001). Gross sown area in the State is 79.46 lakh hectares, while net sown area is 56.03 lakh hectares. There area round 1.04 crore landholdings in the State of which around 83 percent are marginal holdings of size less than 1 hectare(Table 3). With around 90 percent of the total population living in rural areas, agriculture as the primary feeder of rural economy continues to operate not only on margins of land but also on the margins of human enterprise, its productivity being among the lowest in the country. Without increasing returns to these margins, not much can be done realistically to develop the agricultural sector. Thus, agriculture continues to define both the potentialities and constraints to development in Bihar.

Bihar and its adjoining states in eastern region are endowed with ample natural resources comprising fertile and productive soils, good rainfall, high ground water table, bright sunshine throughout the year and enterprising farmers which put together make the region ideal for remunerative cultivation of crops at an intensive scale. As a result, 90% of the population of Bihar lives in rural areas and 77% of state population directly depends on agriculture for livelihood. However, there exists a paradoxical situation that in spite of good resource availability state productivity of major crops is substantially low compared to national average productivity which led to its characterization as a 'high potential low productivity' region. Since irrigated area in the state is only 62% of the net cultivated area, successful and profitable cultivation of crops by and large depends on amount and distribution of rainfall during a particular year. Bihar agriculture is governed by rice cultivation pattern. Moreover, cultivation of crops in a cropping system and cultivation practices *per se* under topography induced specific agro-ecological situations such as *chaur*, *maun* and *diara* lands depend on distribution of rainfall over time and space dimensions. The state annually receives 1234 mm rainfall distributed over the period from June to September in a unimodal pattern which supports widespread cultivation of rice and other *kharif* crops as well. Consequently, rice occupies nearly 36 lakh ha of the 40.5 lakh ha net cultivated area during *kharif,* followed by maize (over 3.0 lakh ha.). Being the most prevalent crop rice affects the livelihood and food and nutritional security of majority of the population of Bihar and contributes significantly to state GDP which still receives almost 45% contribution from agriculture on the whole. Geographically, the state is located between 24°20′ and 27°31′ north latitude and between 830°19′ to 88°17′ east longitude. Agro-climatically the state falls under middle Gangaetic plains region. The state is bounded on the north by Nepal, on the east by West Bengal on the west by Uttar Pradesh and on the south by Jharkhand states. The state lies 53 metres above the mean sea level. The state comprises of two distinct zones, viz. North Bihar and South Bihar. The river Ganges separates the state into North and South. The state is divided into three agro-climatic zones and further in to three agro-ecological sub-zones. These are, Northwest Alluvial Plains (Zone-I), North-East Alluvial Plains (Zone-II) and South Bihar Alluvial Plains (Zone-IIIA and IIIB). In North West Alluvial Plains (Zone-I) there are 12 districts, whereas in North East Alluvial Plains (Zone-II) and in South Bihar Alluvial Plains (Zone-III) there are 9 and 17 districts, respectively. The state has total geographical area of 94.163 lakh ha. Categorization of land according to use is given in Table 3.

Table 4 present a data picture of Share of Rural Non-farm Workers in total Workers in Bihar (Region-wise), table reveals that the involvement of persons in the fields of mining, quarrying livestock, forestry and hunting, etc. have declined constantly over the decades of 1961, 1971, 1981 and 1991 (from 3.39% to 2.44, 2.42 and 1.65) respectively in Bihar. Data were not available for this sector in census 2001. However, data show no remarkable change with regard to male and female participation. The percentages of male participation increased from 83.03 per cent in 1961 to 86.28 per cent in 1991, i.e., over a period of 30 years. On the other hand, female participation declined by 3.25 per cent points during the period. It came down from 16.97 per cent of 1961 to 13.72 per cent in 1991.

TABLE 3

Bihar at a Glance (2001)

Sl. No.	Basic Details	Qty.	Numbers	
			Bihar	India
(1)	(2)	(3)	(4)	(5)
1.	Geographical Area	Sq. km.	94163	3287263
2.	Total Population	Nos.	82878796	102630200
a.	Males	%	50.95	51.90
b.	Female	%	49.05	48.10
3.	Sex Ratio	Female/000 Males	921	933
4.	Population density	Per Sq. km.	880	324
5.	Population living below Poverty line	%	39.38	26.1
a.	Rural Population	%	39.30	25.30
b.	Urban Population	%	23.50	12.50
6.	Literacy	%	47.53	65.37
a.	Males	%	60.32	75.85
b.	Female	%	33.57	54.16
7.	Rural Population	%	89.53	72.22
8.	Urban Population	%	10.47	27.78
9.	Birth Rate	Nos./1000	30.05	25.73
10.	Annual Rates of Employment Growth 1993-94 to 1999-2000	%	1.59	1.07

Source : State Planning Department, Patna and Economic Survey, 2002-03.

TABLE 4
Share of Rural Non-farm Workers in total Workers in Bihar (Region-wise)

Districts	*2001*	*1991*	*1981*	*1971*	*1961*
(1)	*(2)*	*(3)*	*(4)*	*(5)*	*(6)*
Central Bihar					
Patna	23.36	16.67	13.67	11.86	9.93
Nalanda	9.94	6.14	3.82	—	—
Bhojpur	8.77	6.01	4.97	—	—
Rohtas	9.30	5.22	3.81	2.64	—
Aurangabad	7.76	4.80	2.79	—	—
Jehanabad	8.22	4.10	3.02	1.93	—
Gaya	9.24	6.43	4.28	2.69	1.84
Nawada	7.82	6.29	3.17	—	—
Munger	17.15	6.92	3.39	2.81	2.13
Bhagalpur	13.89	5.91	3.86	2.36	—
Sheikhpura	8.81	—	—	—	—
Lakhisarai	10.27	—	—	—	—
Buxar	7.67	—	—	—	—
Banka	9.29	—	—	—	
Kaimur	6.08	—	—	—	—
Jamui	17.15	—	—	—	—
North Bihar					
Saran	7.82	4.76	3.26	2.66	1.82
Siwan	7.02	3.96	2.88	1.97	—
Gopalganj	5.90	3.06	1.93	—	—
West Champaran	6.98	3.98	—	—	—
East Champaran	6.24	3.13	—	—	—
Sitamarhi	7.05	3.75	2.11	—	—
Muzaffarpur	9.72	5.34	3.82	2.47	2.16
Vaishali	8.46	4.43	2.94	1.80	1.94
Begusarai	11.17	6.01	—	—	—
Samastipur	8.23	4.89	3.84	2.03	—

(Contd.)

TABLE (*Contd.*)

(1)	(2)	(3)	(4)	(5)	(6)
Darbhanga	9.04	5.00	3.69	2.16	2.01
Madhepura	4.88	2.93	—	—	—
Purnea	5.69	3.92	2.67	2.03	1.94
Katihar	6.82	4.89	—	—	—
Araria	5.04	3.23	—	—	—
Khagaria	7.65	4.00	—	—	—
Kishanganj	5.64	3.54	—	—	—
Saharsa	6.79	3.39	—	—	—
Madhepura	5.93	3.52	2.91	—	—
Sheohar	4.14	—	—	—	—
Supaul	4.97	—	—	—	—

Source : Department of Statistics, Government of Bihar, Patna.

CONCLUSION

The present study argues that growth in agriculture and manufacturing is important for a broad-based growth of development-induced diversification in the country. Strategies to increase rural non-farm employment growth therefore lay greater stress on these sectors; though other sectors and issues related to non-farm employment in the country are discussed as well. The discussion is more to do with institutional alternatives; and a review of public institutions performances suggest sufficient scope for improvement by making them more accountable. Stakeholder's participation in the management of these institutions and a levy of user's charges for availing of the services of public institutions are some suggestions to make these institutions more accountable. To allay the disadvantages associated with small producers certain innovative institutions like SHGs of producers or cooperatives with corporates as one of the stakeholders have emerged in select parts of the country. Such successful experiments need to be replicated over a wider area. The non-farm sector encompasses a large number of activities, the success of which in a country as diverse as India requires frequent innovations in rural institutions depending on the changed perspective and socio-economic conditions of people.

In direct measures of employment, the rural works programmes are important for at least three reasons, namely, increased marginalization of agriculture land, increased seasonality of employment and the importance of public goods in the rural sector. The study however feels that other employment-generation programmes such as the self-employment generation programme are not less important. This programme primarily caters to the chronic unemployment. Certain anomalies in direct employment generation programmes of such a large magnitude will always be there and such anomalies can be identified with concurrent evaluations of such programmes. With decentralization and effective participation of PRI, certain more frequently cited weakness of employment-generating programmes will be eliminated. The coordination between employment-generating and development programmes will also improve.

References

Acharya, S. and A. Mitra (2000). "The Potential of Rural Industries and Trade to Provide Decent Work Conditions: A Data Reconnaissance in India", SAAT Working Papers, *International Labour Organization*, New Delhi.

Ahluwalia, M. S. (1978). "Rural Poverty and Agricultural Performances in India", *Journal of Development Studies*, Vol. 14, No. 3, April.

Basant, R., B.L. Kumar and R. Parthasarathy (1998). (edited). Non-Agricultural Employment in Rural India: The Case of Gujarat, Rawat Publications, Jaipur.

Bhalla, G.S. and G. Singh (1997). "Recent Developments in Indian Agriculture—A State Level Analysis", *Economic and Political Weekly*, Vol. 32, No. 13 (March 29).

Bhattacharya, B. B., N.R. Bhanumurthy and S. Sakthivel (2004). "Economic Reforms and Structural Changes in Employment: A Comparative Analysis of Gender-Specific Employment Behaviour in Organised and Informal Sector in India", an unpublished report, Institute of Economic Growth, Delhi.

Central Statistical Organization (CSO). Annual Series of Income—Various issues, New Delhi.

———, (2001). Economic Census 1998—All India Report, New Delhi.

Chadha, G.K. (2001). "Impact of Economic Reforms on Rural Employment: No Smooth Sailing Anticipated", *Indian Journal of Agricultural Economics*, Vol. 56, No. 3, pp. 491-97.

Chandrasekhar, C.P. and Jayati Ghosh (2004). "How Feasible is a Rural Employment Guarantee?" *Social Scientist*, No. 4 (July-August): 374-75.

M. Koleswara Rao (Ed.), "Rural Employment: The Non-Farm Sector," Deep and Deep Publications Pvt. Ltd., New Delhi, 2000, p. IX.

Quoted in K. Srinivasa Rao, "Rural Non-Farm Employment and the Residual Sector Hypothesis: A Study in M. Koteswara Rao," *op. cit.*, p. 161.

Jeemol, Unni, "Non-agricultural Employment and Poverty in Rural India: A Review of Evidence," *EPW*, March 28, 1998, p. A43.

Bhalla, G.S. and Hazell, Peter (2003), "Rural Employment and Poverty Strategies to Eliminate Rural Poverty within Generation," EPW, August 16-22, pp. 3475.

Mellor John (1976) "The New Economics of Enterprises in Rural Area," *EPW*, Vol. XXXVIII, No. 23, p. 3482. Acharya, S. (1989), Agricultural Wages in India: A Disaggregated Analysis, *Indian Journal of Agricultural Economics*, Vol.44, pp. 121-39.

Banerjee, A. (1996), Notes Towards a Theory of Industrialization in the Developing World mimeo, M.I.T., Cambridge, Massachussets, USA.

Banerjee, A. and Munshi, K. (2000), Networks, Migration and Investment: Insiders and Outsiders in Tirupur's Production Cluster, mimeo, M.I.T. Cambridge, Massachussets, USA.

Bliss, C., P. Lanjouw, and N. Stern (1998), Population Growth, Employment Expansion and Technological Change, in *Economic Development in Palanpur Over Five Decades* (ed. P. Lanjouw and N. Stern), Oxford: Oxford University Press.

CHAPTER

19

Role of Rural Non-Farm Employment in India

ANGAD SINGH

INTRODUCTION

Generating productive employment for the growing rural-based force remains a formidable challenge for the Indian economy. The capacity of absorbing the incremental rural-based force in agriculture is extremely limited because of : (i) no scope of expansion of the land frontier; (ii) the intensity of cropping has almost reached the limit; (iii) the growth of crop production now depends almost entirely on technological progress resulting in low employment elasticity of output, and (iv) the need for increasing rural-based productivity and reducing unit cost through mechanisation. Recent censuses and rural-based force surveys show a dramatic structural change in the composition of rural-based force in favour of non-farm activities. Doubts, however, continue to persist about the employment generation and growth potentials of the rural non-farm economy due to lack of information on the types of activities, the nature of their operation and the constraints and opportunities.

We distinguish three types of non-farm activities: (i) Manual-based activities, such as self-employment in cottage industries, mechanics, wage employment in rural business enterprises, transport operations and construction-based, (ii) Human capital-based occupations, such as salaried service in public and private sector institutions, teachers, religious leaders, lawyers, village doctors and various typed of personal services (barbers, laundry services, midwives, etc.) and (iii) Physical and human capital intensive activities, such as agro-processing, shop-keeping, peddling, petty trading, medium and large scale trading and contractor services.

INCIDENCE OF RURAL NON-FARM EMPLOYMENT

We begin by examining the incidence of rural non-farm employment at the all-India level. Here we use data for all those years (rounds) at which the NSSO conducted large scale surveys on employment/unemployment. Such data, presented in Table 1, clearly show that the incidence of rural non-farm employment (male and female combined) expanded gradually in all-India during 1972-73 to 1987-88, but the expansion got halted during 1987-88 to 1993-94. The incidence of rural non-farm employment in all-India increased from a low level of 14.3 per cent in 1972-73 to 21.7 percent in 1987-88, which remained almost unchanged at the same level till 1993-1994 (21.6 per cent). However, this again started improving in the period of economic reforms. At the all-India level, the incidence of rural non-farm employment increased to 23.8 per cent in 1999-2000 and further to 27.4 percent in 2004-05. Clearly, the increase has been high during 1999-2000 to 2004-05, the period that is designated as one of agrarian crisis. In any case, the upward trend in the incidence of non-farm employment that was visible in rural India in the early 1990s continued all through the period of economic reforms.

The expansion of non-farm employment in rural India could also be visualised from the absolute figures of non-farm employment at different points of time. Table 1 reveals that the

TABLE 1

Size and Incidence of Rural Non-farm Employment in All-India

NSS Round/Year	*Size of Non-Farm Employment (in million)*			*Percentage of Non-Farm Employment*		
	Male	*Female*	*Person*	*Male*	*Female*	*Person*
27th (1972-73)	25.27	9.96	35.24	16.70	10.30	14.30
32nd (1977-78)	27.06	9.43	36.49	19.30	11.80	16.50
38th (1983)	34.89	11.04	45.92	22.20	12.20	18.40
43rd (1987-88)	41.82	14.14	55.96	25.40	15.20	21.70
50th (1993-94)	48.63	14.56	63.18	25.90	13.90	21.60
55th (1999-2000)	57.00	15.64	72.64	28.70	14.80	23.80
61st (2004-05)	74.16	21.11	95.28	33.50	16.80	27.40

Source : NSSO reports on Employment/Unemployment for various rounds.

size of non-farm employment almost tripled in rural India during 1972-73 to 2004-05 (increasing from 35.24 million on 1st April, 1972 to 95.28 million on 1st January, 2005). Although the absolute size of non-farm employment has been increasing continuously since 1972-73, the rate of increase has been phenomenal in recent years, particularly during 1999-2000 to 2004-05. As shown in Table 2, at the all-India level, the number of rural non-farm employment increased by 4.53 million persons per year during 1999-2000 to 2004-05 as against only 1.58 million during 1993-94 to 1999-2000.

However, if we look at the incidence of rural non-farm employment separately for males and females, it emerges that the increase has been more pronounced in the case of male workers all through the period of 1972-73 to 2004-05. The recent gain in non-farm employment in rural India benefited the male workers much more than the female workers. Table 2 shows that the non-farm employment increased by 3.43 million per year for the males during 1999-2000 to 2004-05 as against only 1.09 million per year for the females.

Another noteworthy point about the rural non-farm sector in India is that the vast majority of its workers actually

TABLE 2
Average Yearly Change of Rural Non-farm Employment in All-India

Average yearly change (in million) Between	*Male*	*Female*	*Person*
1972-73 and 1977-78	0.38	—0.11	0.26
1977-78 and 1983	1.42	0.29	1.71
1983 and 1987-88	1.54	0.69	2.23
1987-88 and 1993-94	1.13	0.07	1.20
1993-94 and 1999-2000	1.40	0.18	1.58
1999-2000 and 2004-05	3.43	1.09	4.53

Source : NSSO reports on Employment/Unemployment for various rounds.

pursued non-farm activities on an enduring basis both during pre- and post-reforms years. As a consequence, at the all-India level, nearly 94 to 95 percent of the rural non-farm workers turned out to be the "Principal status" workers (refer to Table 3). However, some difference between the male and female non-farm workers is noticeable in this regard. While 98 per cent or more of male non-farm workers in rural India belonged to the category of "Principal status" workers, the corresponding figure for the female non-workers varied between 79 to 82 per cent.

TABLE 3
Percentage of "Principal Status" Workers among Rural Non-Farm Workers in All-India

Sex of Workers	*1983*	*1993-94*	*2004-05*
Male	97.91	97.66	98.86
Female	80.71	79.04	82.03
Person	93.78	93.37	95.13

Source : NSSO reports on Employment/Unemployment for 38th, 50th and 61st rounds.

IMPORTANCE OF RURAL NON-FARM EMPLOYMENT IN INDIAN ECONOMY

About 95.28 million workers constituting 27.40% of the total workforce in a country were employed in the non-farm sector as per NSS Survey 2004-05. It plays a vital role in terms of providing employment opportunity to large segment of the working force in the country and contributes to the national product significantly. The contribution of the non-farm sector to the net domestic product.

Thus, rural non-farm sector has a crucial role in our economy in terms of employment and its contribution to the National Domestic Product, saving and capital formation. At present, Indian Economy is passing through a process of economic reforms and liberalization. During the process, merger, integration of various firms within the industry and upgradation of technology and other innovative measures take place to enhance competitiveness of the output both in terms of cost and qualitative to compete in the international market. The low inefficient units either wither away or merge with other ones performing better. In this situation, there is a special need to take care of the interests of the workers by providing them training, upgrading their skills and other measures to enable them to find new avenue of employment, improve their productivity in the existing employment, necessary to enhance the competitiveness of their product both in terms of quality and cost which would also help in improving their income and thereby raising their Socio-economic status. It has been experienced that farm sector could not provide adequate opportunities to accommodate the workforce in the country and rural non-farm sector has been providing employment for their subsistence and survival. Keeping in view the existing economic scenario, the rural non-farm sector will expand further in the years to come. Thus, it needs to be strengthened and activated so that it could act as a vehicle of employment provider and social development.

CONCLUSION

Land, the dominant factor in agricultural production, is

extremely scarce in India. Access of rural households to land has been eroding due to continued growth of population and limited employment generation in the formal industrial and service sector activities. Nearly half of the rural households are 'functionally landless' owning less that 0.2 ha. of land that cannot be a significant source of income. The average size of farm holding has declined. Thus the capacity of agriculture to generate productive employment and provide a decent standard of living is becoming increasingly limited.

The absolute number and incidence of rural non-farm employment expanded rapidly in all-India since early 1990s. The expansion has been particularly high during the recent phase of agrarian crisis (1999-2000 to 2004-05). During this period, rural non-farm employment in all-India increased by 4.53 million per year as against only 1.58 million per year during 1993-94 to 1999-2000. However, the recent gains in rural non-farm employment have benefited the male workers much more than their female counterparts. Thus, while non-farm employment for males increased by 3.43 million per year during 1999-2000 to 2004-05, the corresponding figure for the females is only 1.09 million per year. The recent spurt in rural non-farm employment seems to have occurred due to slack in farm sector employment. In any case, the rural non-farm sector in India appears to have been the sector of 'last resort' to the growing workforce and it would remain so in future too unless there is marked improvement in employment opportunities in the farm sector.

References

Reardon, T. (1997): 'Using Evidence of Household Income Diversification to Inform study of the Rural Non-farm Labour Market in Africa', *World Development*, 25(5) pp. 735-48.

Shand, R. (1986): 'Off-farm Employment in the Development of Rural Asia', Australian National University, Canberra.

Walker, T. and J. Ryan (1990): 'Villages and Household Economics in India's Semi-arid tropics', Johns Hopkins University Press, Baltimore.

Haggblade, Steven, Peter Hazell and Thomas Reardon (eds.) (2007): 'Transforming the Rural Non-farm Economy: Opportunities and

Threats in the Developing World', Baltimore, Maryland, The Johns Hopkins University Press.

Kijima, Yoko and Peter Lanjouw, (2004): "Agricultural Wages, Non-farm Employment and Poverty in Rural India", *World Bank*, Mimeo, Sept. 2004.

Ranis, G. and F. Stewart (1993): "Rural Non-agricultural Activities in Development: Theory and Application", *Journal of Development Economics*, 40(1): 75-101, February.

Lanjouw, Jean O. and Peter Lanjouw (2001): "The Rural Non-farm Sector: Issues and Evidence from Developing Countries", *Agricultural Economics*, 26(1): 1-23, October.

Islam, Nurul (1997): "The Non-farm Sector and Rural Development: Review of Issues and Evidence", FAED Paper No. 22, Washington, D.C., International Food Policy Research Institute.

CHAPTER

20

Empowerment of Rural Labour and Non-Farm Sector—A Case Study of NREGS Workers

P. Tara Kumari, Ch. Masenamma and N. Visalakshmi

INTRODUCTION

India has been mainly a rural economy. About 80% of its population lives in villages. Near about 30% of the rural people are living below the poverty line even today. Rural development is a matter of global concern today because it has the largest reservoir of the human resources. At present, abundant human resources are available in most of the developing countries. But the important aspect of human resources of rural development has not been given its due recognition in these countries.

In developing countries like India most of the population is getting absorbed into agricultural sector and its allied activities. But still the people, who are engaged in agricultural, are lagging behind in socio-economic and cultural development. The development of rural areas is vital for the survival of the third world countries, as majority of the rural people are victims of the vicious circle of poverty.

In India, the National Rural Employment Guarantee Scheme (NREGS) provides social and economic security for the rural poor. The NREG Act is the first tangible commitment by which the poor can expect to earn a living wage, without loss of dignity and demand this as a right.

NREG Act has placed a judicially enforceable obligation on the state governments. Under the provision of the Act, the State Governments are to provide work within 15 days of a person making an application within a radius of 15 kms. from the applicant's residence. Failing this, the state government is to provide an unemployment allowance. Workers are entitled to a statutory minimum wage for their labour, to be paid within seven days after the work is done. Men and Women are to be paid equal wages. This Act is based on the principle of self-selection by focusing on unskilled, manual work.

NEED

Poverty reduction has been an important goal of development policy since the inception of planning in India. No doubt, various anti-poverty, employment generation and basic services programmes have been implemented from time to time in India. The ongoing reforms attach a greater importance to removal of poverty, by addressing specifically the wide variations across the states and the rural, urban divide.

Anti-poverty strategy has three broad components; promotion of economic growth, promotion of human development and targeted programmes of poverty alleviation to address multidimensional nature of poverty. The various programmes aimed at the poor have been streamlined and strengthened in recent years, which include the NREGS.

The NREG Act empowers people to play an active role in the implementation of employment guarantee scheme through Gram Sabhas, social audits, participatory planning and setting up of local vigilance and monitoring committees. It was felt that active community participation was particularly important for ensuring transparency and public accountability. Thus, there is a role for all grass-roots institutions such as worker's associations, local beneficiary committees, self-help groups and user groups in spreading awareness, mobilizing workers, and monitoring the implementation of the scheme.

The Government of Andhra Pradesh has provided employment to 52,25,287 households since the inception of this programme. About 83,28,238 individuals were provided employment of which 46% were men, and 54% were women. Out of the total number of beneficiaries 0.64% were disabled persons. Twenty six percent of person days of employment have gone to members of the scheduled caste and 11% to members of S.T. between April 2007 and March 2008. Some 4,87,623 workers have been initiated between 2006 and March 2009. The works undertaken by NREG schemes mostly include the non-farming activities in the rural areas like road connectivity, flood control and protection, water conservation and water harvesting, drought proofing, desalting of ponds and minor irrigation works.

The NREG Act provides for 100 days of guaranteed employment in one financial year to all rural household members, who are willing to do manual labour. The Act not only guarantees wage employment as a right, but also promotes community monitoring through vigilance and monitoring committees, social audit through Gram Sabha and also makes provision for complete transparency as mandated by the Right to Information Act, 2005.

In Andhra Pradesh, the NREG programme was covered in 3 phases. In the first phase 13 districts were covered. They are Vijayanagaram in coastal Andhra region, Chittur, Kadapa and Anantapur in Rayalaseema region, Mahabubnagar, Medak, Rangareddy, Nizamabad, Warangal, Adilabad, Karim Nagar, Khammam and Nelgonda in Telangana region. In the second phase, 6 districts were covered. They are East Godavari District, Guntur, Prakasham, Nelore and Srikakulam in coastal Andhra region and Kurnool in Rayalaseema region. In the third phase, 3 districts namely West Godavari District, Krishna and Visakhapatnam in coastal Andhra were covered.

Thus, the NREGS gives legal guarantee of providing at least 100 days of wage employment to rural households whose adult members are willing to do unskilled manual labour.

Providing 100 days of employment to the rural-poor especially during the lean agricultural season improves the socio-economic status of the labour force of rural area. In the light of the account give above, an attempt is made in this paper to study the socio-economic characteristics of the NREGS workers in the selected regions. The hypothesis is that NREG schemes work more effectively in the developed region, when compared to under developed regions. The objective of the study is to examine in the efficiency of the NREGS programmes in developed and underdeveloped regions.

METHODOLOGY

For the purpose of the present study two mandals, i.e. one from the developed regions and the other from the under developed regions are purposively selected from Visakhapatnam district. The Bheemunipatnam mandal is considered to be under-developed in view of its longer distance from the district headquarters and other facilities. Similarly the Anandhapuram mandal is considered to be developed which is situated nearer to the town headquarters of Visakhapatnam.

The study is mainly based on primary data. For the purpose of the study 60 households from each selected mandal regions were selected from different categories on the basis of simple random sampling method. Thus, altogether 120 beneficiaries of NREGS were selected and interviewed with the help well-structured questionnaire. Simple statistical techniques like percentages and averages have been employed to analyse the data.

I. Age Structure

One of the important factors that influence the economic development is the age structure in any economy. The statistical information relating to age-wise distribution of respondents is presented in Table 1.

Table 1 presents the age group of the respondents in the two selected mandals. Out of 120 respondents, 47.50 percent are in the age group of 26-40 years, 20.83 percent are in the age group of 21-25 years and 31.67 percent are in the age group of 41-60 years. In Bheemunipatnam mandalam out of 60

TABLE I
Mandal-wise Age Group of the NREGS Workers

Sl. No.	Name of the mandal	Age			
		Below 25 years	26-40 years	Above 41-60 years	Total
1.	Bheemunipatnam	11 (18.33%)	38 (63.34%)	11 (18.33%)	60 (100.00%)
2.	Anandhapuram	14 (23.33%)	19 (31.67%)	27 (45.00%)	60 (100.00%)
	Total	25 (20.83%)	57 (47.50%)	38 (31.67%)	120 (100.00%)

(The figures in brackets are percentages to total).

respondents, 18.33 percent of the respondents are in the age group of 21-25 years, and also the same percentage of respondents in 41-60 years while as high as 63.34 percent are in the age group of 26-40 years. It is noticed that out of 60 respondents in Anandhapuram mandal, as high as 31.67 percent are in the age group of 26-40 years, while 45 percent are in the age group of 41-60 years.

From this it is concluded that as high as 63.34 percentages of the respondents belonging to the age group of 26-40 years in Bheemunipatnam are opting to be in the NREGS while this percentage is relatively lower in Anandhapuram mandal.

2. Educational Background

Education plays a vital role as an ambassador in any society for brining changes in the social status. Education facilitates better knowledge and makes life easier to interact in the various important fields like political, economic, social and others. If one is educated then one will be able to play a successful role in the family and as well as in society.

The statistical information relating to the literacy levels of the NREGS respondents is presented in the Table 2.

Table 2 shows that the percentage of illiterate respondents of NREGS workers is 65 in Bheemunipatnam mandal, while it

TABLE 2

Mandal-wise Educational Status of the Respondents

Sl. No.	*Name of the mandal*	*Illiterates*	*Primary*	*Upper Primary*	*Total*
1.	Bheemunipatnam	39 (65.00%)	16 (26.67%)	5 (8.33%)	60 (100.00%)
2	Anandhapuram	34 (56.67%)	16 (26.67%)	10 (16.66%)	60 (100.00%)
	Total	73 (60.83%)	32 (26.67%)	15 (12.50%)	120 (100.00%)

(The figures in brackets are percentages to total).

is 56.67 percent in Anandhapuram mandal. In both the mandals 26.67 percent of the respondents have primary education. In Bheemunipatnam 8.33 percent of the respondents have upper primary education, while it is 16.66 percent in Anandhapuram mandal.

It may be observed that more than 55 percent of respondents of NREGS are illiterates in each mandal. From the above table it is concluded that though the NREGS schemes are helping both literate and illiterate workers the percentage of illiterate beneficiaries are more in both the mandals.

3. Social Community

Social communities are playing major role in developing countries like India. The information relating to the social community of the respondents of NREGS is shown in the Table 3.

As can be seen from the Table 3 that the majority of the respondents belong to backward castes in both the mandals. It is interesting to note that there are no NREGS respondents in the O.C. category in Bheemunipatnam, and S.T. category in Anandhapuram mandal.

From this it may be concluded that the beneficiaries belonging to B.C. communities are more in number in both the mandals, being the two mandals dominated more by B.C. community.

TABLE 3

Mandal-wise Social Community of the Respondents

Sl. No.	Name of the mandal	Caste				
		S.C.	S.T.	B.C.	O.C.	Total
1.	Bheemunipatnam	5 (8.34%)	8 (13.33%)	47 (78.33%)	0 (0.00%)	60 (100.00%)
2.	Anandhapuram	12 (20.00%)	0 (0.00%)	47 (78.33%)	1 (1.67%)	60 (100.00%)
	Total	17 (14.17%)	8 (6.67%)	94 (78.33%)	1 (0.83%)	120 (100.00%)

(The figures in brackets are percentages to total).

4. Type of the Family

In any society the nuclear family consists of a married couple and their off-spring while the joint families are much larger and consist of members beyond the nuclear families. The family was the most universal and permanent institution of mankind. But today, radical changes are found in the spheres of political, economical, social and cultural life. The changes in socio-economic condition of the people are also reflected in the institution of the family. Now it is of interest to analyse to what extent the family structures are undergoing changes in the study area.

TABLE 4

Mandal-wise Type of the Family by the Respondents

Sl. No.	Name of the mandal	Nuclear family	Joint family	Total
1.	Bheemunipatnam	52 (86.67%)	8 (13.33%)	60 (100.00%)
2.	Anandhapuram	50 (83.33%)	10 (16.67%)	60 (100.00%)
	Total	102 (85.00%)	18 (15.00%)	120 (100.00%)

(The figures in brackets are percentages to total).

It is noticed from Table 4 that in Bheemunipatnam mandal, 86.67 percent of the families are nuclear families and the remaining 13.33 percent are joint families. In Anandhapuram mandal 83.33 percent of the families are nuclear and the remaining 16.67 percent of the families are joint families. Altogether 85 percent of the NREGS sample households are nuclear type and the remaining 15 percent are joint families.

Hence, it may be concluded that nuclear families are predominant in both mandals. From this it may be inferred that the joint family type is fading out giving way for nuclear family type.

5. Main Occupation

One of the important indicators of the socio-economic profile is occupation of the respondents. The statistical information relating to main occupation of the respondents are presented in Table 5.

In Bheemunipatnam mandal, for 46.67 percent of the beneficiaries the main occupation is agricultural labour, while in Anandhapuram mandal for 51.67 percent of the beneficiaries the main occupation is non-agricultural labour activity,

From this it is concluded that the percentage of agricultural labour is more in Bheemunipatnam mandal when compared to Anandhapuram mandal, while the percentage of non-agricultural labour is more in Anandhapuram mandal when compared to Bheemunipatnam mandal. This is attributable to more construction activities undertaken in a developed mandal like Anandhapuram.

6. Average Income Particulars of the Respondents

The data relating to the mandal-wise average income particulars of the respondents is shown in the Table 6.

In Table 6, the data reveals that average income of the respondents is less before joining the NREGS in both Bheemunipatnam and Anandhapuram mandals. Great hikes in average income levels can be observed after joining the NREGS in both the mandals.

TABLE 5

Mandal-wise Main Occupation of the Respondents

Sl. No.	Name of the mandal	Main Occupation					Total
		Agriculture	Agriculture labour	Non-agricultural labour	NREGS labour	Other labour	
(1)	(2)	(3)	(4)	(5)	(6)	(7)	(8)
1.	Bheemunipatnam	5 (8.33%)	28 (46.67%)	16 (26.67%)	4 (6.66%)	7 (11.67%)	60 (100.00%)
2.	Anandhapuram	0 (0.00%)	25 (41.67%)	31 (51.67%)	0 (0.00%)	4 (6.66%)	60 (100.00%)
	Total	5 (4.17%)	53 (44.17%)	47 (39.17%)	4 (3.33%)	11 (9.16%)	120 (100.00%)

(The figures in brackets are percentages to total).

TABLE 6

Mandal-wise Average Income Particulars of the Respondents

Sl. No.	*Name of the mandal*	*No. of house-holds*	*Before joining the NREGS*	*After joining the NREGS*	*Total*
1.	Bheemunipatnam	60	3001.67 (39.74%)	4551.67 (60.26%)	7553.34 (100.00%)
2.	Anandhapuram	60	3617.50 (41.11%)	5181.67 (58.89%)	8799.17 (100.00%)
	Total	120	6619.17 (40.48%)	9733.34 (59.52%)	16352.51 (100.00%)

(The percentages of average income are in the brackets).

From this it may be concluded that, the NREGS schemes are well penetrating into both developed and under developed regions of the districts. This is a welcoming feature.

CONCLUSION

At present the development economists have been arguing that the development process should be job-oriented. In the light of this argument it may be concluded that the NREG schemes provide social and economic security for the rural poor in both developed and under-developed regions. The NREG scheme is tangible commitment on the part of the government to protect the poor, to earn a minimum wage without loss of dignity. As such the NREGS are to be extended more vigourously into the non-farming sector of the rural areas and efforts may also be made to bring in more number of workers into its fold in both developed and under developed regions. Maximizing the number of beneficiaries under schemes like NREGS will ensure the objective of 'inclusive growth'.

PART III

NON-FARM SECTOR : SOME MICRO LEVEL STATE EXPERIENCES

CHAPTER

21

Socio-economic and Working Conditions of Pottery Workers in Tamil Nadu—A Case Study in Thiruvallur District

R. KUMAR

1. INTRODUCTION

Pottery is an age-old craft and a rural tradition in India. Lakhs of people still depend on it for their livelihood. Potters were probably the first engineers in the history of human civilization, the first group to be withdrawn from food production and to be engaged in a full time profession. The potter's wheel was a momentous invention, ranking along with the mastery over fire, as one of those which most influenced the course of human development. Every other invention can be said to be an imitation of nature, but not the continuous rotary motion of the wheel on a bearing.

2. POTTERY IN INDIA

The art of handling of clay called Pottery was one of the earliest skills known to the Indians. From time immemorial, lumps of clay were hand-moulded to form toys and deities of worship. The advent of the Potter's wheel gave man the task of making beautifully shaped pots for his personal use. The movement of the wheel and the pressure exerted by the hands on the clay gives new shapes and forms (www.indian mirror.com).

Pottery is one of those important mediums through which men have expressed their emotions. For thousands of years pottery art has been one of the most beautiful forms of expression. A piece of pottery has a visual message in its shape and colour.

Pottery is the most sensual of all arts. In India, we have had a great tradition of pottery making. In fact, being an agricultural country, pots for storage of water and grains were in demand. The real beginning of Indian pottery began with the Indus Valley Civilization and the art of shaping and baking clay articles as pottery, earthenware and porcelain has continued through the ages. While pottery and earthenware are definitely utilitarian and often decorative, porcelain and studio pottery belong to the sphere of art.

There is proof of pottery making, both handmade and wheel-made, from all over India. In the Harappan civilization potter's place was quite an important one in society. The craft was well advanced. Rectangular oven for firing the product were in use. Seals and grain and water containers were made that were put to use effectively. The potter occupies a unique position in the craft traditions of India. India is home to more than a million potters. The potters are wonderful masters of their trade (www.culturalindia.net).

Pottery in Indian villages presents a wonderful amalgamation of concept, design and execution. Potteries can be both hand-made and wheel-made, which is practised all over India. There are mainly, two types of pottery found in the Indian villages, namely the unglazed pottery and glazed pottery. The unglazed pottery is considered the finest pottery in Indian villages and it has a wide range. There are three different styles in unglazed pottery, namely, the paper-thin, scrafito and highly polished. Black pottery is another famous form of unglazed pottery in Indian villages and it resembles the Harappan pottery style. In the paper-thin pottery, the biscuit coloured pottery is decorated with incised patterns. Very fine paper-thin pottery is produced in the villages of Kutch, Kanpur and Alwar in India. Alwar is well known for a particular type of paper-thin pottery called Kagzi. The villages of Kanpur are also famous for making thin pottery with incised designs (www.indianetzone.com).

Today, in the Indian villages, around 15 lakh potters with traditional skills are plying their profession. About 95 per cent of them are engaged in the traditional red or local clay pottery work. The extent of employment of outsiders in the village pottery activities is about 9 per cent. The rest 91 per cent are potters family members who assist the potters in various operations, from preparation of clay to baking the raw products in the kiln.

The socio-economic conditions of the traditional potters too are far from satisfactory. Most of the potters families have no landed property, own source of availability of raw materials, working shed, furnace, etc. Thus, even if any financial organization offers them assistance they are unable to avail of it as they do not have property for mortgage against such assistance.

Objectives

1. To highlight the socio-economic conditions of potters.
2. To show the cost-benefit of pottery production.

Methodology

The study is based on exclusively primary data. The data have been collected through a sample survey. The simple random sampling technique has been used to select the respondents from the universe. The study includes 133 respondents living in the cluster of villages' viz., Kosavanpatti, Edapalayam and Vydalampaddu in Thiruvallur District, Tamil Nadu.

3. RESULT AND DISCUSSION

Habitation

In Tamil Nadu, potters usually live in settlements in interior villages. Each settlement is related to a particular potter community (caste). Most of these potter settlements are located in river basins or near to paddy fields. This is obviously related to raw material availability. The work area is mostly located adjacent to the homes. In several places the interior of the houses are also used to keep the pots either baked or raw (for drying). In certain places, a few potters were found living in their kiln shed itself. They use the same place as the work area as well.

Most of the settlements have a shrine or temple, which is common to all the families. Most of the potters worship Maariamma or Kattaree. A male member from within their own community functions as the temple priest also. Potters keep their raw material and fuel in the courtyard or nearby places where space is available. In almost all houses there will be an extension area from the main house/building. This place is used for pottery production work.

It is located that potter groups are found in all the 14 districts of Tamil Nadu. Thrunveli district has the maximum number of potter colonies and Coimbatore the least. All the Grama Panchayaths where the potters are found have been listed. In several Panchayaths there are two or more potter groups. The population of these communities may be 20 to 25 times higher than the number of persons engaged in pottery work because in several villages only less than 5 per cent households of potter communities are engaged in pottery.

Community

Potters are living all over Tamil Nadu, belonging to various communities. They are known by different caste names in different places. There are three communities viz., Kusavan, Andhra Udhyar and Telugu Kusavan. They have no inter-marital relationship with other communities. In some places Telugu cultured Kusavan are known as Mannudayan. The revenue authorities issue caste certificate as Kusavan. So they have accepted 'Kusavan' as their caste name. They still believe that there is no caste called Kusavan.

Educational Status

Potters are educationally and economically backward irrespective of the place of residence. They are considered more backward among the 'Most Backward Communities'. One of the reasons for their educational backwardness is said to be child labour that prevailed in their community. In early days children used to help their parents in their traditional occupation and it resulted in their dropping out from school. Among females 41.86 per cent reached high school and 19.38 per cent reached PDC and 1.55 per cent became graduates. It shows that the 50 to 60 age group, 36.7 per cent are illiterates, 51.7 per cent dropped out in primary classes, and only 6.7 per cent reached high school. Among females, 72 per cent are illiterates, 22.8 per cent dropped out in primary classes, and only less than 1 per cent reached high school. The trend has changed among the youngsters in the 20 to 30 age group. 49.29 per cent reached high school and 9.15 per cent reached PDC and 2.11 per cent got graduated.

The potter parents encourage their children to help them in their traditional occupation, but on the other hand they have neither the time nor the capability to help their children in their study because of their illiteracy. Even the female members of the potter family are usually busy with assisting their men folk in their occupation.

Social Status

The Social status of potter communities is very low in comparison with other backward communities. Persons employed in government sector, or even organized private sector, are very few. Most of the dropouts from their traditional occupation are engaged in manual labour, agricultural labour, masonry work, construction work, etc., and some of them are engaged in low paid menial jobs in shops and other institutions. Among the children, only very few students have reached higher education or professional courses. Very few persons belonging to these communities are living in reasonably good living conditions and have reasonable economic status. Many persons belonging to these communities live in thatched houses. Most of them live in tiled houses, but they are not well furnished. Terraced roofs are rarely found among these communities.

Among the potter communities, Udhyar community is the most backward one, educationally and economically. We could not find any professionally qualified person in this community. Graduates are very rare in this community. Most of the students from this community stop their studies at the primary level. The living conditions of these people are also poor. Some Udhyar families live under their kiln roof and they have no separate shelter for living. Most of these families depend on public wells, public taps and neighbour's well for drinking water. Alcoholism among the male members of the Udhyar community is one of the reasons for the poor living conditions. The situation is similar in all the communities and all the colonies we visited. We met some inebriated persons in the morning itself, during our field visits. Even though the communities are included in the OBC list for public services, their share in the government jobs is very poor.

Deviation from the traditional occupation is very high in the Udyar community. Only 3 out of 100 Kusavan households are engaged in the traditional occupation but it is 78 in the case of Udhyar of Coimbatore. Modernisation has not yet reached this artisan community. Very few Velan potters use motorised wheel. In the case of kiln, most of them are using traditional kilns. We also found the round shaped improved kiln, but its fuel consumption is very high in comparison with traditional kiln. No individual potter uses the pug mill. They pug the clay using their foot and hand, which needs considerable physical effort. They have no technology at hand to test the clay quality, temperature of the kiln, etc. They adopt traditional methods in colouring and polishing Kumbarans use red earth for colouring the products, and cloth pieces for polishing them.

4. ECONOMIC STATUS

Production : Demand for Pottery

The demand condition for pottery shows two distinct features of regular and seasonal demand. The regular demand for pottery arises from the household demand as a vessel for cooking, storing and for other purposes. The flower pot, terracotta etc., create regular demand for this production.

The large scale demand for pottery products arises during Pongal Festival and Karthigai Deepam festival in Tamil Nadu. The religious ceremonies of marriages and death also create sizeable demand for earthen were in the state. The seasonal demand for pottery results in significant proportion of production of these goods. Only during this season these people can get remunerative price for their products. They used to work for a very long hours inorder to meet the increased demand.

Potters are mainly dependent on their traditional occupation for their income. Very few have other sources of income like government or private employment. Among 133 families are engaged only in pottery. Out of this, 86 males and 47 females are engaged in pottery work. None of these potter families is happy with the earnings from the pottery profession. Table 1 shows monthly income of potters in study area.

TABLE 1

Monthly Income of Potters

Income level	*Number of households*	*Percentage*
Below 500	22	16.54
500-1000	49	36.84
1000-1500	32	24.06
1500-2000	21	15.79
Above 2000	9	6.77
Total	133	100.00

Seasonal income of potters is given in Table 2.

It is reveals that 45.9 per cent of the respondents have earned Rs. 20000-30000 during the festival season. 22.6 per cent of the sample potters replied the income receipt of Rs. 10000-20000 and 17.2 per cent of the potters have received rupees less than 10000, 9.02 per cent of them have reported the income of

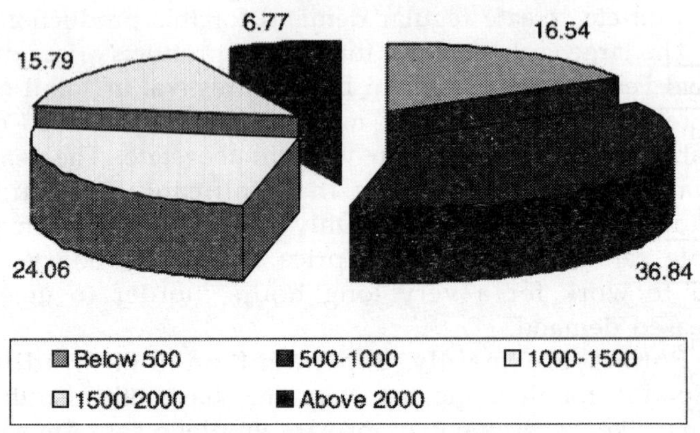

TABLE 2

Seasonal Income of Potters

Festivals	*Number of households*	*Percentage*
Below 10000	23	17.20
10000-20000	30	22.60
20000-30000	61	45.90
30000-40000	12	9.02
40000-50000	7	5.27
Total	133	100.00

Rs. 30000-40000 and 5.27 per cent of the respondents have got Rs. 40000-50000 during the festival season. It implies that these people can earn sizable income only during the period of seasonal in the study area.

The living condition of the potters is not generally good. In certain villages some potters are living under the kiln roof. Even though some of them are living under the tiled roofs, these facilities are very poor. In certain areas 30 per cent houses have thatched roofs. In both cases the potters are working in the area attached to the houses. Type of households is shown in Table 3.

TABLE 3
Type of Households

Type of house	*Number of household*	*Percentage*
Thatched	23	17.29
Tiled	61	45.86
Terraced	49	36.84
Total	133	100.00

It is interested to note that there is hardly any other mode of supplementary income to these communities. Land holding of members of these communities is also very small. Most of them have less than 10 cents of land. Only very few families are holding agricultural land. In lean seasons they depend on manual labour for living. A few of them opined that they do not know agricultural work, so the landowners do not like to employ them, otherwise they could have gone for agricultural work in the lean seasons. Most of them complained that they have to depend on money lenders for money to purchase clay and fire wood. Now-a-day these potters have to pay for clay to the land owners. The amount varies from place to place. In olden days they did not have to pay for clay. Cost of clay is Rs. 500 to 1000 per bullock cart in different places. Along with this they have to pay for transportation charge. Most of them collect 2 bullock cart of clay per year. 6 to 10 labours are required to collect the clay. It varies according to the depth of clay source and quality of clay. Total cost of clay may be up to Rs. 8000 per load, according to the distance of the clay source. During production they have to spend a considerable amount of money for firewood, husk and hay. More than 85 per cent of these potters are using traditional kiln so the fuel consumption is very high. In olden days firewood was collected from various places free of cost but now-a-days its cost has gone up considerably. That has resulted in the hike of production cost. Apart from these, another problem is the high damage of the products in the kiln during firing. The damage happens when most of the work is completed and the damage is 5 to 50 percent at certain times. In such cases, potters lose up to 50 per cent of the products before the completion of the manufacturing process. They have to work for 10 to 30 days to prepare the pottery articles for a kiln firing. It varies according to the size of the kiln and the number of workers. Most of them fire the kiln twice a month. A few of them have changed their traditional wheel, from stone to ball bearings. The use of this type of primitive technology is a major reason for their poor plight. These problems compel them to accept other occupations, which are inferior in social status to pottery, like agricultural labour, and unskilled labour, etc. Table 4 indicates a rough calculation of the income of a typical four member potter family (man and wife and two children) a typical cost calculation for terra cotta production is indicated below.

TABLE 4

Cost Benefit of Pottery Production

Clay quantity required for a year	2 bullock cart
Cost of clay at source (2 × 100)	Rs. 200
Transportation (2 × 400)	Rs. 800
Labour (5 mandays)	Rs. 500
Fire wood per year (2100 kg)	Rs. 3400
Production per day	25 pots
Interval of each kiln firing	25 days
Number of pots per kiln	350 nos.
Damage during firing	10 per cent
Total pots per year	3150
Average rate per pot	15
Total revenue of 3150 pots @ Rs. 15	Rs. 47250
Income after expenditure	Rs. 32700

Table 4 brings out the cost-benefit analysis of pottery production in Tiruvallur district of Tamil Nadu. The table shows that the input cost of producing 3150 pots in a year is Rs. 14550. The total revenue receipts of selling these pots at

Rs. 15 per pot would be 47250. The benefit from the sale of these pots is Rs. 32700. The table suggests that the cost of production is only 30.79 per cent of the total revenue of the output. Income is excess of expenditure is nearly 70 per cent of the total revenue. It is evident from the above discussion that the production of pottery is a profitable viable non-farm activity in rural Tamil Nadu.

The major works, pugging, making of pots on wheel, finishing of the pots, placing of pots in kiln, and kiln firing are the Potter men's work. In pottery work irrespective of the communities and colonies, women participation is higher than in the other traditional jobs. Polishing, drying and helping to place the pots in kiln are women's work, irrespective of community. Among Kusavan community, potter women help to make moulded products. Udyar women collect the fuel-firewood and paddy straw from villages.

Products and Processes

Products

In olden days potters used to make earthen wares for cooking and also for storage of grains and other items. Pots were also used for rituals in temples by all communities. These products were used commonly in all villages. Vessels used for cooking, pots used for gardening and, mattoms used for toddy drawing are the major products currently in this field. Water storing pots, different types of stoves, tiles, irrigation pipes, products for indoor and outdoor decoration and fancy products are also produced by the potters. In some districts, Telugu Kusavan are making kitchen vessels, flowerpots, and some studio pottery items.

Techniques

There are three types of pottery items currently made in Tamil Nadu. They are: Wheel made pottery, handmade pottery and Mould made pottery. In all the districts the potters are using these three methods, but there is some difference in the tools or equipment, in different areas and in different communities. The Udhyar communities of Kosavanpatti, Edapalayam and Vydalampaddu are using 'spoked wheel'.

Only very few people are manufacturing pots these days. For ceiling tile manufacturing, these people use dies and machines, which can be operated by hand. In all these cases specially made stones are used for finishing and polishing of pots and red mud is used for colouring. In institutions, pug mill, motorised wheel and moulds are widely used. Motorised wheels are rarely used by individual potters.

Shrinkage

During drying the water is driven off from a clay body. The removal of water film and absorbed water results in shrinkage. The last traces of mechanically combined water in clay can be removed by heating the clay above the boiling point of water (100°C). Clays differ greatly in their shrinkage. The degree of variation of drying shrinkage of clays is similar to the variation in the amount of water necessary to develop their plasticity. Greater the plasticity of clay, greater will be the water absorbed and greater will be the thickness of water film and greater will be the shrinkage when it is dried. Likewise, clays which contain more clayey substance and less non-plastic clays will show high shrinkage. Because of high shrinkage some clay are difficult to get dried without undergoing cracking. The outside of a bigger and larger piece tends to shrink faster than the inside and this is the reason how surface cracks take place. Non-plastic and non-shrinking material addition can decrease the magnitude of shrinkage and also the tendency to crack. Sand is usually added to plastic clay to reduce shrinkage in the manufacture of bricks. Lesser the shrinkage of a clay better it is for the pottery, of course the required plasticity has to be maintained for workability.

Problems Faced by Potters during Production of Pots

Preparation of clay requires good skill and high physical effort. Women and children also have to help in this work. The separation of stone from the clay is drudgery. It consumes considerable time and physical effort of the potter's family. The potters family women are rotate these wheels, high physical effort is needed; they have to repeat this for each pot. In this case they engage one person to rotate the wheel and one person to make the pots. Sometimes the women folk of the household rotate the wheel. In both the cases this work consumes considerable effort. Another problem is the cracking of pots before firing. While drying in the sun the pots may crack due to the poor quality of clay. Damage to the pots is very high in the traditional kiln. It is 5 to 20 percent in different areas. It varies according to the pot placing skill of potters. Breakage during firing is the main problem faced by some potters.

Marketing Strategies

Most of the members of the kusavan colonies market their products by head load vending. Usually potter women take the products to the houses of neighbouring areas in the early morning. Till recently they exchanged their products for rice. But now they prefer payment in cash. Potters of some places take their products to distant places, mostly to other districts, by vehicles. From there they take the products to the neighbouring houses to sell. For this purpose these people stay in those areas for a few days, and take the products to the markets by vehicles. They usually stay in the markets for weeks in the rooms hired by them from Panchayat, or private parties. Here also women are dominant in marketing. In certain places of Thiruvallur district the potters sell their products at site. In these areas agents or merchants from distant places come and purchase the products. In this case also potters get advance money from the buyer. In institutions, the products—mainly flower pots are sold to other firms in government and private sector, and to individual buyers. Some of the institutions sell the damaged products and broken pieces at reduce rates to fill the floor before concreting. Studio pottery manufacturing institutions sell their products to institutions like interior decorators and individual buyers. Some times they get export orders, for which they depend on other firms which hold export licenses. Most of the potters complained that there is no demand for their products. Some members of this potter community are engaged in pottery trading only.

5. CONCLUSION

The socio-economic condition of potter in the study area is very poor. They used to get very low level of regular income. The analysis shows that the seasonal demand for pottery generates sizable income to the potters. To conclude pottery as a non-form activity, provide only a seasonal employment to the rural people.

References

http://www.indianmirror.com/crafts/cra5.html

http://www.culturalindia.net/indian-crafts/indian-pottery.html

Gupta, K.C., 1988, Progress and Prospects of Pottery Industry in India: a Case Study of U.P., Mittal Publications, Delhi, India.

http://www.indianetzone.com/41/pottery_indian_villages.htm

Barber, Edwin Atlee, *The Pottery and Porcelain of the United States*. New York: G.P. Putnam's Son's, 1893.

Hiller, Bevis, *Pottery and Porcelain 1700-1914*. New York: Meredith Press, 1968.

CHAPTER

22

Determinants of Inter-Mandal Variations in Rural Non-Farm Employment

S. RATHNAKUMARI AND J. SREENIVASULU

The generation of productive and gainful employment, with decent working conditions, on a sufficient scale to absorb our growing labour force must from a critical element in the strategy for achieving inclusive growth. Past record in this respect is definitely inadequate and the problem is heightened by the fact that the relatively higher rate of growth achieved during the last decade or so is not seen to generate a sufficient volume of good quality employment.

The basic weakness in our employment performance is the failure of the Indian economy to create a sufficient volume of additional high quality employment to absorb the new entrants into the labour force while also facilitating the absorption of surplus labour that currently exists in the agricultural sector into higher wage, non-agricultural employment. A successful transition to inclusive growth requires migration of such surplus workers to other areas for productive and gainful employment in the organized or unorganized sector.

In the process of economic development, agriculture and the modern formal Industrial sector are not able to absorb the growing labour force in the developing countries. Since the agricultural sector in the rural areas is over-saturated, the growth of rural non-farm sector (RNFS) is seen as a promising sector for creation of sustainable employment and as the solution to the problem of poverty in rural areas. Until recently, the commonly held view was that the RNFS was a non-productive sector, which is expected to within away with development. However, there recently much interest has been created in the nature, characteristics structure, operational problems, policies and factors affecting the growth of the RNFS. At present, it is argued that in a situation of rural workforce growing faster than the employment potential of agriculture, the RNFS has positive role to play in promoting growth and welfare. So, it is argued that the RNFS :

1. Can lower unemployment by absorbing growing labour force;
2. Reduce rural-urban migration and thereby contribute in easing urban congestion and lowering pressures on scarce urban infrastructure facilities;
3. Employ women and provide seasonal employment of residual employment for those left out of agriculture and for the poor who cannot find sustenance in agriculture since they are small landholders or landless;
4. Use more appropriate technology, particularly in small-scale rural manufacturing and thus generate grater income from available productive inputs;
5. Improve household security through diversification;
6. Contribute to national economic growth;
7. Produce commonly consumed goods at lower prices which are mostly demanded by the poor; and
8. Promote a more equitable distribution of income (Lanjouw and Lanjouw, 1995).

The rural non-farm sector was neglected by economists who were mainly working in the two sector growth models (Lewis, 1972; Ranis-Fei, 1961). In the two sector model it is assumed that employment in the modern industrial sector will expand and absorb the supply of labour from agriculture. This assumption has proved to be unrealistic. Lewis and Ranis-Fei both refer to two sectors as agriculture and non-agriculture, irrespective of whether these are rural or urban. However, the logic of their approach does not exclude the RNFE. Indeed, (as in China, Taiwan and Korea) much early industrial growth was in the RNFE.

An explicit effort to incorporate RNFS into a growth paradigm was Mellor's (1976) topic of discussion of agricultural growth linkages (AGL): forward and backward and consumption linkages from agriculture to the non-farm sector. Forward linkages include those to processors and distributors in the non-farm sector (such as processing, transport and marketing of agricultural activities). Such linkages are frequently quite strong and the strength of such forward linkages depends crucially on the choice and location of the processing technology involved. The backward production linkages from agriculture to RNFE (rural input suppliers) are also crucial to agricultural performance and there may also be backward linkages to those who supply and maintain produced goods. Finally, consumer demand linkages are generated as a result of increasing farm incomes. Consumption linkages from agricultural income to non-farm sector are normally defined as one type of forward linkages. Push and pull factors are both important in influencing the level and growth of the RNFE. In some rural areas non-farm activities may be increasing because of agricultural growth and in some other areas, the workforce might have been pushed out of the farm sector because of lack of farm growth. The AGL model emphasizes the pull factors.

In this paper an attempt is made to identify the factors determining the rural non-farm employment in Kurnool district of Andhra Pradesh. The data used in this study relate to year 2007-08. This paper seeks to identify the determinants of inter-Mandal variations in the shares of rural non-farm employment (RNFE) in Kurnool district (AP). The focus is on non-farm main workers i.e., workers who had worked for the major part of the year and whose principal work was in RNFE. A study of the determinants of non-farm employment can help to interpret the structure and functioning of the labour markets. This paper attempts to identify these determinants and examines their inter-relationship with the help of Mandal-wise data in the Kurnool district.

The data on share of RNFE in total employment among the different Mandals revealed wide variations in RNFE. The average rural non-farm employment worked out to be 19.3 in 2007-08. Among the Mandals the rural non-farm employment was less than 15% in 17 Mandals, in between 15 to 30% in 20 Mandals, more than 30% in 7 Mandals. These variations are explained using Simple Linear Regression method. The percentage of RNFE is related to each of the following variables:

1. Rural workers in non-farm activities as a percent of total workers
2. Commercialization defined as percentage of non-food crops to total crops
3. Irrigation ratio defined as percentage of irrigated area/total cropped area
4. Average land holding size in hectares
5. Literacy rate(percentage)
6. Percentage of villages with 1000 population and above
7. Percentage of Banks that are rural in total number of banks.
8. Operated area per worker.

These independent variables are explained below.

Commercialisation (x1): Proportion of area under commercial crops may influence non-agricultural employment. For instance, in areas where non-food commercial crops predominate, employment opportunities in non-farm activities like processing, grading, and marketing, may be more widespread. The hypothesis is that, since commercial crops are mostly market-oriented, a large area under commercial crops implies more opportunities for non-farm employment.

Irrigation (x2): An increase in the irrigation ratio leads to changes in cropping patterns from less to more remunerative crops and to improvements in factor productivity. The net effect in increased-value added, which eventually increases incomes in the agricultural sector. The hypothesis is that irrigation increases incomes in agriculture and that this will lead to an increase in the demand for non-farm activities through production and consumption linkages, thereby increasing RNFE.

Land Holding Size (x3): The share of RNFE is hypothesized to vary inversely with the average size of the operational holding. The limited absorptive capacity of their operational holdings compels the poor, landless labourers and land poor farmers to go for RNFE. On the other hand, the agricultural output or income creates greater demand for RNFS and generates surplus for investment in the RNFS. A substantial improvement will be to count irrigated land or the crop twice. Therefore it is argued that farm size may have positive or negative relationship with RNFE.

Levels of Literacy (x4): Generally the impact of literacy on RNFS is expected to be positive (Chadha, 1991) but there is little evidence. However, non-farm activities are of two types, traditional and modern. In the case of traditional non-farm activities, literacy may be a deterring factor to participation or employment. In contrast it might have a favorable impact in the modern RNFS.

Villages Size (x5): The proxy variable used for measuring the agglomeration size is the percentage of villages with more than 1000 people in a Mandal. This variable mainly operates on the production side, specifically in manufacturing activities. It is supposed that regional industrial agglomeration/clusters provide extra benefits for RNFE, possibly due to transfers of technology, the availability of special inputs and a better or more organized business atmosphere.

Percentage of Bank Branches in villages (x6): As proxy variable for the infrastructure we have used percent of bank branched in villages. Rural infrastructure influences the supply costs (farm and non-farm) and expected to have positive effect on RNFE. However, in the case of Maharashtra, a negative relation was observed. (Shukla, 1991) He notes influences the articulation of supply and demand we can said that investment in infrastructure a rural area enhances the position of agriculture at the cost of RNFE, though this is not productive counter to RNFE.

Operated Land per workers (x7): one obvious reason as to why people take up non-agriculture operation is the absence of adequate land. If land base is too small not many people could get gainful employment in it. So as operated land for worker decrease, the proportion of non-agricultural workers in the family may increase. On the other hand, the agricultural output or income creates greater demand for RNFS and generates surplus for investment in the RNFS. Therefore, farm size may have positive and negative relationship with RNFE.

THE DETERMINANTS OF INTER MANDAL VARIATIONS IN THE SHARES OF RNFE

In order to understand the nature of relationship between dependent and independent variables a correlation matrix of

TABLE I

Means and C.Vs. on Dependent and Independent Variables

Sl. No.	*Variables*	*Mean*	*Std. Diviation*	*C.V.*
1.	% of non-farm sector/ total workers	19.29	10.10	52.35
2.	% of non-food crops/ total cropped area	38.83	23.39	60.23
3.	% of irrigated area/ total cropped area	30.75	26.29	85.50
4.	Literacy Rate (%)	48.78	8.5	17.42
5.	% of villages/1000 and above population	82.00	13.50	16.46
6.	% of Rural Banks/ total Banks	58.69	32.00	54.52
7.	Operated area/ total no. of Holdings	1.78	0.36	20.22
8.	Operated area/total no. of workers	2.70	12.50	462.96

Source : Primary data.

all the variables was obtined. This is helpful to check whether dependent variables are related to each other or move together. The correlation matrix do not show abnormally high correlation between the dependent variables. Hence, there is no problem of multi-co-linearity in this model.

The estimated regression coefficient for the selected variables are presented in the Table 2.

The regression results are presented in Table 2. The explanatory power of the regression equation is indicated by the value of R-square. The equation explain 69 percent of the inter-Mandal variation in RNFE as a share of total employment. With the exception of the coefficient associated with commercial crops, all other coefficiences have the expected signs. The relationship between the commercialization of agriculture of rural non-farm employment is the opposite of that predicted, but that is not significant. This could imply that Mandals with rapid agricultural growth had better absorption of labour in the farm sector itself and thus less spill-over effects

TABLE 2
Regression on Coefficients

Sl. No.	*Variables*	*Regression coefficient*	*t-test*	*Sig.*
1.	% of non-fam sector/ total workers	Constant	-0.80	0.43
2.	% of non-food crops/ total cropped area	-0.04	-0.32	0.74
3.	% of irrigated area/ total cropped area	0.14	1.42	0.17
4.	Literacy Rate (%)	0.39	2.99	0.05
5.	% of villages/1000 and above population	0.20	0.20	0.83
6.	% of Rural Banks/ total Banks	-0.18	-1.81	0.79
7.	Operated area/total no. of Holdings	-0.11	1.14	0.25
8.	Operated area/ total no. of workers	0.76	8.30	0.00

R-square=0.69, F-value=14.22.

in to the RLFS. However, the negative sign may be due to the fact that Rural towns are not included in the analysis.

Irrigation is expected to be positively related with the rural RNFE. The regression coefficient for irrigation is 0.15 and is not statistically significant in explaining variations in Mandal level FNFE Share. The hypothesis that growth linkages from irrigated agriculture lead to RNFE is rejected. The regression coefficient for average land holding size is 0.38 and is significant at 5 percent level. The RNFE elasticity to literacy is 0.2 the hypothesis that literacy has positive on RNFE is proved. The operated land per worker has positive relationship and significant at 1 percent level.

Thus, it can be concluded that all the selected variables except commercialization are positively related to RNFE. The average land holding size and operated land per worker are significantly related to RNFE.

References

Rural Non-farm Growth (Sign of Farm Success of Failure?), Dr. M. Prasada Rao.

Rural Employment (the Non-farm Sector), M. Koteswara Rao.

Economic Survey, 2008-09.

Kurnool District—Hand Book Statistics, 2007-08.

Indian Rural Development Report, 1999.

Eleventh Five Year Plan, 2007-12.

CHAPTER

23

Women Empowerment in Rural Development Through NGOs

MALLIKARJUN G. NAIK AND ASHOK R. RATHOD

INTRODUCTION

Rural development is an important sector of national development for any developing economy. Rural development includes poverty alleviation, poverty reduction coupled with agriculture development, rural industrial development and women empowerment. Poverty is often referred as a severe failure of basic capabilities and often related to inadequate incomes. Poverty as a concept encompasses many aspects of wants and disadvantages. Chambers (1995) recognizes that lack of assets: physical weakness, isolation, vulnerability, and powerlessness are the five clusters of disadvantages characterize the poor, particular to women in the rural areas.

Even though considerable advance has taken place in gender equality in recent years, gender discrimination still remains persistently in many dimensions of life, worldwide.

Gender gaps are widespread in access to and control of resources, in economic opportunities, in power, and political voice. Women and girls bear the largest and most direct costs of these inequalities. For these reasons, gender equality and empowerment of women is recognized globally as a key element to achieve progress in all areas. Gender equality is a core development issue—a development objective in its own right. It strengthens countries abilities to grow, to reduce poverty and to govern effectively. Hence, promoting gender equality is thus an important part of a development strategy that seeks to enable all people—women and men alike—to eradicate poverty and improve their standard of living.

The concept of empowerment is fundamental aspect in the gender literature. Srilath Batliwala, an NGO activist puts it, "I like the term empowerment because no one has defined it clearly yet, so it gives a breathing space to work it out in action terms before we have to pin ourselves down to what it means".

The empowerment of women is one of the central issues in the process of development of all developing countries in the world. Historically, women in India are deprived socially and economically compared to men. Disparities between men and women prevails here in education, health, employment and income opportunities, control over assets, personal security and participation in the political process that make women disadvantaged and less empowered, which limits the country's ability to achieve its full potential. It is well established that women have less access than men to investments in skills, knowledge and lifelong learning. The empowerment of women is an essential prerequisite for the exclusion of world poverty and the upholding of human rights.

Empowerment results from control over resources and the power to take decisions on all major issues concerned. "The empowerment process encompasses several mutually reinforcing components but begins with and supported by economic independence". (Ranjana Kumari) Empowerment would consist of greater access to knowledge and resources, greater autonomy in decision-making and free them from shackles imposed on them by custom, belief and practice.

Backwardness among rural women is such more severe to various socio-economic and cultural reasons. International,

national and local government organizations and non-government organizations are involved in various experiments of women empowerment in backward regions.

NGOs AND RURAL WOMEN EMPOWERMENT

NGOs can frame many activities for women empowerment. Historians of feminism have long back noted that nineteenth century philanthropy offered a pathway for women in western societies to move from the private to the public sphere. Often denied access political participation and barred from donors, volunteers and organizational entrepreneurs nonetheless left their imprint on national legislation and institutions in a variety of countries. Through their philanthropic contributions of time, money and possessions carved out a public niche for themselves in diverse religions, political and economic regimes. In India, several NGOs are involved in women empowerment programmes in general and rural women empowerment in specific.

OBJECTIVES

The study was carried out with the following specific objectives:

- To assess the impact of micro-credit on employment and income generation of women empowerment in the selected areas.
- To ascertain the factors related to empowerment of women.
- To give suitable suggestion in respect of NGOs and empowering women in society.

METHODOLOGY

Data Collection and Selection of Villages: The primary data have been collected from the selected NGOs revealed that it is working in 30 villages. Out of those, three villages were selected randomly on the basis of following criteria, highest number of SHGs organized by NGOs, highest programmes and kinds of training programmes carried out. In this regard, three villages of Bijapur district (Karnataka) are selected, namely; Minchinal, Aheri and Ainapur.

Selection of the Beneficiaries: Complete list of all the programme beneficiaries were prepared for all the selected villages. From the listed villages by employing the following criteria. Beneficiaries should have knowledge about SHGs, beneficiaries of NGOs and its staff consideration of this criteria randomly selected beneficiaries village-wise, as follows.

TABLE I

Village-wise Distribution of Beneficiaries Selected for the Study

Village	*Total No. of Beneficiaries*	*Selected No. of Beneficiaries*	*Percentage*
Minchinal	92	45	40.90
Aheri	95	45	40.90
Ainapur	110	20	18.20
Total	227	110	100

NGOs AND MICRO-FINANCE (SHG's)

To bridge the gap between demand and supply of funds in the lower rungs of the rural economy, the formal sector took the initiative to develop a supplementary credit delivery mechanism by encouraging institutional arrangements outside the financial system. With the launching of NABARD's pilot scheme, microfinance—the development buzzword of the ninetees to cure the illness of rural poverty gained visibility in the Indian development landscape.

Micro-finance is being referred to as small scale financial services provided to people who work in agriculture and allied sectors; operate small and micro-enterprises; provide services; work for wage and commission and other individuals and groups at the local levels of developing countries under rural areas. With the SHG-linkage programme introduced in 1992, the NGO sector has been recognized as a crucial partner. Recognizing the strengths of the NGOs in organizing the community and the potential in savings and credit programmes (both under the linkage programme and other credit delivery innovations), NABARD also started associating with them increasingly. Mainly, their services in the area of capacity building, training and promotion of SHGs and to a limited extent as financial intermediaries also. NGOs are building social capital in the form of groups that can generate a sound base for their members to develop their credibility as borrowers and encourage financial institutions to develop confidence in establishing a lending relationship with groups.

NGO has trained and developed the SHGs to a level where it can do business on equal terms with a bank. Social and communication barriers often make it necessary for the NGOs to introduce the group to the bank and to poster the relationship for some time. Large number of NGOs have responded to this new 'Market', since it is a powerful method of empowering their clients and making them independent of future assistance.

According to NABARD data, by the end of the 1990s about 800 NGOs were participating in its SHG Bank-linkage programme. (E. Shunmugam, 2000) Several others have developed microfinance function either with assistance from international donors or exclusively on the strength of individual savings. The loans are also provided to NGOs for on lending to the self-help group members. The process proceeds as follows :

- NGOs evaluate the SHGs by using the format devised by the bank.
- For those SHGs, which quality for financial assistance, NGOs assesses loan requirements of each member separately.
- Then, the NGO prepares a detailed proposal indicating SHG-wise and member-wise loan requirement.
- This proposal is submitted to the bank. The bank scrutinizes the proposal by evaluating the NGOs and

10.15 percent SHGs are selected randomly. The bank has not taken security for providing loans to these NGOs.

SABALA has made a pioneering effort in this region, in community organization and monitoring the self help group in which rural women are largely involved. Women participation in the savings and the credit movement has paved the way for the speedy socio-economic development of rural women.

SABALA organizes SHGs with following objectives :

1. To include self-help activities among women folk.
2. To develop collective leadership.
3. To enhance effective women's participation in their development programmes.
4. To promote savings habits and develop an indigenous banking systems within the village among the women folk.
5. To federate these SHGs under one umbrella or apex body not only for credit purpose but also to promote women's solidarity and eventually women's empowerment.

While traditional women's associations look up for help from outside, the SHGs help themselves with their savings and resources. Hence, helping themselves becomes the prime motto of SHGs.

SABALA has organized 330 SHGs in its area of operation. There are 5010 total members in various groups. Hence, an attempt has been made to understand the different aspects of SHGs. The study of savings details, loan details, impact on employment generation, impact on generation of income etc., has been analyzed. The study of these independent variables helps us to know not only the status of rural women but also to understand the changes in the life of members.

PURPOSE OF BORROWINGS

In order to meet unforeseen expenditure of family, taking loan is common. In the case of rural women in the present study who are members of SHG, who are economically weak, loan becomes inevitable in order to provide some facility to the family members. The purposes of borrowing are given in Table 2.

TABLE 2
Purpose of Borrowing

Purpose	*Frequency*	*Percentage*
Agriculture	18	16.36
Petty Business	19	17.27
Medical Facility	10	9.09
Children's Education	08	7.27
To clear old Dept.	12	10.90
Household Expenditure	43	39.09
Total	110	100

Table 2 indicates that 18 (16.36 percent) respondents have borrowed for the purpose of agriculture, 19 (17.27 percent) respondents have for petty business, 10 (9.09 percent) respondents taken to meet out the medical expenses, 8 (7.27 percent) have borrowed for the sake of their children's education, 12 (10.90 percent) taken to clear old debts and 43 (39.09 percent) have borrowed for the sake of fulfillment of household expenditure.

The data clearly shows that 16.36 percent of the respondents have borrowed to take up agriculture such as, seed, fertilizers, pesticides, etc., and 17.27 percent of the respondents have borrowed to take up petty business such as small kirani shops, poultry, tailoring, cloth business, hotel, etc., which brings additional income to the family. And also SHGs are meeting different needs of the respondents. The inability of institutions to deal with the credit requirements of the poor effectively has led to the emergency of microfinance or micro credit system as an alternative credit system to the poor. Usually credit institutions provide finance for productive purposes but sometimes poor people need money for consumption or for emergency purpose. In this study also majority of the respondents have taken loan to meet out their household expenditure, medical facility, etc.

AMOUNT OF DEBT

Table 3 gives information about the amount of debt. It indicates that 36 (32.72 percent) of the respondents have borrowed Rs. 2,000 to 4,000, 39 (35.45 percent) have borrowed Rs. 4,001 to 6,000 and 35 (31.81 percent) have borrowed Rs. 6,001 and above.

TABLE 3
Amount of Loan from SHG (Debt)

Amount	*Frequency*	*Percentage*
Rs. 2,000 to 4,000	36	32.72
Rs. 4,001 to 6,000	39	35.45
Rs. 6,001 and above	35	31.81
Total	110	100

The data indicates that the number of respondents increases with increasing amount of debt. Basically the respondents who are poor borrow large amount with low rate of interest as compared to other private money lenders in rural areas. The respondents are also conscious that, the interest they pay goes to their own group and shared by all the members of group.

Savings

The basic objective of SHGs is to develop saving habits among the poor sections of the society, particularly among rural women, which in turn reduces dependency on financial institutions and develops self-relience. Hence, an attempt has been made to collect information about savings of members. The data presented in Table 4.

Table 4 indicates that 25 (22.72 percent) of the respondents save Rs. 30 to 40, 28 (25.45 percent) respondents save Rs. 41 to

TABLE 4

Monthly Savings of Respondents

Saving Amount Rs. Per Month	*Frequency*	*Percentage*
Rs. 30 to 40	25	22.72
Rs. 41 to 60	28	25.45
Rs. 61 to 80	36	32.72
Rs. 81 to 100	21	19.09
Total	110	100

60, 36 (32.72 percent) respondents save Rs. 61 to 80 and 21 (19.09 percent) respondents save the amount between Rs. 81 to 100.

The analysis of the above data clearly shows that a significant number 36 (32.72 percent) of the respondents save Rs. 61 to 80 and 28 (25.45 percent) save Rs. 41 to 60. It shows that SHG has developed the saving habit among the rural women with their limited income. Moreover, the benefits of self-help groups are based on co-operation rather than competition. It follows the real principle of "contribute and save according to your ability and extract according to your need". Regular savings by each member is an indication of the members' commitment to the group and to personal growth and progress. There is an aphorism in English that the little drops of water makes a mighty ocean and this has made true among SHGs.

The source of saving is an important fact. The respondents source of saving shows Table 5. It indicates that 29 (26.36 percent) respondents have saved from wages, 10 (9.09 percent) have saved from tailoring, 17 (15.45 percent) respondents saved from pretty business, 16 (14.54 percent) have saved from dairy, 30 (27.27 percent) have saved from craft work and only 8 (7.27 percent) respondents have saved from other source.

The Table 5 indicates that majority of the respondents saving source is craft work (income) and their employment wages, some others save from dairy and petty business, etc. It indicates that sources of saving are increasing trend.

TABLE 5
Saving Source of the Respondents

Source	*Frequency*	*Percentage*
Wage	29	26.36
Tailoring	10	9.09
Petty Business	17	15.45
Dairy	16	14.54
Craft Work	30	27.27
Others	08	7.27
Total	110	100

EMPLOYMENT GENERATION

No doubt any financial assistance, if utilized properly, generates gainful employment opportunities in the rural economy. It was observed in the field survey that the sampled some members also got gainful employment opportunities as shown in Table 6. The data shows that on an average, the loans received generated 175 person days of employment per

TABLE 6
Impact on Employment Generation

Financing Activity	*Frequency*	*Total Employment Generated*	*Average Employment Generated*
Agriculture	18	2,304	128
Dairing	04	400	100
Tailoring	10	2,200	220
Cloth Business	03	870	290
Poultry	04	480	120
Hotel Business	02	600	300
Kirani Shops	03	840	280
Total	44	7,694	175

household. In all, 7,694 person days of employment was generated for 44 household members. It was noted that non-farm activities generated higher number of person days of employment in the sample village. Hotel business and cloth business generated 300 and 290 person days of employment, tailoring generated 220 person days of employment. On the contrary, agriculture could generate 128 person days of employment on an average per household followed by 120 person days of employment by poultry and 100 person days by dairying.

A look at this data makes us infer that the loans provided by SHGs are productive and efficient in generation of employment to rural farm and non-farm workers in general.

GENERATION OF INCOME

There is a symbolic relationship between generation of income and employment opportunities and the potential of employment can be judged by the amount of income generated in any activity. Table 7 makes an attempt to explain the impact of loans provided on generation of income. It seems the loans provided by SHGs had a favorable impact on generation of

TABLE 7

Impact on Generation of Income

Financing Activity	*Frequency*	*Total Income Generated*	*Average Income Generated (In Rs.)*
Agriculture	18	2,16,000	12,000
Dairing	04	20,000	5,000
Tailoring	10	66,000	6,600
Cloth Business	03	18,000	6,000
Poultry	04	24,000	6,000
Hotel Business	02	10,000	5,000
Kirani Shops	03	18,000	6,000
Total	44	3,72,000	8,455

income in the selected villages. On an average, each selected family could get an income of Rs. 8,455 which is sufficient to bring the poor families above the poverty line. No doubt, the income generation varies from activity to activity and each activity has its own capacity to generate income.

The data also reveals this fact income generated in the selected activities shows that it varies from Rs. 5,000 per annum in the case of daring and hotel business, Rs. 12,000 in the case of agriculture. Highest amount of income generation is seen from agriculture, tailoring proved an efficient activity as it generated an average income of Rs. 6,600 per household, followed by Rs. 6,000 in cloth business, poultry and kirani shops.

FINDINGS

1. Maximum of 35.4 percent beneficiaries joined the SHG's for saving purpose, 30 percent joined SHG's for income earning purpose, 10.9 percent were joined for employment purpose and 23.63 percent beneficiaries were joined SHG's for loan purpose.
2. Amount of loan from SHG's, majority of 32.7 percent of the beneficiaries have taken Rs. 4,001 to 6,000 and 31.8 percent have taken Rs. 6,001 and above. The number of beneficiaries increases with increasing amount of loan. Basically the respondent who are poor, borrow large amount with low rate of interest as compared to other. The majority of beneficiaries saving sources are wage, petty business, tailoring, craft work income, dairy and others. It indicates that sources of saving are increasing trend.
3. These SHG's are mobilizing thrift deposits and receiving timely matching and revolving funds to generate employment activities to earn their livelihood.
4. Private moneylenders in rural areas.
5. The loans provided by SHG's had a favorable impact on generation of income in the selected villages on an average, each selected family could get an income of

Rs. 8,455, which is sufficient to bring the poor families above the poverty line.

CONCLUSION AND SUGGESTIONS

The results of the study provides sufficient evidence that micro-credit program is contributing to some extent in women empowerment in rural particularly selected areas. Though it shows that the overall status of empowerment of the respondent women is poor, however, the women who are actively involved with micro-credit programmes their status of empowerment are relatively better than the individual housewives. Thus, it can be concluded that micro-credit programmes have significant impact in empowering the rural women in selected areas.

SUGGESTIONS

(1) In the study area is scarcity of infrastructure facilities. Therefore, organization should involve in providing of infrastructure to the rural poor.

(2) SABALA organization has been providing loans to the SHGs. But it is not sufficient. Therefore, there is required more loan facilities for the betterment and improvement of the people.

(3) Organization should be developed to build harmonious relations to rural people, particularly the grass-root level.

(4) Organization should conduct more seminars, meetings and moral education to the village people.

References

Bhatia, Anju (2000), "Women Development and NGOs". Rawat Publication, New Delhi.

Meenu Agrawal, Shobana Nelasco (2009), "Empowerment of Rural Women in India", Kanishka Publishers, New Delhi.

Ram Naresh Thakur (2009), "Rural Women Empowerment in India" in *Empowerment of Rural Women in India*, Kanishka Publishers, New Delhi.

Shobana Nelasco and Junofy Antorozarina (2009), "Rural Women Empowerment through Self Help Groups" in *Empowerment of Rural Women in India,* Kanishka Publishers, New Delhi.

Lipi (2009), "Women Empowerment: Globalization and Opportunities" in *Empowerment of Rural Women in India,* Kanishka Publishers, New Delhi.

Batliwala, Srilatha (1994), "The meaning of Women's Empowerment: New Concepts from Action", in *Population Policies Reconsidered: Health, Empowerment and Rights.* G. Sen, A. Germain, and L.C. Chen, eds. Cambridge, MA: Harvard University Press.

Bisnath, Savitri and Diane Elson (1999), *Women's Empowerment Revisited.* Background Paper, Progress of the World's Women. UNIFEM.

Sadik, N., "Women, the Center of Development", *Development,* 1988, Vol. 1.

CHAPTER

24

Role and Importance of Women Entrepreneurship in Informal Sector in U.P. : With Special Reference to Allahabad City

J. NARAYAN AND NIHARIKA SRIVASTAVA

I

"When women move forward, the family moves, the village moves and the Nation moves."

—Pandit Jawahar Lal Nehru

"Women" as Entrepreneurs in India: Women owned businesses are highly increasing in the economies of almost all countries. The hidden entrepreneurial potentials of women have gradually been changing with the growing sensitivity to the role and economic status in the society. Skill, knowledge and adaptability in business are the main reasons for women to emerge into business ventures. 'Women Entrepreneur' is a person who accepts challenging role to meet her personal needs and become economically independent.

Defining Women Entrepreneur,[1] "An enterprise owned and controlled by a woman having a minimum financial interest of 51% of capital and giving at least 51% of the employment generated by the enterprise to women."—Government of India. "A women entrepreneur can be defined as confident, innovative and creative women capable of achieving self-economic independence individually or in collaboration, generates employment opportunities for others through initiating, establishing and running the enterprise by keeping pace with her personal, family and social life."—Kamal Singh.

A strong desire to do something positive is an inbuilt quality of entrepreneurial woman, who is capable of contributing values in both family and social life. With the advent of media, women are aware of their own traits, rights and also the work situations. The glass ceilings are shattered and women are found indulged in every line of business from Pappad to power cables. The challenges and opportunities provided to the women of digital era are growing rapidly that the job-seekers are turning into job creators. They are flourishing as designers, interior decorators, exporters, publishers, garment manufacturers and still exploring new avenues of economic participation.

In India, the term informal sector has not been used in the official statistics or in the National Accounts Statistics (NAS). The terms used in the Indian NAS are 'organized' and 'unorganized' sectors. The organized sector comprises enterprises for which the statistics are available from the budget documents or reports, etc. On the other hand, the unorganized sector refers to those enterprises whose activities or collection of data is not regulated under any legal provision or do not maintain any regular accounts. In the unorganized sector, in addition to the unincorporated proprieties or partnership enterprises or enterprises run by cooperative societies, trust, private and limited companies are also covered. The informal sector can therefore, be considered as a sub-set of the unorganized sector.

In the light of definition of informal sector encompassing private unincorporated enterprises as mentioned above, NSS 55th round, 1999-2000 also covered non-agricultural enterprises in the informal sector in India. As per survey, there were 44.35 million enterprises and 79.71 million workers employed thereof in the non-agricultural informal sector of the economy. Among these 25.01 million enterprises employing 39.74 million workers were in rural areas whereas 19.34 million enterprises with 39.97 million workers in the urban area. Among the workers engaged in the informal sector, 70.21 million are full time and 9.5 million part time. Percentage of female workers to the total workers is 20.2 percent.

II

THE GLOBAL WORKFORCE : A STATISTICAL PICTURE

In 2002, the WIEGO network analyzed available national statistics on the informal economy (broadly defined) for an ILO publication called Women and Men in the Informal Economy: A Statistical Picture. The available evidence shows that, in developing countries, informal employment comprises one-half to three-quarters of non-agricultural employment in developing countries and, if informal employment in agriculture is measured, a higher percentage still of total employment. Self-employment comprises a greater share of informal employment than wage employment. And informal employment comprises a greater share of women's employment than of men's employment. What follows is a summary of the main statistical findings of the 2002 ILO publication, including data on non-standard work in developed countries:

Comparative International Size of the Informal Economy[2]

Informal employment broadly defined comprises one-half to three-quarters of non-agricultural employment in developing countries: specifically,

47 per cent in the Middle-East and North Africa;
51 per cent in Latin America;

71 per cent in Asia; and
72 per cent in sub-Saharan Africa.

Comparative International Composition of the Informal Economy[3]

As noted earlier, the informal economy is comprised of both self-employment in informal enterprises (i.e., small and/or unregistered) and wage employment in informal jobs (i.e., without secure contracts, worker benefits or social protection). In developing regions, self-employment comprises a greater share of informal employment outside of agriculture than wage employment. More specifically, self-employment represents—

70 per cent of informal employment in sub-Saharan Africa;
63 per cent in the Middle-East and North Africa; and
60 per cent in Latin America; and
57 per cent in Asia.

Relative Importance of Women and Men in the Informal Economy[4]

Informal employment is generally a larger source of employment for women than for men in the developing world. Other than in the Middle-East and North Africa where 42 per cent of women workers are in informal employment, 60 per cent or more of women workers in the developing world are in informal employment (outside agriculture). Here are the comparative figures for informal employment as share of non-agricultural employment by sex and region: in sub-Saharan Africa, 84 per cent of women workers compared to 63 per cent of men workers; in Latin America 58 per cent of women workers in comparison to 48 per cent of men workers; and in Asia 73 per cent of women workers in comparison to 70 per cent of men workers. Although women's labor forces participation rates are lower than men's, the limited data available point to the importance of women in home-based work and street vending in developing countries:

30-90 per cent of street vendors (except in societies that restrict women's mobility);

35-80 per cent of all home-based workers (including both self-employed and home workers);
80 per cent or more of home workers (industrial outworkers who work at home).

Although women's labor force participation rates are lower than men's, women represent the vast majority of part-time workers in many developed countries. In 1998, women were 60 per cent or more of part-time workers in all OECD countries reporting data. Women's share of part-time work for specific countries was as high as 98 per cent in Sweden, 80 per cent in the United Kingdom, and 68 per cent in both Japan and the United States.

Importance of Women Entrepreneurs

50% of the population is traditionally outside the domain of economic activities. They must be made part of the economic development, because it will not only ensure the economic and social development of the women along with providing more human resources to strengthen the economy of the country. The economic status of women is now accepted as an indicator of a society's stage of development. Therefore, Women's participation has become important in economic activities. They are doing jobs as a teacher, clerk, etc. they have also been interested in entrepreneurship.

III

HYPOTHESIS, OBJECTIVES AND METHODOLOGY OF THIS STUDY

The hypothesis of this study is *"Participation of Women as a Women Entrepreneur is increasing in normal life of the country."* The above hypothesis shall be tested on the basis of primary survey of Allahabad city. The main objective of this study is as follows:

- To identify factors which are relatively more important in providing empowerment to women.

- To find out, how much contribution of women is in economic activities.
- To assess the problems and obstacles which becomes hindrance in the entrepreneurial activities of women.
- To predict new areas or methods that can help to promote women entrepreneurship in modern era in Allahabad city.

We propose Allahabad city (Urban part of the Allahabad city) as an area of survey for conducting the study of above-mentioned objectives. Allahabad, the highest populated district in U.P. and is a native land of various important personalities of the country. Allahabad city has religious importance on the religious map of the world, important from the point of view of largest Hindu Fair "Kumbh Mela" at the interval of every 12 years. Allahabad city having a population of approximately 14 lacs. The survey was performed through close-ended questionnaire in this study. Survey are divided into three parts—

- Those women who are running business now (30 respondents).
- Those women who are in jobs (35 respondents).
- Those women who do not engage in any economic activity (35 respondents).

Thus, the sample size is 100 and all women above 20 years old either middle-Income class or upper middle-Income class constitutes the universe of the proposed survey. For studying, so-called objectives, Allahabad city has been chosen purposively. Now at the second stage, Allahabad city has been divided into different strata level (according to the income specifically) for choosing middle-income group and upper middle-income group, strata for women were prepared. Among these selected strata group, 100 women are selected. Statistical tools—namely Mean, Mode, Correlation Coefficient; Rank Correlation Coefficient has been worked out and they were tested by Chi-square and t-tests. Pie diagram is also used.

IV

ANALYSIS AND INTERPRETATION OF THE SURVEY

After analyzing the data it was found that Women Entrepreneurship depends on education of women, family income, personal interest, family interest for job, age, marital status, etc. Among them age, marital status, education and family income are identified most important factors to determine women entrepreneurship. Mathematically, we can write it as follows :

$$W.E.P. = f\{E, Y_f, A, M, I_p, I_f\}$$

where *W.E.P. stands for Women Entrepreneurship Preference, E for Education,* Y_f *for Family Income, A for Age, M for Marital Status,* I_p *for Personal Interest* I_f *for Family Interest.*

On the basis of survey (See in Table 1) we come to this conclusion that 60% women preferred business and 55% of women entered into the entrepreneurship around the age of 38 and above. This result is consistent to another study in which it was found that women entrepreneurs started their business or venture at the age of 35 (Hisrich and Brush, 1986) and 80% of women at the age of 39 entered into the entrepreneurship (Rajesh Kumar Shastri and Avantika Sinha).[5] This gives an impression that women entrepreneurs by the age of thirties

TABLE I

Mean/Mode, Standard Deviation and Standard Error Mean of Women Entrepreneurship

Factors	*N*	*Mean/Mode*
Preferences	100	Business (60%)
Age	60	38.3
Education	60	13.15
Marital Status	60	Married (66.7%)
Spouse Income	60	8083.33
Promoted by	60	Personal Interest (58.33%)

become independent of their family responsibilities and left with no work. So, this is the time in which they can think of some innovative ideas or to give certain direction to their lives. They try to spend their time into some productive work, which came into the form of starting some business. Almost every woman opted for the business related to beauty-parlor and law. Probably this is the profession, which they could easily be able to manage and work efficiently. Educational background of the women entrepreneurs also seems to be quite impressive, as most of them have completed their studies up to graduation.

If we see the relationship between family income and women entrepreneurship we find that very low-income group family women and upper middle-income group family women are generally low risk bearer. They are not very sound economically. Some have no need, so they give smaller preferences to business but Middle-income group and upper middle-income group women (77%) mostly prefer entrepreneurship because they are high risk bearer, having family support and personal interest. (See Table 2 and Chart 1)

TABLE 2

Preferences of Women Entrepreneurship by Family Income Level

Income level	*Preferences (in %)*
Less than 5000	20.0
5000-10000	56.7
10000-15000	18.3
15000-20000	1.7
20000 and Above	3.3
Total	100.0

And Income and Education factor analysis highlights that high income and low educated women prefer entrepreneurship. Middle income group and average educated women also like entrepreneurship. (See Tables 1 and 3)

FIG. I

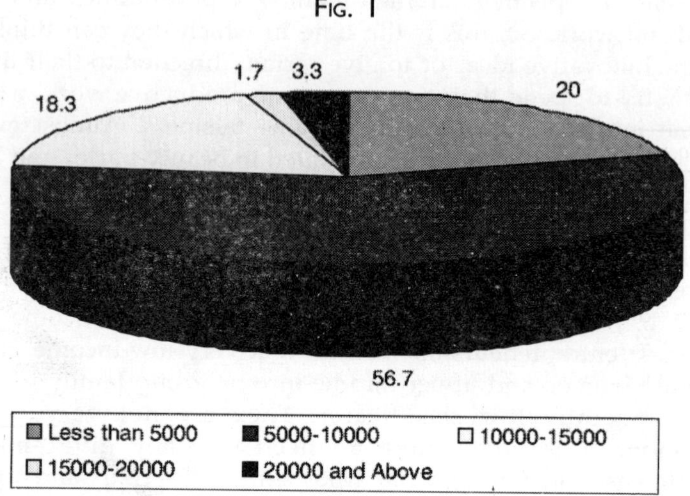

TABLE 3

Preferences of Women Entrepreneurship by Income and Education Level-wise (In %)

Income	*Business Preferred*			*Service Preferred*		
	Up to inter-mediate	*Grad-uate*	*Post-Grad-uate*	*Up to inter-mediate*	*Grad-uate*	*Post-Grad-uate*
Less than 5000	29.41	8.70	25.0	—	20.0	53.85
5000-10000	17.65	86.96	55.0	30.0	80.0	30.77
10000 or Above	52.94	4.35	20.0	70.0	—	15.38
Total	100	100	100	100	100	100

Correlation Coefficient and their testing between different factors are shown in the Table 4. Summary of this table is that Preferences for Women Entrepreneurship is positively Correlated with Education while it is negatively correlated with the Age-group-wise distribution. Income and Women Entrepreneurship is positively Associated to each other (on the basis of t-test, there is no correlation) and Personal interest and

TABLE 4

Correlation Coefficient, Rank Correlation Coefficient and their Testing Results

Sl. No.	*Particulars*	*Results*
(1)	*(2)*	*(3)*
1.	Rank Correlation Coefficient between women preferences for business and age-group-wise distribution	-1, perfectly negatively correlation
2.	T-Test results for testing Correlation Coefficient between women preferences for business and age-group-wise distribution	$t_{(cal,1,.05)}$ =infinite $t_{(tab,1,.05)}$ =12.706 Ho : r=o H_1 : r≠o H_1 is accepted.
3.	Rank Correlation Coefficient between women preferences for business and Educational level distribution	0.9, highly positive correlation
4.	T-Test results for testing Correlation Coefficient between women preferences for business and educational level	$t_{(cal,3,.05)}$ =3.61 $t_{(tab,3,.05)}$ =3.182 Ho : r=o H_1 : r≠o H_1 is accepted.
5.	Rank Correlation Coefficient between women preferences for business and Income level distribution	0.8, highly positive correlation
6.	T-Test results for testing Correlation Coefficient between women preferences for business and Income level	$t_{(cal,3,.05)}$ =2.3067 $t_{(tab,3,.05)}$ =3.182 Ho : r=o H_1 : r≠o H_0 is accepted.
7.	Karl Pearson Correlation Coefficient between Educational level and age-group-wise distribution	0.28, positive correlation
8.	Karl Pearson Correlation Coefficient between Income and Educational level	-0.01, negative correlation

(Contd.)

TABLE 4 (*Contd.*)

(1)	(2)	(3)
9.	Rank Correlation Coefficient between women preferences for business and promotional aspects	1, perfectly positively correlation
10.	X^2-test between women preferences for business and promotional aspects	$X^2_{(cal,2,.05)}$ =24.10 $X^2_{(tab,2,.05)}$ =5.991 Ho: All promotional aspects have equal preferences. H_1: All promotional aspects have not equal preferences. H_1 is accepted.

family interest are also positively related to women entrepreneurship.

Tailoring and Garments, Jewelers, Daily needs product shop is that business in which generally women enter or start with her husbands or relatives. Coaching classes, Painting and Dance classes, Tailoring, Beauty parlor is low investment business that can be managed or financed either by self or by borrowing from friends, relatives, and family members. These can be run in their own premises and without any support. These are full time but 8 to 10 months jobs. Someone can earn between Rs. 5,000-10,000 per month through these jobs. It depends on investment, location, charges and quality of service.

Some businesses like Garments and Tailoring, Jewelers, School and crèche are high investment business (more than Rs. 10,000) and returns are also in the same range but they cannot run without the support of family. They may run in own premises too and the job expectation is highly probable income. It is also full time and 8 to 12 months job. Supporting staff (1 to 4) will be hired round about wages to the tune of 5,000 in crèche, Garments and Tailoring, Jewellers.

In all the number of workers, they had between 3 to 5. This implies that due to having small-scale enterprises and limited earnings women preferred to keep up to 10 workers.

This result is consistent to another study in which it was found that women entrepreneurs had 5-10 workers (Rajesh Kumar Shastri and Avantika Sinha).

The number of hours, which they spent on their business, was around 8 or 12 h. This amount of time is quite satisfactory and sufficient according to the market timing of Allahabad where markets open around 11 O'clock and close at 9 O'clock. Similarly, the amount of time, which they spent with their family, is more than 12 h. Thus, it shows that they make a balance between the work and family. They are quite committed to both and know about their responsibilities and how to manage them. They keep their children in the childcare facility center as both the husband and wife work and most of them having a nuclear family. Though the women are not willing to keep their children in a child care centers but are forced due to safety reasons.

Some problems are also faced by women entrepreneur like—*Marketing* [in all], *Power* [Beauty parlor, School and crèche, PCO and Daily Need products shop], *Stock maintaining* [Garments and Tailoring, Necessary product shop, Jewellers], *Finance* [School and crèche, Garments, Jewellers]. Refer to Table 5.

The findings of this paper for Allahabad City as well as the findings of the paper, 'The Social Cultural and Economic Effect on the Development of Women Entrepreneurs' wrote by Rajesh Kumar Shastri and Avantika Sinha for Allahabad District as a whole are more or less similar. And since Allahabad District includes both Rural and Urban segment, therefore, the findings of Urban Allahabad regarding the Women Entrepreneurship (Non-Form Sector) are important and will focus some light and rays of directions to the policy-makers for non-farm sector of the economy.

V

FACTORS INFLUENCING WOMEN ENTREPRENEURS AND FOR NEW AREAS

Factors influencing women entrepreneurs' Economic independence Establishing their own creativity Establishing

TABLE 5

Summary of Running Women Enterprises and Problems

Particulars	*Coaching Classes*	*Beauty Parlor*	*Tailoring*	*School and Crèche*
(1)	(2)	(3)	(4)	(5)
Entry/New or Self	New or Self	New or Self	New or Self	Entry
Investment	Very low	5000-10000	Up to 5000	High
Place	Own premises	Own premises rental building	Own premises	rental building
Time Duration (in hours)	6-6	8	8	8
Month Duration	6-12	8-12	10-12	10-12
Job Expectation	Low	Pros pours/Average	Pros pours/Average	Pros pours
Other person requirement (except family member)	No need	May be Maximum 4	Yes Maximum 4	Yes Many
Payment	—	—	Below 5000	High
Problems	Marketing	Marketing and Power	Marketing, Stock Maintaining	Marketing, power, Finance
Earning	Up to 5000	5000-15000	5000-8000	High

(*Contd.*)

TABLE 5 (Contd.)

Particulars	*Painting and Music*	*Garments*	*PCO and Necessary Product Shop*	*Jewellers*
(1)	(6)	(7)	(8)	(9)
Entry/New or Self	New or Self	Entry	Entry	Entry
Investment	Very low	5000-25000	5000-10000	Very high 15000 and above
Place	Own premises	Own premises/ rental building	Own premises/ rental building	Own premises/ rental building
Time Duration (in hours)	4-6	8	8	8
Month Duration	8-8	10-12	12	10-12
Job Expectation	Low	Pros pours	Average	Pros pours
Other person requirement (except family member)	No need	Yes 1-4	No need	Yes more than 3
Payment	—	Below 5000	—	Below 5000
Problems	Marketing	Marketing, Finance, Stock Maintaining	Stock Maintaining, Power	Marketing, Finance, Stock Maintaining, Raw-material
Earning	Up to 5000	5000-25000	5000-15000	Very High

their own identity Achievement of excellence, Building confidence, Developing risk-taking ability, Motivation, Equal status in society, Greater freedom and mobility. Highly educated, technically sound and professionally qualified women should be encouraged for managing their own business, rather than dependent on wage employment outlets. The unexplored talents of young women can be identified, trained and used for various types of industries to increase the productivity in the industrial sector. A desirable environment is necessary for every woman to inculcate entrepreneurial values and involve greatly in business dealings. The additional business opportunities that are recently approaching for women entrepreneurs are:

- Eco-friendly technology
- Bio-technology
- IT enabled enterprises
- Event Management
- Tourism industry
- Telecommunication
- Plastic materials
- Vermi culture
- Mineral water
- Sericulture
- Floriculture
- Herbal and health care
- Food, fruits and vegetable processing

Empowering women entrepreneurs is essential for achieving the goals of sustainable development and the bottlenecks hindering their growth must be eradicated to entitle full participation in the business. Apart from training programs, Newsletters, mentoring, trade fairs and exhibitions also can be a source for entrepreneurial development. As a result, the desired outcomes of the business are quickly achieved and more of remunerative business opportunities are found. Henceforth, promoting entrepreneurship among women is certainly a short-cut to rapid economic growth and development.

Notes

1. www.authorstream.com/Presentation/aSGuest14954-164241-presentation3-entertainment-ppt-powerpoint.
2. www.wiego.org/stat_picture, p. 1.
3. www.wiego.org/stat_picture, p. 2.
4. www.wiego.org/stat_picture, pp. 3-4.
5. Shastri, Rajesh Kumar and Sinha, Avanika (2010), "The Socio-Cuiltural and Economic Effect on the Development of Women Entrepreneurs (With Special Reference to India)", *Asian Journal of Business Management*, 2 (2), pp. 30-34.

References

Abraham Gladis, Dr. (2003), "Female Workforce Participation: Need for Empowerment," *Yojana*, Vol. 47, No. 5, May.

Allen, S. and Truman, C. (Eds.) (1993), "Women in business: perspectives on women entrepreneurs", London and New York, Routledge.

Carter, S. and Cannon, T. (1992), "Women as entrepreneurs: A study of female business owners, their motivations, experiences and strategies for success", London: Academic Press.

Chauhan Sonia (2003), "Sex Imbalances in India: 1991-2001," *Yojana*, Vol. 47, No. 5, May, p. 42.

Ghose Jaynti (2006), "Self-Employment Opportunities: Private Investigates," *Employment News Weekly*, Vol. XXX, No. 52, March, 25-31, p. 1.

Goel Mridula (2003), "Gender Bias and Economic Development in U.P. with reference to Female Education," unpublished thesis, Giri Institute of Development Studies, Lucknow.

Himachalam, D. Dr. (1990), "Entrepreneurship Development in Small Scale Sector," *Yojana*, Vol. 34, No. 3, February.

Jothi Arul P., "A Theoretical Respective of Women Entrepreneurship," Department of Economics, Bharathidasan University Tiruchirapalli.

Kumar A. (2007), "Women Entrepreneurship in India", Regal Publications, New Delhi.

Lekhi, K.R., "ISC Economics," Kalyani Publishers, New Delhi, Noida, Vol. II, pp. 643-4.

Mishra Sabita and Das Baran Ashok (2006), "Women Entrepreneurship for Self-Employment," *Employment News Weekly*, Vol. XXXI, No. 9, May, 27-June, 2, p. 1.

Naik Sumangala (2003), "The Need for Developing Women Entrepreneur," *Yojana*, Vol. 47, No. 7, July, pp. 36-7.

Prof. Narayan, Jagdish and Dr. Purwar, Sristi (2008), "Scenario of Education and its Impact on Growth and Development Process for India with special reference to employability for women", paper presented in 91th *Annual Conference of Indian Economic Association.*

Sethi, Priya (2009), "Women Entrepreneurship in Small Scale Industries", *UPUEA Economic Journal,* Vol. 5, Conference No. 5, p. 351.

Shanmugasundaram, Yasodha (1999), "Data File on Women in India," Mother Teresa Women's University Press, Kodarikanal.

Shastri, Rajesh Kumar and Sinha, Avanika (2010), "The Socio-Cultural and Economic Effect on the Development of Women Entrepreneurs (With Special Reference to India)", *Asian Journal of Business Management,* 2 (2), pp. 30-34.

The Indian Economic Association, Part II, 91st Annual Conference Volume.

The Journal of Entrepreneurship, January-June (2002), Vol. II, Nos. 1 and 2.

UPUEA Economic Journal, Vol. 3, Conference No. 3

UPUEA Economic Journal, Vol. 4, Conference No. 4

Various Census of India.

Various Economic Surveys.

www.articlesbase.com/entrepreneurship-articles/women/as/entrepreneurs-in-india-212759.html

www.authorstream.com/Presentation/aSGuest14954-164241-presentation3-entertainment-ppt-powerpoint

www.wiego.org/stat_picture

CHAPTER

25

Women Role in Agriculture and Industry Sectors—A Special Reference of Andhra Pradesh

M.V. Narasimha Sarma, K. Nageswara Rao
and Bh. Anjaneya Reddy

The women are the backbone of agricultural workforce but worldwide her hard work has mostly been paid. She does the most tedious and back-breaking tasks in agriculture, animal husbandry and homes.

Women constitute almost half of the country's population. Woman is said to be the best of all creations. In our religious treatises and Vedas also the glory of women has been acclaimed. Every year all over the world, on 8th March, International Women's day is celebrated. Women contribution towards society in deed is commendable. There is field left where she hasn't shown her dexterity. She is brushing her shoulders with her men counterparts in every field, be it—

Politics, Social Welfare, Engineering, Business, Defence, Administrative Services and Sports. Due to various social and economic changes, process of urbanization and industrialization, there has been shift from traditional joint family to nuclear family. Women are also taking up new roles in the market sector. Now, I am going to deal with the women's productive position in Agricultural sector, Industrial Sector as well as in Panchayatiraj Institutions.

F.A.O., 1991 justifies that *"Most Farmers in India are Women"*. Where women are absolute decision-makers in families where the male members migrate to cities for better earning purpose.

NATIONAL RESEARCH CENTRE FOR WOMEN IN AGRICULTURE (NRCWA)

The National Research Centre for Women in Agriculture (NRCWA) has been functioning at Bhubaneshwar, Orissa, for developing methodologies, for identification of gender implications in farming systems approach and to develop women specific technologies under different production systems.

According to population census of India 2001, there are about 402.5 million rural workers of which 127.6 million are cultivators and 107.5 million are agricultural labourers (Table 1). In other words, pure agricultural workers constitute nearly 58.4 per cent of the total rural workers, of which 31.7 percent are owner cultivators and 26.7 percent are mainly agricultural wage earners (Agriculture Statistics at a Glance, sourced from Registrar General of India, New Delhi 2001). The latest available agricultural census data (Govt. of India, Agricultural Census Division, Ministry of Agriculture 2002) also reveal that about 78 percent of operational holdings in the country are marginal and small, having less than 2 hectares. About 13 percent holdings have 2 to 4 hectares and 7.1 per cent have 4 to 10 hectares of land. (Haque 2003)

The relatively large holdings above 10 hectares number only about 1.6 percent of the total operational holdings. However, these 1.6 per cent of the large holdings occupy about 17.3 per cent of the total area, while 78 percent of holdings which are less than 2 hectares, operate only about 32.4 percent of the total area. This speaks of inequality in the distribution of operational holdings. Also there is inequality of income between agricultural and non-agricultural workers, which is evident from the fact that percentage share of agriculture in current total GDP is only 24.2, while the percentage share of agricultural work force to total work force comes to about 60 percent.

TABLE I

Year	*Rural Population*	*Cultivators*	*Agricultural Labourers*	*Other Workers*	*Total Rural labourers*
1951	298.6 (82.7)	69.9 (49.9)	27.3 (19.5)	42.8 (30.6)	140 (100.0)
1961	360.3 (82.0)	99.6 (52.8)	31.5 (16.7)	56.6 (30.5)	188.7 (100.0)
1981	523.9 (76.7)	92.5 (37.8)	55.5 (22.7)	96.6 (39.5)	244.6 (100.0)
1991	628.7 (74.3)	110.7 (35.2)	74.6 (23.8)	128.8 (41.0)	314.1 (100.0)
2001	741.7 (72.22)	127.6 (31.7)	107.5 (26.7)	167.4 (41.6)	402.5 (100.0)

Source : Registrar General of India, New Delhi, 2001.

ROLE OF WOMEN IN AGRICULTURE

Since the dawn of civilization in which women played a significant role or participated in large number, is Agriculture. Women played an important role because agriculture is largely a household enterprise. Recent studies highlight that women in India are major producers of food in terms of value, volume and number of hours worked. Nearly 63% of all economically active men are engaged in agriculture as compared to 78% of women. Almost 50% of rural female workers are classified as agricultural labourers and 37% as cultivators. About 70% of framework was performed by women. Thus, it aptly justifies "Most farmers in India are Women".

Out of the total Indian population, 31.39% are main workers and 3.49% are marginal workers. The percentage of female main workers and female marginal workers is 16.8 and 6.26 respectively. Of the female workers 34.55% are cultivators, 43.56% are agricultural labourers and 4.65% are engaged in live- stock, forestry and fisheries. Remaining 17.62% female main workers and 9.02% female marginal workers are in non-agricultural sector. Thus in India women forces are overwhelmingly engaged in agriculture both as main and marginal workers and played a pivotal role from ancient times.

Women perform a variety of tasks such as weeding, hoeing, fertilizer application, plant production, harvesting, threshing, winnowing and also participate in activities like sowing, transplanting, irrigation, threshing, providing fodder to the milk animals, preparing food for labour and also in the

TABLE 2

Table Showing Distribution of Women in Agriculture as Per their Activities and Socio-economic Groups in India

Agricultural activity	*Lower socio-economic group*	*Lower middle socio-economic group*	*Upper middle socio-economic group*	*High socio-economic group*
Sowing	55.22	24.24	—	—
Transplanting	100.00	84.85	—	—
Weeding	100.00	100.00	100.00	—
Harvesting	100.00	100.00	25.00	—
Winnowing	100.00	90.91	66.67	—
Threshing	26.87	24.24	—	—
Providing fodder to the milk animals	4.48	33.33	37.50	—
Preparing food for labour	—	100.00	100.00	85.71
Caring poultry and cattle	46.27	75.76	87.50	7.14

Source : Women in Agriculture—A Case Study.

important allied agricultural acitivities like—entire management of dairy animals, caring, poultry and cattle, etc.

Women performing agricultural activities on their own farm where categorized as upper socio-economic group. Women owning small land holdings and work as a wage labourers on other farms constituted the lower middle class. The landless women wage labourers were grouped under Socio-economic class.

The Table 2 depicts the fact that 55% performed sowing; poultry care was undertaken by about 46% of women in this category. Majority of women in this group are performing winnowing (90%) and transplanting (85%) and 75% of them were engaged in poultry care.

Women supervises farm activities and in decision-making on farm activities. Women are generally consulted more for selected agricultural decisions like amount of grain to be used, stored and sold, getting credit and its repayment, buying new equipment and selling and buying of land and property, etc.

WOMEN EMPLOYMENT IN ORGANISED SECTOR

Employment of women constitute an important component of the organized sector employment in the country. Employment of women in the organized sector has gone up steadily over the last 10 years. It comprises 16.3% of the total women employment in the organized sector in the country. An overview of women employment is given in Table 3.

It may be observed from the Table 3 that women employment in the organized sector went up from 33.40 lakhs in 1986-87 to 45.27 lakhs in 1995-96 in India as a whole.

WOMEN IN INDUSTRIAL SECTOR

Women are not only working in agricultural sector, but are also working in Industries and services like hunting, forestry, fishery, mining, quarrying, manufacturing, electricity, gas, water, construction, wholesale and retail trade, restaurants, hotels, transport, storage, communication, financing, real estate, business, community, social and personal services.

TABLE 3

Employment of Women in Organised Sector

Year Ended 31st March	*Women Employment in Lakhs*	*Percentage Change Over Previous Year (+)*	*Total Employment in Lakhs*	*Percentage of Women Employment in Total Employment*
1987	33.4	2.20	253.88	13.20
1988	34.36	2.90	257.12	13.40
1989	35.45	3.20	258.97	13.70
1990	36.46	2.80	263.53	13.80
1991	37.81	3.70	267.33	14.10
1992	39.08	3.40	270.56	14.40
1993	40.26	3.00	271.77	14.80
1994	41.54	3.20	273.75	15.20
1995	43.04	3.60	279.87	15.40
1996	45.27	5.20	277.80	16.30

Souce : Based on *Employment Review*, 2006-07.

The distribution of women's employment by major Industries as on 31st March, 1995 and 1996 is given in Table 4.

The details of some industries which employed more females than males like processing, processing of edible nuts, manufacturing of beedies, cotton spinning, weaving, coir textiles, manufacturing of matches, welfare service organizations and domestic services, are shown in Table 5. From the figures in Table, it can be inferred that the women employment rate is higher than male employment rate in industrial sector also. The female percentage rate is high (69%) in manufacturing of match industry and beedi industry. Women participation rates are 67% of edible nuts, 65% of coir fibres, 64% of textiles, 63% of welfare services and 58% of domestic services.

ROLE IN PANCHAYAT RAJ INSTITUTIONS

India is perhaps the first country to recognize this social

TABLE 4

Table Showing Women Employment in Industries : 1995-96

Nic Code	*Industry*	*Women's Employment in Thousands as on 31st March*		*Percentage Change 1996/ 1995*
		1995	*1996*	
0	Agriculture, hunting, forestry and fishery	489.20	544.70	11.35
1	Mining and quarrying	78.60	75.50	-3.94
2 and 3	Manufacturing	760.20	877.60	15.44
4	Electricity, gas and water	39.00	39.80	2.05
5	Construction	65.20	67.30	3.22
6	Wholesale and retail trade and restaurants, hotels	39.70	41.90	5.54
7	Transport, storage and communication	161.50	170.50	5.57
8	Financing, insurance, real estate and business services	224.10	215.70	-3.75
9	Community, social, personal services	2446.80	2493.50	1.91
	Total	4304.30	4526.50	5.16

Source : Based on *Employment Review*, 2006-07, Ministry of labour.

fact underlined by Lenin on the International Working Women's Day in 1921, and to have concrete measures to draw women into leadership positions and thereby into politics by giving them 1/3rd reservation in what may now be called the third tier of governance—the Panchayatiraj. The constitutional 73rd Amendment Act, 1992, Article 243(3) reads

"Not less than 1/3rd (including the number of seats reserved for women belonging to Scheduled Castes and Scheduled Tribes) of the total number of seats to be filled by

TABLE 5
Industry which Employed more Females than Males : All India, 1991

Sl. No.	*Description*	*Number of workers*		*% Female*
		Female	*Male*	
1.	Processing, canning and preserving of fish crustacia and similar foods	14215	13938	50
2.	Processing of edible nuts	61761	31092	67
3.	Manufacturing of beedi	1454224	648316	69
4.	Cotton spinning other than in mills	150159	109718	58
5.	Preparatory operation (including cording and combing) on coir fibres	17975	9471	65
6.	Spinning, weaving and finishing of coir textiles	53936	30974	64
7.	Manufacturing of matches	91638	40816	69
8.	Welfare services rendered by organizations operating on a profit basis for the promotion of welfare of the community such as relief societies, cretches, homes for the aged and physically handicapped, etc.	170177	99921	63
9.	Domestic services.	562396	410527	58

Source : Based on *Employment Review,* 1995-96.

direct election in every panchayat shall be reserved for women and such seats may be allotted by rotation to different constituencies in Panchayats."

And clause (4) of the Act has the following provision :

". . . not less than $1/3^{rd}$ of the total number of offices of chair persons in the Panchayats at each level shall be reserved for women"

"Provided also that the number of offices reserved under this clause shall be allotted by rotation to different panchayats at each level".

The constitutional amendment providing 1/3rd representation to women in elected bodies as well as reserving 1/3rd of the offices and chair persons for them will have far-reaching consequences in Indian political and social life. This is indeed a welcome, though delayed, gesture without which we cannot democracy meaningful in a traditional society like ours.

Despite dynamic constitutional and legal provisions and various other measures adopted by the Governments from time to time to transfer India Society to make it gender just and to make a way for empowerment of women. The basic problems among women are illiteracy and ignorance about their rights and duties. If these problems are removed, .hey could perform their role effectively as a member of Panchayatiraj Institutions no less than their male counterparts.

WOMEN IN POVERTY ALLEVIATION PROGRAMMES

The share of women in employment generated under poverty alleviation programmes are shown in Table 6.

TABLE 6
Women in Poverty Alleviation Programmes

Year	*Total number of families assisted in self-employment*	*Percentage share of women*	*The number of man days generated*	*Percentage share of women*
1990-91	3.10	32.29	873.80	24.64
1991-92	2.80	35.20	809.20	24.01
1992-93	2.30	35.22	782.10	24.69
1993-94	2.80	35.47	1075.30	22.82
1994-95	2.50	35.42	1225.70	22.25
1995-96	2.30	34.16	1239.40	29.67
1996-97	1.70	31.39	730.10	30.52

Source : Based on *Employment Review,* 2005-06, Ministry of Labour.

State response to women issues has been mixed and ambivalent. It has been both a process of progression and retrogression. The Table 5 shows that the programmes of the state are ensuring positive results in the poverty alleviation of rural women. Even though, for the last 20 years the State Policies ostensibly trying to improve the situation of women, simultaneously we can observe their inability to effectively address the issues of equality between men and women, the manner of approaching women issues within them. The gap in the implementation raises doubts about the State's intentions to actually improve the status of women in the programmes which are meant to empower women as they not only unable to provide minimum service conditions in the form of working hours and salaries, but also do not provide security to women working under the programmes. Hence, the gap between the State rhetoric and the actual State Policy has not been noticed and it calls for more struggles to bridge this gap.

As a woman welfare measure, the year, 2001 is observed in India as the year of "*Women Empowerment*". This veritably presupposes the nation's urge to hoist the status of women in the Society. Policies for rural women should be devised so as to meet the practical and strategic needs of the women after taking into consideration, the rural topography of the area and their social cultural factors as well. It is important to understand that in providing technological innovation to rural women and to mitigate their problems, the physical capacity of women and their social, psychological factors have to be considered in the proper perspective and it should come from the women themselves.

References

Alka, Janaki and Archana, Programme for Agriculture, Indian Social Institute, New Delhi, women in Agriculture and the various Government programmes, *Women's Link*, Jan.-Mar. 1997, Vol. 3, No. 1.

Dr. Anil, P. Joshi, Dr. Kiran Rawat, Choice of Technologies for Women, *Kurukshethra*, October, 2001.

Bhagirathi Dash, Role of Women in Agriculture, *Yojana*, November, 2000.

Chander Bhan, Raj Singh, Women's Empowerment for Gender Equality, *Kurukshethra*, August, 2001.

Dr. B.K. Dattanaik, Gender Economic Empowerment and Rural Poverty Alleviation, *Kurukshetra*, December, 2001.

Dr. I. Sobha, Women in Agriculture : A Case Study, *Yojana*, July, 2001.

R.K. Ojha, Women's Employment in Organized Sector, *Yojana*, May, 2001.

Dr. P. Ravishankar, M. Pargunan, Women's Development Policy and Gender Equality : Myth and Reality, *Kurukshethra*, January, 2002.

Sweta Mishra, Women and 73rd Constitutional Amendment Act—A Critical Appraisal, *Social Action*, Vol. 47, January-March, 1997.

CHAPTER

26

Access to Basic Health Care Services by Rural SCs and STs : A Case Study of Two Villages of Balangir District of Orissa

SACHITA NANDA AND G. SRIDEVI

I. INTRODUCTION

Health is a fundamental human right and a worldwide social goal. Health is necessary for the realization of basic human needs and to attain the status of a better quality of life (WHO, 1979). The issue of health is of great importance both from the point of view of individuals and nation as well. The WHO declaration on "Health for all" in 1979, has led most of the governments including developed and less developed countries to give much more priority to their health care system through higher allocation and better utilisation of resources in order to improve the quality of health care services. India too has been attempting towards this end. The major hindrances that India is facing are inadequate allocation of resources for health sector, rapid population growth, inefficient use of resource allocated and lack of public consciousness about their own health. Health being a state subject in Indian federal system, different states in the country have been trying to meet the WHO health goal through utilization of resources. In case of Orissa the health care services are one of the most backward and inadequate in this region. Orissa is one of the most backward states in India and still 47 percent of its population live below poverty line (Economic Survey, 2009). Regarding the various health indicators, Orissa is being the worst performer state in India especially in Infant mortality rate (around 79), Death rate (10.73%), Life Expectancy (Rural-56% and urban-64.7%), malnutrition (underweight above 50%), maternal mortality (367/100000 population), morbidity (higher than the national average) and accessibilities and utilization of public health care facilities (<20%) (Human Development Report, 2004 and Economic Survey, 2009, Orissa).

The above mentioned health indicators are not same in all over the state and it is more severe in the inlands (south western and KBK region—Kalahandi, Balangir and Koraput) region of Orissa. In 2001, five out of the eight KBK districts had an IMR of more than 100 (IIPS estimates based on census 2001). The people of this region specifically SCs and STs are in a worst condition due to poor health services. Needless to say, due to their long years of isolation and inhabitation in the inaccessible forest areas in a primitive economic set-up of food gathering and subsistence farming, the tribal in the schedule area are yet to be integrated with the mainstream development process of post-independent India (*ICMR Bulletin*, 2003). Also in the exploitative and oppressive caste system of Hindu society the SC comprising the ex-untouchable population of India are yet to be brought at par with the general category population in their socio-economic development status (Baraik, V.K. and P.M. Kulkarni, 2006). Till today health services provided to people have continued to be urban-oriented and neglect the rural area.

2. REVIEW OF LITERATURE

This chapter gives a brief overview of selected studies on availability and accessibility of basic health care services with a view to understand the issue involved and identify the appropriate methodology for this study. These studies are related to various aspects of health like social and economic inequality in health and nutrition by different sections of the society, maternal and child health, household and government health spending and lastly utilization of health care services in India. Some studies are related to health conditions of the people and provision of health services in Orissa. The literature on the issue of health care accessibility is very limited and those available mostly address the theoretical and policy aspects of accessibility to health care services. Klinoubol (1997) empirically found that the close location of health centre and cheaply available health care services enhances the accessibility of rural health care services to the rural masses. The quality of health care in terms of infrastructure, personal, basic adult and child health services has large and significant effect on demand. Regarding the choice between price and quality, the choice is more pronounced with quality than with distance or price (Alderman, H. and Victor Lavy, 1996). From the review, for example, [Roy, T.K. and *et al.* (2004), Sen, G. and *et. al.* (2002), Joe, and Mishra *et. al.* (2008), Radhakrishna R. and C. Ravi (2004), Sambhamoorthi, and Thimothy, *et. al.*, (2004), Pandey, and Roy, *et. al.* (2004)] focus on social and economic inequality in health and nutrition by different sections of the society. Other studies [Das R.K. and Dasgupta *et. al.* (2000), Kulkarni M.K. (1992)] concentrated with the aspects of maternal and child health. Some other studies [Sanyal S.K. (2000), Berman P. and Rajeev Ahuja (2008), Bhat R. and Nishant Jain (2004)] explains household and government health spending, and studies like Purohit B.C. and Tasleem A. Siddiqui (1994), Lindelow M. (2008), Srinivasan, K. and S.K. Mohanty (2008), Alderman H. and Victor Lavy (1996), Berman, and Sisler *et. al.* (1989), Nayar K.R. (1997), Bhat R. (2000), Baluch M. and Saima shahid (2004) look into the utilization aspects of health care services. There are some literatures which deal with different health-related problems in Orissa by Swain, S. (2008), and

Padhi, S. (2001). It is known that earlier researchers had dealt with the issue of supply side of health services and studied the various health-related problems and indicators for the different social and economic sections and neglected the demand side, particularly the accessibility of health care services of the SCs and STs. The present study attempts to explore the factors which are responsible for the poor accessibilities of health care and for the poor health status of the SCs and STs in the backward region of Orissa.

The present study attempts to deal with accessibility of health care services and health status of SC and ST communities from the two villages of Gudvela block of Balangir district in Orissa. This place is selected as the study area because no research work has yet been done on health-related issues here. The specific objectives dealt with in the study are:

(1) To determine the factors responsible for poor health status of SC and ST households in two villages of Gudvela block of Balangir district in Orissa.

(2) To look into the accessibility of basic health care services provided to infant and mother of SC and ST households of the two villages.

Hypothesis

(1) Low income, illiteracy and lack of health care facilities are the major determinants of poor health standard of SC and ST households of this region.

(2) The SC and ST households have very low accessibility to basic health care services in rural areas of the region under study.

Data and Methodology

To articulate the above objective the present study would like to use both secondary and primary data. Keeping the level of accuracy in mind, maximum dependence will be on primary data, from the field through household survey. Secondary data is collected from Department of Health and Family Welfare (GoO), Economic Survey of Orissa, Statistical Handbook of

Balangir District, District at a Glance (GoO), and Official websites of Orissa and Balangir District.

A Logit model will be used in order to determine the health status of these people. Here health status being taken as the dependent variable and the factors which are effecting are taken as independent variables or control variables. The independent variables are income of the household, education of the head of the family, medical facilities within the locality and Assets holding of the households (in terms of Land) others which include demographic and socio-economic factors.

Health status (HS) = F {Income (TI), Education (ED), Medical set-up (MF), Assets holding (AS), and others (U)}.

3. HEALTH STATUS IN ORISSA

Orissa is one of the poor states in the country, where economic growth in recent years has not been adequate to bring a significant reduction in poverty. Still, 47.15 percentage of the people lives below the poverty line which is highest in the country (Orissa Human Development Report, 2004). Orissa has one of the worst set of health indicators in the country as per the WHO reports. It has rated as one of the miserable state so far the health care is concerned. Public expenditure on health is not so impressive, while the number of 'national programmes is large, the financing is not; neither does the state spend anywhere close to the required amounts for its health services. Regarding the state expenditures on health, family welfare, water supply, and sanitation in 1999-2000 prices (per capita) as ratios of GSDP and total expenditure has decreased over the years (Orissa Human Development Report, 2004).

As per the Register General of India source, the estimated birth rate in Orissa lies much below as compared to other bigger states. In 2007, the estimated birth rate in Orissa was 21.5 percent as against 23.1 percent at all India level. The birth rate in rural and urban Orissa was 22.4 percent and 16.1 percent respectively. The death rate in Orissa, during 2007 was highest among all the states and Union Territories, i.e. 9.2 as against 7.4 at all India level (Economic Survey, 2007-08, GoO). Though the infant mortality in Orissa has declined considerably from 83 in 2003 to 71 in 2007, it is still highest at all India level. The infant mortality rate in rural Orissa was 73 while in urban Orissa it was 52 (Economic Survey, 2008-09, GoO). The situation is particularly worse-off in the tribal dominated districts (which includes the KBK districts) of the state. In 2001, five out of the eight KBK districts had an IMR of more than 100 (IIPS estimates based on census 2001). In 1997, the state had an MMR of 367, which was lower than the MMR for the country as a whole (408). But it was changed in 2001-03; the maternal mortality in India during this period was 301 per 100000 live births as against 358 in Orissa (Economic Survey, 2008-09, GoO). There is an improvement in case of Life expectancy at birth from 58.6 years for male and 58.7 years for females in 1999-2003 to 60.3 for male and 62.3 for female during 2001-05. The projected level of life expectancy at birth in Orissa is 64.3 for male and 67.3 for female during 2011-15 as against 67.3 for male and 69.6 for female at all India level (Economic Survey, 2008-08, GoO). In general, the rates of morbidity and mortality are high in the state (GoO, 2005).

The high levels of IMR and MMR in the state are caused by two important factors: low access to effective health care facilities and high levels of malnourishment (GoO, 2004). As per the 58th round of National Sample Survey Organisation (NSSO) survey on village-level facilities, about 63 percent of the villages in the state are more than 5 km away from a sub-centre/dispensary (as compared to 36 percent at the national level). Despite this problem of access, there has been no addition to health facilities (allopathic) in the last ten years. Apart from long distances to the health facilities that limit access, the status of infrastructure, staff, equipment and supply of drugs therein is one of the worst in the country. The low access partially results from a severe shortage of allopathic doctors in the state. The ratio of doctor population in the state is as low as 1:7440 relative to the national average of 1:2000 (Economic Survey 2008, GoO). Additionally, the low access to health facilities is coupled with high levels of malnourishment in the state. The situation is further worsened by the poor state of water supply and sanitation. Health care accessibility is thus relevant to the study because it provides a background that should be reflected in analyzing the factors which are responsible for the poor health status of SCs and STs and the relative accessibility of health care services by these people. In short due to lack of health care facilities, socio-economic constraint and lack of government intervention, the health status of the people's like SCs and STs remain poor in the district and also in the State.

4. HEALTH STATUS OF SC AND ST HOUSEHOLDS IN THE STUDY AREA

The objective of the present study is to analyse the health status of the SC and ST households having children below five years of age, and for this purpose household is taken as the sample unit. In the present study, two villages have been taken, they are Buromal village from Samra Panchayat which is close to health centre and another is Laitara village from Jamut Panchayat which is very far away from the health centre having maximum number of SC and ST households. All the variables like income, health expenditure, food expenditure, unless otherwise specified, are calculated and listed on annual basis.

Modeling on Health Status

As stated earlier, the main aim of the study is to understand the factors responsible for the poor health status of SCs and STs in the study area. In this section an attempt is made to estimate the health status of SCs and STs in the two villages from the Gudvela block of Orissa. Here we have an attempt to look into the health care access of SCs and STs by considering the health status of the sample households. For this we have used the Binomial Logit model, the dependent variable is health status. Logit model is used when the dependent variable is dichotomous. The model is as follows:

Model :

$$L_i - l_n\left(\frac{P_i}{1-P_i}\right) = \beta_1 + \beta_2 TI_i + \beta_3 ED_i + \beta_4 MF_i + \beta_5 AS_i + U_i$$

Where P_i = 1, if the health status is good and
0, if it is bad.

β_1 = intercept term
β_2 = coefficient of the total income (annual) of the family
β_3 = coefficient of the educational qualification of the head of the household
β_4 = coefficient of health care facilities
β_5 = coefficient of assets holding (in terms of land) of the families
U_i = error term

Here Total income (TI), and assets holding (AS), Education level (ED) and health care facilities (MF).

Education (ED)

Here educational qualification is divided into literate and illiterate, if the head of the household is literate then the score as one and if illiterate then zero.

ED = 1 = if literate
0 = otherwise

Health Care Facilities (MF)

Like education, Health care facilities are grouped into two types like good and bad. If the health care facilities are good then the score will be one and if it is bad then zero.

MF = 1 = if it is good and easily accessible
0 = if it is bad.

Before estimating the health status model, we have to be first clear about the score allocated to the dependent variable. Here health status is given the score as one or zero based on different indicators like income of the household, education of the head of the household, physical distance of the health centre, number of visits to health centre, number of times of anti-natal care, provision of iron and folic acid tablets, type of anti-natal care, post-natal care, number of medical supervisions of delivered mother, place of delivery, complete set of immunisation, birth distance, number of still-born babies, type of removing umbilical cord and sanitary facilities. By looking into the above factors score is given. Out of the thirty households from Buromal village eighteen households have got the score one that means they have better health status and other twelve have got score zero. On the other hand, in case of Laitara village out of thirty families, only six households have got score one and rest households have got zero score. The above analysis shows that people from Buromal village are somehow better in health status in comparison to Laitara village.

Estimated Results of the Health Status Model

TABLE 1

Estimated Results of Health Status Model

Variable	*Coefficient*	*z-Statistic*	*Prob.*
C	-11.0023	-3.3141	(0.0009)***
TI	0.0003	2.4786	(0.0132)***
MF	3.313	3.0197	(0.0025)***
ED	2.2789	2.0626	(0.0391)**
AS	6.97E-06	0.7079	0.479
LR statistic	49.6592		
Prob (LR stat.)	0.0000		
Mc Fadden R-squared	0.6148		

*** denotes significant at 1% level.

** denotes significant at 5% level.

* denotes significant at 10% level.

C = Constant

TI = Total income (annually)

MF = Health care facilities

ED = Educational status of the head of the household

AS = Assets holding of the family (in terms of land)

Table 1 represents the estimated values of the above variables. It can be seen that all the Logit variable coefficients are positive which show the positive effects of these variables on the health status. But, all the variables are not significantly affecting the health status of the people. It can also be seen that only total income of the household, the educational status of the head of the household and Health Care facilities are significantly affecting the health status the people in the study area. Here in the health status model, the total income and medical facilities are significant at 1% level and educational status is significant at 5% level. The estimated LR statistic is highly significant with very less probability value 0.0000, indicates that together all the variables are significantly determining the health status of people in the study area. Here the Mc Fadden R-squared is 0.6148, which is significant in the health status model. Only the assets (AS) variable is not significantly affecting the health status of the people but it has some positive impact as its coefficient is 6.97 which is positive. So from the above analysis, it is clear that medical facilities, income of the households and educational status of the head of the households are significantly affecting the health status of the families in the study area as a whole.

Thus the people from high income category, literate people and village closest to the health centre have better access to health facilities then people from low income category, illiterate people and village distance to the health centre. So from the above analysis it is seen that, more than 60 percent of the households belonging to SC and ST communities are in the bad health status condition which is very high as compared to the other communities of the population. Thus, the health status of the households as a whole is very poor in the Laitara and Buromal village and the staple reason for this are: illiteracy, low income and lack of health care facilities.

5. HEALTH CARE ACCESSIBILITY OF CHILDREN AND WOMEN IN THE STUDY AREA

This section deals with the accessibility of health care services to females and infants in the Gudvela block of Balangir District in Orissa. If the mother and children of a household are having a good access to basic health care services then it will indicate that the overall health status of the household is also good. There are various health services like Antenatal care and types (number and timing of antenatal care visit), Tetanus

Toxoid vaccination and iron and folic acid tablets, Condition of delivery and birth, Immunisation of children, post-natal care, medical supervision during pregnancy, etc. The present study made an attempt to show the health care accessibility of children and women in the study villages based on the above components of maternal and child health services. The analysis is based on two steps, first village-wise accessibility of health care services to children and women and second, overall accessibility of health care services as a whole in the study area. First we will look into the accessibility of health care services to children and women in Buromal village (which is close to the medical centre). Table 2 shows that, in Buromal village out of 15 SC households, 11 households are from Rs. 10000-Rs. 20000 income group (annual), two households are from Rs. 20000-Rs. 30000 income group and the other two households are above Rs. 30000 income group. We will look into the different indicators of health care accessibility by the 15 SC households and then the same is done for other 15 ST households.

From the detail analysis of health care accessibility of SC community, it is seen that four households (each group two households) from the last two income level (Rs. 20000-Rs. 30000 and Rs. 30000 and above) have cent percent access in all the components of health services except medical supervision but it was very low in case of the Rs. 20000 to Rs. 30000 income group. Out of 11 households from this category, only three households have three and more times ANC, three households have two times ANC and no ANC was availed by rest five households, for TT, iron and folic acid consumption, five households have completed the required doses and the rest have not completed the required doses although they have taken the tablets from the health servant. In case of medical supervision during pregnancy, only three households have consulted to the physician and the rest have not consulted any physician and for post-natal care, only three households availed two and more times PNC services, two households have availed one time PNC services and no PNC was availed by rest six households. Regarding the place of delivery and child immunisation, four household's delivery took place in health facilities and rest seven households at home and parent's home and for immunisation five households

TABLE 2

Health Care Accessibility of Children and Women in Buromal Village (Both SCs and STs)

Income (Rs.) Level	No. of Families	Ante-natal Care				TT, I and FA		MS		Post-natal Care			Delivery		Immunisation	
		1	2	≥3	No	(C.)	Not (C.)	Yes	No	1	≥2	No	Safe	Unsafe	(C.)	Not (C.)
(1)	(2)	(3)	(4)	(5)	(6)	(7)	(8)	(9)	(10)	(11)	(12)	(13)	(14)	(15)	(16)	(17)
10000 to 20000	11	—	3	3	5	5	6	3	8	2	3	6	4	7	5	6
20000 to 30000	2	—	—	2	—	2	—	1	1	—	2	—	2	—	2	—
30000 and above	2	—	—	2	—	2	—	2	—	—	2	—	2	—	2	—
Sub-Total (SC)	15	—	3	7	5	9	6	6	9	2	7	6	8	7	9	6
10000 to 20000	9	—	2	6	1	8	1	4	5	—	6	3	7	2	7	2
20000 to 30000	4	—	—	3	1	3	1	2	2	—	3	1	4	—	3	1
30000 and above	2	—	—	2	—	2	—	—	2	—	1	1	2	—	2	—
Sub-Total (ST)	15	—	2	11	2	13	2	6	9	—	10	5	13	2	12	3
Total (SC+ST)	30	—	5	18	7	22	8	12	18	2	17	11	21	9	21	9

Source : Field Survey, May 2009.
I and FA = Iron and Folic acid tablets

* 1 = One time AC
* 2 = Two times AC
*≥3 = Three and above three times
*≥2 = Two and above three times

(C.) = Completed
Not (C.) = Not completed
No = No AC

have completely immunized their children and the remaining six families have not completely given the required amount of vaccines to their children. Similarly for ST community, out of 15 households nine households are from Rs. 10000 to Rs. 20000 income group, four households are from Rs. 20000 to Rs. 30000 income group and the rest two households are from Rs. 30000 and above income group.

Now we will look into the accessibility of all the components of health services for ST community. For Ante-natal care out of 15 ST households eleven households availed ANC services for three or more than three times, two households availed two times ANC and no ANC was availed by the other two households. Similarly for Tetanus Toxiod (TT), iron and folic acid consumption, 13 households have completed the required amount of TT, iron and folic acid consumption and the other two households have not completed the required amount. In case of medical supervision during pregnancy, six households have consulted to the physician and the other nine households have not consulted any physician and for post-natal care ten households availed two and more times PNC services, and no PNC was availed by rest five households. Regarding the place of delivery it is found that in case of 13 households delivery of the children takes place in health centre while in case of other two households delivery has taken place at home. So far as child immunisation is concerned, it is found that, twelve households have completely immunised their children and rest three households have not completely given the required amount of vaccines to their children.

Now we will look into the health care accessibility of child and women in Laitara village (which is far away from the health centre). From the Table 3, it is known that out of 15 SC households, three households are from below Rs. 10000 income group, 9 households are from Rs. 10000-Rs. 20000 income group, two households are from Rs. 20000-Rs. 30000 income group and one household is above Rs. 30000 income group. We will look into the different indicators of health care accessibility by the SC and ST households. In case of ante-natal care services, out of 15 SC households, one household availed the ANC for three and more than three times, no ANC was availed

TABLE 3

Health Care Accessibility of Children and Women in Laitara Village (Both SCs and STs)

Income (Rs.) Level	*No. of Families*	*Ante-natal Care*				*TT, I and FA*		*MS*		*Post-natal Care*			*Delivery*		*Immunisation*	
		1	*2*	*≥3*	*No*	*(C.)*	*Not (C.)*	*Yes*	*No*	*1*	*≥2*	*No*	*Safe*	*Unsafe*	*(C.)*	*Not (C.)*
(1)	*(2)*	*(3)*	*(4)*	*(5)*	*(6)*	*(7)*	*(8)*	*(9)*	*(10)*	*(11)*	*(12)*	*(13)*	*(14)*	*(15)*	*(16)*	*(17)*
Below 10000	3	2	1	—	—	0	3	2	1	2	—	1	—	3	NA	3
10000 to 20000	9	—	5	—	4	0	9	2	7	4	—	5	1	8	1	8
20000 to 30000	2	—	—	—	2	—	2	—	2	—	—	2	—	2	—	2
30000 and above	1	—	—	1	—	1	—	1	—	—	1	—	1	—	1	—
Sub-Total (SC)	15	2	6	1	6	1	14	5	10	6	1	8	2	13	2	13
10000 to 20000	3	1	1	—	1	—	3	—	3	—	—	3	1	2	—	3
20000 to 30000	10	—	4	3	3	4	6	3	7	—	3	7	2	8	4	6
30000 and above	2	—	—	2	—	2	—	2	—	—	2	—	1	1	2	—
Sub-Total (ST)	15	1	5	5	4	6	9	5	10	—	5	10	4	11	6	9
Total (SC+ST)	30	3	11	6	10	7	23	10	20	6	6	18	6	24	8	22

Source : Field Survey, May 2009.
I and FA = Iron and Folic acid tablets.

* 1 = One time AC
* 2 = Two times AC
*≥3 = Three and above three times
*≥2 = Two and above three times

(C.) = Completed
Not (C.) = Not completed
No = No AC

by six households, six households availed two times ANC services and the remaining two households availed one time ANC services. For the consumption of TT, Iron and Folic acid tablets, only one household has completed the required amount of iron and folic acid consumption and the rest fourteen households have not completed the required amount.

In case of medical supervision during pregnancy, five households have undergone medical supervision and the other ten households do not have any medical supervision. In case of post-natal care services, one household availed PNC for two and above two times, no post-natal care was availed by eight households and the rest six households availed only one time PNC services. Regarding the delivery status, only two households have safe delivery and rest 13 households have non-institutional delivery such as at own home, parents home and others. In case of immunisation status, two households have completely immunised their child for all the dangerous diseases and the rest 13 households have not given the required amount of vaccines to their children.

On the other hand in case of the ST category, out of 15 households three households are from Rs. 10000 to Rs. 20000 income group, ten households are from Rs. 20000 to Rs. 30000 income group and rest two households are from Rs. 30000 and above income group. Now we will look into the accessibility of all the components of health services for the ST category. For Ante-natal care out of 15 ST households, five households availed ANC services for three and more than three times, another five households availed two times ANC, only one household availed one time ANC and no ANC was availed by the rest four households. Similarly for TT, iron and folic acid consumption, 6 households have completed the required iron and folic acid consumption and the rest nine households have not completed the required doses. In case of medical supervision during pregnancy, five households have consulted to the physician and the rest ten households have not consulted to any physician and for post-natal care ten households availed no PNC services, and the rest five households availed two times and above two times of PNC services. Regarding the place of delivery and child immunisation, only four households opted for having delivery at health centre and the rest eleven households chose the home as the right place for delivery and for immunization six households have completely immunised their children and the rest three households have not completely given the required amount of vaccines to their children.

Table 4 shows the overall accessibility of health care services to children and women in the two villages as a whole. Similarly in the village-wise health care accessibility, here also the total income of the household and the components of health care services are divided identically as mentioned above. The data shows that, in the study area out of 60 households (consisting of both SCs and STs from the two villages) only three (5 percent) households are from below Rs. 10000 income group, 32 (53 percent) households are from Rs. 10000 to Rs. 20000 income group, eighteen (30 percent) households are from Rs. 20000-Rs. 30000 income group and rest seven (around 12 percent) households are above Rs. 30000 income group.

From the above analysis of health care accessibility, it is known that seven households from the Rs. 30000 and above income level, eight household from Rs. 20000 to Rs. 30000 income level and the rest nine households from Rs. 10000 to Rs. 20000 income level have better access in all the components of health services except medical supervision. But most interesting thing is that, out of this 24 households (whose accessibilities are good) from all the income categories, 18 households are from Buromal village and the rest six are from Laitara village. It is seen that the health care accessibilities in all the components of maternal and child health are relatively less as compared to the non-accessibility of the above health services. The SC and ST households in Buromal village are better than that of Laitara village so far as the components of health care services are concerned. But as a whole in the study area, out of 60 households 36 households (60%) have no access and the rest only 24 households (40%) have access to the basic health service. From the above study, it is known that health care accessibility of children and women are very poor or low in the study area as a whole.

TABLE 4

Health Care Accessibility of Children and Women in the Study Area (Both SCs and STs)

Income (Rs.) Level	*No. of Families*	*Ante-natal Care*				*TT, I and FA*		*MS*		*Post-natal Care*			*Delivery*		*Immunisation*	
		1	*2*	*≥3*	*No*	*(C.)*	*Not (C.)*	*Yes*	*No*	*1*	*≥2*	*No*	*Safe*	*Unsafe*	*(C.)*	*Not (C.)*
(1)	*(2)*	*(3)*	*(4)*	*(5)*	*(6)*	*(7)*	*(8)*	*(9)*	*(10)*	*(11)*	*(12)*	*(13)*	*(14)*	*(15)*	*(16)*	*(17)*
Below 10000	3 (5)	2	1	—	—	0	3	2	1	2	—	1	—	3	—	3
10000 to 20000	32 (53)	1	11	9	11	13	19	9	23	6	9	17	13	19	13	19
20000 to 30000	18 (30)	—	4	8	6	9	9	6	12	—	8	10	8	10	9	9
30000 and above	7 (11.6)	—	—	7	—	7	—	5	2	—	6	1	6	1	7	—
Total (SC+ST)	60	3	16	24	17	29	31	22	38	8	23	29	27	33	29	31
Percentage	100	5	26.66	40	28.33	48.33	51.66	36.66	63.33	13.3	38.3	48.3	45	55	48.3	51.6

Source : Field Survey, May 2009.
I and FA = Iron and Folic acid tablets.

* 1 = One time AC
* 2 = Two times AC
*≥3 = Three and above three times
*≥2 = Two and above three times

(C.) = completed
Not (C.) = Not completed
No = No AC

CONCLUSION

The study shows that the health status of SC and ST households in the study area is very poor and the staple reason for this are: illiteracy, low income and lack of health care facilities. In this section first we will look into the major findings and then the policy recommendations of the present study.

Findings

- As regards the village--wise accessibility of health services Buromal village is in a better position than Laitara village.
- Regarding the accessibility between SCs and STs as a whole, the accessibility of health care services by the STs are more than the SCs in the study area as a whole.
- But as a whole in the study area, 60 per cent households have no access and the rest 40 per cent have access to the basic health services. From the above study, it is found that health care accessibility of children and women are very poor or low in the study area as a whole.
- Health care facilities are also very poor in the study area. Health institution is situated at a mean radial distance of more than 20 km from that of another. At the same time there is also severe shortage of Government health service providers in this area.
- Income of the households, education level of the head of the households and medical facilities are significantly affecting the health status of the households. So the factors we hypothesized in our study having greater influence on the poor health standard come true.

Policy Recommendations

- Plan efforts both at government and organizational

level should be combed out to strengthen the educational and health infrastructure system in the rural areas.

- Education should be made free to the SC and ST people till higher secondary level.
- More employment opportunities should be created. To increase the employment opportunities in the study area the NREGS programme should be made broad-based and implemented properly.
- Medical facilities should be provided to the people at the doorsteps so that morbidity and mortality figures are brought under control.
- There should be emphasis on the Public Private Partnership (PPP) approach so that the large gap in the field could be filled up.
- Different agencies should be involved to raise awareness among the illiterate people of the region.
- Health expenditure as percentage of State Domestic Product should be increased so that per capita availability to the people in terms of health care facilities will give them better access in this region.

References

Ager, A. and Katy Pepper (2005), "Patters of health services utilization and perceptions of needs and services in Rural Orissa" *Journal of Health Policy and Planning,* Vol. 20, No. 3, pp. 176-84.

Alderman Harold and Lavy Victor (1996), "Household responses to public health services: Cost and Quality Tradeoffs" *The World Bank Research Observer,* Vol. 11, No. 1, pp. 3-22.

Banerji, D. (1986), "Technocentric Approach to health: Western approach to Alma Ata", *Economic and Political weekly,* Vol. XXI, 28, pp. 1233-34.

Barik, V.K. and P.M. Kulkarni (2006), "Health status and access to health care services—disparities among social group in India", Indian Institute of Dalit Studies, New Delhi, Working paper series, Vols. 1, 4.

Berman, P., Daniel G. Sisler and Jean-Pierre Habicht (1989), "Equity in public sector primary health care: The role of service organization in Indonesia", *Economic Development and Cultural Change,* Vol. 37, 4, pp. 777-803.

Das, K. Sujit. (1992), "Public Health: A Question of Entitlement", *Economic and Political Weekly,* Vol. XXVII, 39, pp. 2095-96.

Das, R.K. and Purnamita Dasgupta (2000), "Child Health and Immunization: A Macro-Perspective", *Economic and Political Weekly*, Vol. 35, No. 8/9. pp. 645-55.

Government of Orissa (2004), Human Development Report, 2004, Orissa, Planning and Coordination Department, Bhuvaneswar.

Guruswamy, M., Sumit Mazumdar and Papiya Mazumdar. (2008), "Public Financing of Health Services in India: An analysis of Central and State Government Expenditure", *Journal of Health Management*, Vol. 10, No. 1, pp. 49-85.

Joe, W., U.S. Mishra and K. Navanneetham (2008), "Health inequality in India: Evidence from NFHS-3", *Economic and Political Weekly*, Issue: Vol. 43, No. 31 August 02 to August 08, 2008.

Klinoubol, K., "*Public Health Development and Administration*", Deep and Deep Publication, New Delhi, 1997.

Kulkarni, M.K. (1992), "Universal immunization programme in India: Issues of Sustainability", *Economic and Political Weekly*, Vol. 27, No. 27, pp. 1431-37.

Meher, R. (2007), "Livelihood, Poverty and Morbidity: A study on Health and Socio-economic Status of Tribal Population in Orissa", *Journal of Health Management*, Vol. 9, No. 3, Sept.-Dec. 2007, pp. 343-67, Sage Publications.

Narayana, K.V., 1997, "*Health and Development*", Rawat Publications, Jaipur.

Padhi, S. (2001), "Infant and child survival in Orissa", *Economic and Political Weekly*, Vol. 36, Issue No. 34, 2001.

Pandey and *et al.* (2004), "Maternal Health Care Services: Observations from Chhattisgarh, Jharkhand and Uttaranchal" *Economic and Political Weekly*, Issue: Vol. 39, No. 07 February 14 to February 20, 2004.

Purohit, C.B. and Tasleem A Siddiqui (1994), "Utilization of Health Services in India", *Economic and Political Weekly*, Vol. 29, No. 18, pp. 1071-80.

Puttaswamaiah S and Shashanka Bhinde (2008), "Government Spending on selected public Health Services in India: Centre, State and Local Governments", Monographs, No. 13, pp. 1-114, ISEC, Bangalore.

Radhakrishna, R and C. Ravi., (2004), "Malnutrition in India: Trends and Determinants", *Economic and Political Weekly*, Vol. 39, No. 07, pp. 671-76.

Ramalingaswami, P. (1987), "Women's Access to Health Care", *Economic and Political Weekly*, Vol. XXII, Issue 27, pp. 1075-87.

Rout, S.H. (2008), "Socio-economic Factors and Household Health Expenditure: The case of Orissa", *Journal of Health Management*, Vol. 10, No. 1, pp. 101-18.

Roy, T.K., Sumati Kulkarni and Y. Vaidehi (2004), "Social inequalities in health and nutrition in selected states", *Economic and Political Weekly*, Vol. 39, No. 07, pp. 667-83.

World Health Organisation (1979), "Global Strategy for Health for all by the year 2000", Health For All Series 2, Geneva.

World Health Organisation (1981a), "Global Strategy for Health for all by the year 2000", Health For All Series 3, Geneva.

World Health Organisation (1981b), "Global Strategy for Health for all by the year 2000", Health for All Series 4, Geneva.

World Health Report (2000), "Health System: Improving Performance", World Health Organisation, Geneva.

CHAPTER

27

Levels of Rural Development in Punjab

SUKHJEET K. SARAN, MINI GOYAL
AND PRATIBHA GOYAL

INTRODUCTION

Rural development is a choice influenced by time, space and culture. The term rural development connotes overall development of rural areas to improve the quality of life of rural people. In this sense, it is a comprehensive and multi-dimensional concept, and encompasses the development of agriculture and allied activities, small and cottage industries, socio-economic infrastructure, community services and facilities and above all, human resources in rural areas of a country and its states. Structural changes started taking place in the villages of Punjab after green revolution. Katcha houses were progressively converted into Pucca houses and at present almost every village has Pucca houses. All the villages were electrified as road links were developed in almost all the villages. Irrigation has covered 97.7 percent of the net sown area during the period 2007-08 (Govt. of Punjab, 2008). Simultaneously, credit facilities for farm mechanization and other inputs were extended. But the level of rural development in each region and state is not the same in the country. Some regions in the country, because of historical or location factors would always enjoy certain advantages *vis-à-vis* other regions and this ultimately leads to disparities. It is increasingly being recognized that the identification of regions, according to their differential levels of development, typology and dimensions of backwardness is critical for implementing plans (Chaudhary, 1992). Based on this notion, an attempt has been made to identify the level of rural development in the different districts of Punjab state from 1997-98 to 2007-08 and to see the extent of rural development during these two periods of time.

MATERIALS AND METHODS

Punjab is an important state of India. It lies between 29° 30′ to 32° 32′ N latitude, 73° 55′ to 76° 50′ E longitude and with an average altitude of 240 meters above sea level. The total geographical area of state is 53076 kilometers and the population is more than two crore (Census 2001). Nearly 70 percent of the population lives in rural area.

The spread of the green revolution has considerably changed the profile of the Punjab farmers. The green revolution not only brought prosperity to the Punjab farmers but also changed their psyche, mindset and pattern of living. As a result, Punjab was considered to be one of the states of India coming under the category of highest level of human development as the value of HDI was 0.822 whereas India with a HDI value of 0.612 (Bhargava, 2010). Considering the overall development that has taken place so far, this state was chosen for this study.

The rural people suffer deeper levels of poverty than their urban counterparts and have much more limited access to basic social services such as sanitation, safe water, health services, primary education, etc. (Phougat *et. al.* 2007). Thus, the following fifteen economic indicators were selected for rural areas for measuring rural development in Punjab:

(I) Percentage of literates to the total population, 2001
(II) Percentage of agricultural workers to total workers
(III) Intensity of cropping
(IV) Percentage of net irrigated area to net sown area
(V) Number of tractor per 1000 hectares of net sown area
(VI) Livestock population per veterinary institution
(VII) Percentage of villages with drinking water to total inhabited village
(VIII) Population served per medical institution
(IX) Population served per primary school
(X) Population served per high/senior secondary school
(XI) Percentage of inhabited village with pucca roads
(XII) Number of co-operative societies per thousand of population
(XIII) Population served per commercial bank
(XIV) Area served per market committee per square kilometer
(XV) Population served per post office

The above mentioned indicators are very important and affect the level of rural development in the state. High percentage of literates to total population represents the attainment of education in rural Punjab. Low percentage of agricultural workers to total workers denotes the higher level of mechanization and thus higher level of rural development. Percentage of net irrigated area to net sown area represents the specific level of irrigation. The dynamics of rural development were measured at two points of time spaced ten year apart, i.e. 1997-98 and 2007-08. Once the homogeneously developed districts are identified and delineated and the main factors contributing of disparities in the level of development analyzed, strategies for more development at micro-level may be suggested.

For measuring the relative score of various attributes of rural development in different districts of Punjab state, standard score technique (Kellong, 1963) has applied (Z score):

$$Z_i = \frac{X_i - \bar{X}}{S.d.}$$

where Z_i = Standard score for the ith observation
X_i = Original values of observation
$\bar{X}$ = Mean for all the value of X
S.d. = Standard deviation of X

Further the results of standard score obtained for different indicators were aggregated. On this aggregation, again standard score technique has been applied to calculate the composite standard score so that the regional disparities in the level of development of districts may be obtained on a common scale.

The composite standard score may be explained as (Smith 1977):

$$T_j = \sum_i^m Z_{ij}$$

i = 1

where j = the districts
m = set of indicator
Z_{ij} = Standard score of the ith to jth indicators

In order to classify the districts, according to magnitude of development, the composite score were divided into the three classes as follows (Bhatia, 1966):

Class	*Range of composite standard score*
1. High	> 1.0
2. Medium	1.0 to –1.0
3. Low	< –1.0

The data regarding the rural development was collected from the statistical abstracts of Punjab for the years 1997-98 and 2007-08.

RESULTS AND DISCUSSION

The process of development is going on in all the district of Punjab, but the extent of utilization resource potential and the levels of development accrued vary from district to district. The seventeen districts of the state (newly framed districts were merged into their origins to facilitate the comparison at two points of time) were grouped under the three categories i.e. high, medium and low for two points of time i.e. 1997-98 and 2007-08, on the basis of composite standard score of the fifteen indicators for rural development as under:

(i) *Districts with high level of rural development (above + 1.0)*: This category consists of four districts namely Ludhiana, Ferozepur, Faridkot and Patiala during the period 1997-98. Ludhiana and Patiala fall in the central plain zone whereas Ferozepur and Faridkot fall in the south-western zone on the basis of topography in Punjab. However in 2007-08, Ludhiana is replaced by SBS Nagar from the top position and to occupy the position in the second category and Patiala and Faridkot maintained their positions in the first category. Ferozepur slipped from first category in 1997-98 to second category in 2007-08 while Hoshiarpur gained the top position during 2007-08 (Table 2) and this district fall in the sub-mountainous zone.

(ii) *Districts with medium level of rural development (+ 1.0 to -1.0)*: Nine districts of the state fall in this category during the period 1997-98. Out of these nine districts, four districts, namely Mukatsar, Moga, Bathinda and Mansa fall in the south-western zone of the state. Central plain zone comprising of Jalandhar, Sangrur and Fatehgarh Sahib also fall in this category and only two districts of sub-mountainous zone i.e. Gurdaspur and SBS Nagar are in the category of medium level of rural development. However, major shuffling has taken place in this category during the study period and in 2007-08 out of eleven districts comprising the second category, Ludhiana and Ferozepur were earlier in the first category (Table 1)

TABLE I

Standard Scores of Various Economic Indicators (I-XV)* of Rural Development in Punjab, 1997-98

District	I	II	III	IV	V	VI	VII	VIII	IX	X	XI	XII	XIII	XIV	XV	CSS**
(1)	(2)	(3)	(4)	(5)	(6)	(7)	(8)	(9)	(10)	(11)	(12)	(13)	(14)	(15)	(16)	(17)
Gurdaspur	0.61	-0.49	-0.73	-1.81	-1.08	-0.84	-1.60	2.26	-0.85	0.99	0.38	-1.47	0.82	-0.12	0.50	-0.75
Amritsar	-0.42	-2.21	0.73	0.16	-0.88	1.16	-0.11	-0.61	-0.17	0.33	-0.56	-0.98	-1.26	-0.30	-0.67	-1.18
Kapurthala	0.60	0.41	0.83	0.86	0..21	0.11	-1.81	-0.47	-0.85	-0.51	-3.78	0.14	-1.19	-0.32	-0.73	-1.46
Jalandhar	1.07	-0.64	-1.21	0.83	0..21	0.13	-1.32	0.53	-0.47	-0.84	0.38	1.18	-1.17	-0.54	-1.13	-0.87
SBS Nagar	1.05	0.20	-0.43	-0.95	-0.63	-1.17	0.76	-1.40	-0.53	-0.70	0.30	2.30	-0.34	-0.52	-1.28	-0.74
Hoshiarpur	1.71	-0.75	-2.47	-1.72	-1.18	-0.29	-0.20	-1.05	-0.98	-0.60	0.22	1.06	0.19	0.97	-1.34	-1.30
Roopnagar	0.97	-1.20	-1.02	-2.11	-0.78	-0.88	0.55	-0.42	-1.05	-0.56	-0.72	-0.23	-0.69	2.87	-0.48	-1.17
Ludhiana	0.54	-0.33	1.22	0.87	1..31	-0.09	-0.15	1.74	-0.02	-1.16	0.38	1.52	-0.56	-0.40	-0.65	1.53
Ferozepur	-0.93	1.89	0.63	0.66	0..98	1.88	0.10	-0.01	-0.49	1.86	0.38	-1.08	1.30	0.71	0.13	1.29
Faridkot	-0.86	1.17	0.54	0.44	1.86	0.82	0.94	0.44	0.59	-0.52	0.38	-0.21	0.55	-0.35	0.94	1.08
Mukatsar	-1.06	1.32	-0.53	0.37	0.81	-1.52	1.02	-0.80	0.81	-0.87	0.38	0.92	-0.09	1.56	-0.33	0.22
Moga	-1.89	-0.87	1.51	0.85	1.61	-0.78	1.10	-0.24	2.24	-0.75	0.38	0.41	-1.89	-1.34	-0.75	-0.21
Bathinda	-1.16	1.14	-0.43	0.31	-0.63	-1.03	1.10	-0.48	1.87	-0.06	0.38	-0.23	0.04	-0.31	1.51	0.22
Mansa	-1.80	1.27	-0.53	-0.56	-0.88	0.04	1.10	-0.12	1.00	2.35	0.38	-0.46	1.83	0.08	1.36	0.78
Sangrur	-0.98	0.97	0.73	0.74	0.01	0.04	0.90	0.06	0.70	-0.03	0.38	-0.31	0.60	1.43	1.06	1.00
Patiala	-0.22	0.03	0.73	0.21	-0.33	2.09	0.50	1.88	-0.67	1.33	0.38	-0.77	0.64	-0.14	1.89	1.23
Fatehgarh	0.75	0.01	0.44	0.87	1.36	0.34	-1.34	-0.25	-1.12	-0.26	0.38	0.62	1.19	-0.47	-0.05	0.31

* Economic indicators of rural development mentioned in methodology part, and ** composite standard score.

TABLE 2

Standard Scores of Various Economic Indicators (I-XV)* of Rural Development in Punjab, 2007-08

District	I	II	III	IV	V	VI	VII	VIII	IX	X	XI	XII	XIII	XIV	XV	CSS**
(1)	(2)	(3)	(4)	(5)	(6)	(7)	(8)	(9)	(10)	(11)	(12)	(13)	(14)	(15)	(16)	(17)
Gurdaspur	0.40	-1.17	-0.51	-1.33	-0.63	-0.94	-4.07	3.12	-0.85	1.26	0.4	-1.45	0.77	-0.43	0.89	-1.06
Amritsar	-0.29	0.18	0.88	0.66	-1.35	1.26	0.42	0.02	-0.04	0.92	0.4	-0.97	-0.60	0.15	-0.66	-0.01
Kapurthala	0.35	-0.25	0.60	0.66	0.07	0.09	-0.23	-1.02	-1.03	-1.15	0.4	0.46	-1.33	0.25	-1.12	-0.78
Jalandhar	0.61	-0.97	-1.21	0.65	-0.18	-0.18	0.15	-0.89	-0.70	-1.34	0.4	1.05	-1.62	-0.11	-1.38	-1.27
SBS Nagar	0.72	-1.39	-0.10	-0.72	0.47	-1.98	0.33	-2.15	-0.72	-1.13	0.4	3.01	-1.80	0.43	-1.78	1.46
Hoshiarpur	1.00	-1.11	-2.60	-2.37	-1.27	-1.01	0.48	-0.85	-1.04	-0.64	0.4	0.80	-1.09	0.05	-1.16	2.32
Roopnagar	0.63	-1.18	-0.86	-1.75	0.81	-1.11	-0.33	0.10	-0.96	-0.16	0.4	-0.02	-0.23	2.66	0.19	-0.47
Ludhiana	0.53	-1.01	0.32	0.66	1.07	0.05	0.38	-0.41	0.03	-1.11	-3.6	1.08	0.17	-0.15	-0.62	-0.65
Ferozepur	-0.59	1.31	0.32	0.65	-1.72	1.38	-0.21	0.43	-0.38	2.69	0.4	-1.02	1.34	-0.13	0.35	0.94
Faridkot	-0.49	1.10	1.16	0.65	0.92	0.48	0.48	1.12	0.85	-0.16	0.4	-0.13	-0.47	-0.40	1.10	1.33
Mukatsar	-0.69	1.69	-1.21	0.12	-0.13	1.78	0.48	0.41	1.07	0.61	0.4	-0.76	0.54	-0.03	-0.38	0.75
Moga	-0.23	0.76	1.16	0.66	1.12	-0.50	0.48	0.24	1.43	0.75	0.4	0.10	-0.23	-0.69	-0.59	0.95
Bathinda	-0.57	0.72	0.18	0.50	-0.33	-0.51	0.46	0.01	1.77	0.10	0.4	-0.16	0.06	-0.61	1.19	0.60
Mansa	-1.07	1.16	-0.79	-0.59	-1.32	0.59	0.45	0.22	1.17	0.45	0.4	-0.50	1.84	-0.39	0.98	0.47
Sangrur	-0.49	0.75	0.88	0.66	0.34	0.86	0.27	0.03	1.17	-0.22	0.4	-0.32	0.91	-1.01	0.96	0.95
Patiala	-0.13	0.31	1.16	0.65	0.67	0.87	0.48	0.50	-0.73	1.18	-1.6	-0.25	0.49	0.005	1.73	1.05
Fatehgarh	0.49	-0.61	-0.24	0.66	1.66	0.87	-0.35	-0.29	-1.17	0.14	0.4	0.41	1.17	0.05	-0.22	0.55

* Economic indicators of rural development mentioned in methodology part, and ** composite standard score.

and were at the first and second top positions respectively. Amritsar, Kapurthala and Roopnagar shifted from third category to second category during the period from 1997-98 to 2007-08. So the change has witnessed in all the three zones, other districts i.e. Mukatsar, Moga, Bathinda, Mansa, Sangrur and Fatehgarh Sahib remained in the same category, however, positions have interchanged.

(iii) *Districts with low level of rural development (below –1.0):* Four districts come under this category, namely Amritsar, Kapurthala, Hoshiarpur and Roopnagar in 1997-98. Amritsar and Kapurthala fall in central plain zone whereas Hoshiarpur and Roopnagar come under sub-mountainous zone according to topography of Punjab. In 2007-08, Gurdaspur and Jalandhar shifted to third category from the medium level of rural development in 1997-98. These districts fall in the sub-mountainous zone and central plain zone respectively.

Thus, it is observed that there is a change in the spatial pattern of districts in the level of rural development over the entire period under study. Out of nine districts, five districts, i.e. Amritsar, Kapurthala, SBS Nagar, Hoshiarpur and Roopnagar have shown an improved level of rural development from the period 1997-98 to 2007-08. On the other hand, four districts namely Gurdaspur, Jalandhar, Ludhiana and Ferozepur slipped down from their previous category. All other districts maintained their category level though their positions have interchanged. Districts having high level of rural development are less in number, out of which SBS Nagar and Hoshiarpur fall in sub-mountainous zone, Faridkot fall in south-western zone and Patiala fall in central plain zone, having all the advantages required for the speedy development. These include better access to new technology, proximity to well developed urban areas, more workforce, high cropping intensity, education, medical facilities, existence of cooperative societies and communication facilities. Medium level of development is found in more number of districts and out of eleven districts, Amritsar, Kapurthala, Ludhiana, Fatehgarh and Sangrur belong to central plain zone. The entire south-western zone accepts Faridkot has medium level of rural development. These districts have all the facilities to have moderate level of rural development but are scoring low on some indicators like labour force, cropping intensity, irrigation, machinery, veterinary institutions, education, roads, banks and post offices. Only one district from sub-mountainous zone i.e. Roopnagar is in the second category of rural development. This may be due to high rate of literate persons, more villages attached to pucca roads, tractorization and establishment of market committees as compared to rest of the zone and it is having medium level of rural development.

The low level of development is found only in two districts, namely Gurdaspur and Jalandhar falling in sub-mountainous and central plain zones respectively. As the name suggests, topography of Gurdaspur is sub-mountainous, which hampers the development process. Low level of irrigation affects the yield and area under high yield varieties which is the cause of low agricultural development in Gurdaspur. Shortage of agricultural workforce and drinking water are the other reasons which lead this district into the third category of development. So far as Jalandhar district is concerned, lack of education, banks, post offices and area under crops are the factors which lead to low level of rural development in this district. The development of infrastructure-rural roads and bridges, irrigation facilities, flood control, power supply, education, health, agriculture research and extension services etc. is essential for accelerated economic development of rural areas (Saldi, 2005). With an avowed objective of stimulating development in backward areas and to reduce regional disparities, the Government of India has already launched the backward district initiative scheme for rural development in 2003 (Krishan, 2005). In the backdrop of the latest developments that are taking place in information and communication technologies, there is a need to develop an 'Info-structure' for rural development, information by bringing together various institutions as a network for resource sharing and dissemination (Raju, 2001).

CONCLUSION

The general conclusion which may be derived from the analysis of the levels of rural development is that rural development has not been uniform in the state; inequalities do exist among the districts in the development of rural area. A change has witnessed overtime in the districts regarding development. The districts i.e. Ludhiana and Ferozepur were leading earlier are now lagging behind. The districts i.e. Patiala and Faridkot have higher level of rural development; do not need any special effort. SBS Nagar and Hoshiarpur districts slipped up in the first category from second and third category respectively, a great achievement of Hoshiarpur district. In the districts which have medium level of development, a little push up is needed in some areas. In Gurdaspur, public sector should provide irrigation facilities to the farmers. In Jalandhar and Kapurthala districts medical facilities, senior secondary schools, banking facilities and communication facilities should be taken into account. Tractor intensity must be increased in Amritsar, Hoshiarpur, Ferozepur and Mansa districts. Kapurthala, Jalandhar, Hoshiarpur, Roopnagar, SBS Nagar and Fatehgarh Sahib are lacking behind on account of schooling of the children. The Government should take initiative to provide primary and secondary education to the children of the villages of these districts. But what need priority at the moment are the districts having low level of rural development. These need a wholesome strategy with major thrust on providing irrigation facilities which will lead to increase in area under high yield varieties and higher level of tractorization so that their level of development may be raised. Moreover, other facilities related to medical, veterinary, education, banking and drinking water must be provided; only then it can be called that Punjab has made a rural development in the right sense.

REFERENCES

Bhargava, Pushpa M. (2010), Measuring Human Development, *The Sunday Tribune*, Chandigarh, July 4, 2010:12.

Bhatia, S.S. (1966), Spatial Variations, Changes and Trends in Agricultural Efficiency in Uttar Pradesh, *Indian Journal of Agricultural Economics*, 207(4).

Chaudhary, P. (1972), Reading in Indian Agricultural Development, Geogre Allen Unwin Ltd., London.

Census of India (2001), Government of India, New Delhi.

Government of Punjab (2008), Statistical Abstracts of Punjab, Chandigarh.

Kellong (1963), Interactions in Agricultural Development, Science, Technology and Development, World Food Congress, Washington, D.C.

Krishan, Gopal (2005), Map Series I—Backward District Initiative Scheme, *Man and Development*, 27(2): 164-68.

Phougat, Sunil and Surender Kumar (2007), Rural Development under WTO Regime, *Kautilya, Haryana Economic Journal*, 27(1-2): 116-120.

Raju, K.A. (2001), Developing Partnership for a Rural Info-Structure, *Journal of Rural Development*, 20(4): 673-91.

Standard Scores of Various Economic Indicators (I-XV) of Rural Development in Punjab, 2007-08.

Saldi, Rattan (2005), Priorities for Rural Development, *Kurukshetra*, 55(7): 33-35.

Smith, D.M. (1977), Human Geography : A Welfare Approach, Edward Arnold (Publishers) Ltd., London.

CHAPTER

28

Sectoral Distribution of Non-Farm Workers in India

K. GOVINDARAJALU AND C. MYTHILI

INTRODUCTION

The key to India's development lies in the development of its rural areas. There are as many as six lakh villages where about 70 per cent of the population lives. The agricultural sector thus occupies a pivotal place in the national economy both in terms of its contribution to the gross domestic product and employment generation. However, segmenting rural employment growth into the farm and non-farm sectors would demonstrate that non-farm employment growth had been significantly higher than farm sector employment growth throughout the period 1972-73 to 2004-05. Rural-Non-Farm-Sector (RNFS) includes all economic activities viz., household and non-household manufacturing, handicrafts, processing, repairs, construction, mining and quarrying, transport, trade, communication, community and personal services etc. in rural areas. Rural-Non-Farm-Activities (RNFAs), thus, play an important role to provide supplementary employment to small and marginal farm households, reduce income inequalities and rural-urban migration. Though, agricultural sector has played a very significant role for generation of rural employment in the Asia and Pacific region, its contribution to the overall economy has greatly reduced in the recent past (Asian Productivity Organization, 2004). Therefore, development of various non-farm-activities can effectively be exploited as a potent stimulator for further economic growth offering rural communities better employment prospects on a sustainable basis.

IMPORTANCE OF RURAL NON-FARM SECTOR

The non-farm sector, particularly in rural areas is being accorded wide recognition in recent years for the following reasons:

- Employment growth in the farm sector has not been in consonance with employment growth in general.
- A planned strategy of rural non-farm development may prevent many rural people from migrating to urban industrial and commercial centers.
- When the economic base of the rural economy extends beyond agriculture, rural-urban economic gaps are bound to get narrower along with salutary effects in many other aspects associated with the life and aspirations of the people.
- Rural industries are generally less capital-intensive and more labour-absorbing.
- Rural industrialization has significant spin-offs for agricultural development as well.

SIZE OF THE RURAL NON-FARM (RNF) SECTOR

Table 1 reveals a much higher rate of increase in non-farm employment among male workers than their female counterparts during 1977-78 to 2004-05. The percentage of males rose from 16.7 per cent in 1972-73 to 33.5 per cent in

2004-05 whereas that of females registered a rise from 10.3 per cent to no more than 16.7 per cent. The increase in both categories was noticeably sharper during the period 1999-00 to 2004-05 than in earlier periods. Quite significantly, the male workers increased by 4.9 percentage points while the females by only 2.1 percentage points.

Table 1 reveals a much higher rate of increase in male workers than the female counterparts in non-farm employment during the period 1977-78 to 2004-05. The percentage of males rose from 16.7 per cent in 1972-73 to 33.5 per cent in 2004-05 whereas that of females registered a rise from 10.3 per cent to no more than 16.7 per cent. The increase in both categories was noticeably sharper during the period 1999-00 to 2004-05 than in earlier periods. Quite significantly, the male workers increased by 4.9 percentage points while the females by only 2.1 percentage points.

TABLE I

Distribution of Rural Workforce in Non-Farm Activities

(in percentages)

Year	*Persons*	*Males*	*Females*
1972-73	N.A.	16.7	10.3
1977-78	16.6	19.3	11.8
1983	18.5	22.2	12.5
1987-88	21.7	25.4	15.3
1993-94	21.6	26.0	13.8
1999-00	23.7	28.6	14.6
2004-05	27.6	33.5	16.7

Source : NSSO Reports Various issues.

Given the wide class-wise and region-wise differences in the level, rate of growth and pattern of rural employment, the question which naturally arises is the degrees to which this diversification of rural employment is a result of distress-push as opposed to demand-pull factors. We hold the view that it is not feasible to answer the question on the basis of data on the aggregate level. But the all-India and state-level data are bound to point to tendencies that are indicative of which one of these two sets of predominantly influence employment growth in the rural sector.

The aggregate all-India figures on non-farm employment since 1993-94 can, however, be taken to suggest that demand-pull has played a greater role in the more recent periods. Since the sharp rise in the last two of these sub-periods occurred in a context of extremely high rates of expansion of the manufacturing and services sectors, this is partly seen as being driven by demand-pull factors, though prevailing unemployment and underemployment meant that this was not accompanied by substantial improvements in the quality of and earnings associated with such non-farm employment.

INDUSTRIAL DISTRIBUTION OF THE WORKFORCE

An analysis of the industrial division of the workforce could help us assess whether this indication of a greater influence of demand-pull factors in the 1990s and after is valid. Table 2 essentially reviews the compositional importance of the sub-sectors in rural India at the one-digit level. The Table 2 shows that while agricultural activities continue to be the mainstay for the rural workers, their relative importance declined substantially by nearly 11 percentage points between 1977-78 and 2004-05 starting from 83.4 per cent in 1977-78. Their proportional share constituted 72.7 per cent of the workers in 2004-05. A closer perusal of the table reveals that primarily, it is the withdrawal of male workers (14.2 percentage points) rather than the female workers (5 percentage points) from agricultural activities that was responsible for the fall in the share of agriculture in rural employment.

In the non-farm sector, manufacturing is the largest source of non-farm employment in rural India. Its proportion rose from 6.2 per cent in 1977-78 to 8.1 per cent in 2004-05. This kind of diversification does seem to tally with conventional expectations of diversification away from agriculture to more productive manufacturing, supporting the role played by demand-pull factors representing a degree of dynamism.

Until 1999-00, the second largest non-farm employment source was other services sector. While this could include some

TABLE 2
Sectoral Distribution of the Workers in Rural India : 1977-78 to 2004-05

(in percentages)

Sectors	*1977-78*	*1983*	*1987-88*	*1993-94*	*1999-2000*	*2004-05*
Rural Persons						
Agriculture and Allied	83.4	81.5	78.3	78.4	76.3	72.7
Mining and Quarrying	0.4	0.5	0.6	0.6	0.5	0.5
Manufacturing	6.2	6.8	7.2	7.0	7.4	8.1
Electricity, gas and water	0.1	0.1	0.2	0.2	0.2	0.2
Construction	1.3	1.6	3.3	2.4	3.3	4.9
Secondary Sector	8.0	9.0	11.3	10.2	11.4	13.7
Trade, hotels and restaurants	3.3	3.4	4.0	4.3	5.1	6.1
Transport and communication	0.8	1.1	1.3	1.4	2.1	2.5
Other Services	4.5	4.9	5.1	5.7	5.2	5.0
Tertiary Sector	8.6	9.4	10.4	11.4	12.4	13.6
All	100	100	100	100	100	100
Rural Males						
Agriculture and Allied	80.7	77.8	74.6	74.0	71.4	66.5
Mining and Quarrying	0.5	0.6	0.7	0.7	0.6	0.6
Manufacturing	6.4	7.0	7.4	7.0	7.3	7.9
Electricity, gas and water	0.2	0.2	0.3	0.3	0.2	0.2
Construction	1.7	2.2	3.7	3.2	4.5	6.8
Secondary Sector	8.8	10.0	12.1	11.2	12.6	15.4
Trade, hotels and restaurants	4.0	4.4	5.1	5.5	6.8	8.3
Transport and communication	1.2	1.7	2.0	2.2	3.2	3.9
Other Services	5.3	6.1	6.2	7.1	6.1	5.9
Tertiary Sector	10.5	12.2	13.3	14.8	16.1	18.1
All	100	100	100	100	100	100

Rural Females						
Agriculture and Allied	88.2	87.5	84.7	86.2	85.4	83.3
Mining and Quarrying	0.2	0.3	0.4	0.4	0.7	0.3
Manufacturing	5.9	6.4	6.9	7.1	7.6	8.4
Electricity, gas and water	—	—	—	—	—	—
Construction	0.6	0.7	2.7	0.8	1.1	1.5
Secondary Sector	6.7	7.4	10.0	8.3	9.4	8.3
Trade, hotels and restaurants	2.0	1.9	2.1	2.1	2.0	2.5
Transport and communication	0.1	0.1	0.1	0.1	0.1	2.0
Other Services	3.0	2.8	3.0	3.4	3.7	3.9
Tertiary Sector	5.1	4.8	5.2	5.6	5.8	8.4
All	100	100	100	100	100	100

Source : NSSO Reports, Various issues.

modern services, many of these activities may be in the nature of residual opportunities exploited when employment in the commodity producing sectors is not growing fast enough. The relative share of these services in total employment stood at 5.0 per cent in 2004-05 against 4.5 per cent in 1977-78. However, it is notable that their employment contribution registered a decline of 0.7 percentage points during the nineties, from 5.7 per cent in 1993-94 to 5.0 in 2004-05. That is when the proportion of employment in manufacturing sector is growing, that of other services is stagnant or on the decline supporting the perception of a degree of dynamism.

The share of the next most important sector trade, hotels and restaurants rose consistently from 3.3 per cent in 1977-78 to 6.1 per cent in 2004-05. Construction and transport, storage and communications were also emerging as important sectors in the provision of non-farm employment particularly in the nineties. These are sectors in which the growth of employment at the aggregate may be the result either of distress-push or demand-pull. On the other hand, employment in mining and quarrying and electricity, gas and water was static at a low level of employment.

Table 2 also shows the employment distribution for male and female workers. It is likely that separate male-female distribution of workers may correspond in some of the non-farm sectors. The common sectors for both gender groups in terms of quantum and trends are as indicated below. The manufacturing sector employment is crucially important both for male and female workers. The proportions were more than 7 per cent in 2004-05 for both the gender groups. Other services absorbed the majority of the rest of the workers though the relative position of this sector lessened later for the male workers. The proportions were 5.9 and 3.9 per cent respectively for male and female workers in 2004-05. Employment in mining and electricity, gas and water remained nominal at less than one per cent throughout the period.

In rural areas, the average decline in regular women workers' real wages over the first five years of this decade was by 32 per cent. Even if we ignore the outlier, viz. women with primary education, the decline in real wages in other categories was substantial. Illiterate women workers in regular employment in rural areas faced average wage cuts of 20 per cent, while those who had secondary and higher secondary education faced average cuts of nearly 30 per cent. More than 66 per cent of all rural women workers were illiterate. The second explanation could be that a decline in real earnings in the principal source of employment in agriculture could be forcing rural women to take up non-agricultural activities to sustain family income. This does seem to be of relevance, especially since a major form of non-agricultural employment growth has been an increase in self-employment.

Finally, a third and complementary explanation could be that women's employment in non-agricultural has risen in recent times because the earnings from these activities is far below the reservation wage for men, leaving these sources of income open to women. The gender-wise trends in rural non-agricultural employment seem to point to the role of distress-driven increases in such employment, especially among women.

STRENGTHS AND WEAKNESSES OF NON-FARM SECTOR

Non-farm activities either keep the poor falling into deeper poverty or are advantageous in lifting the poor above the poverty line. Keeping this in view, it becomes imperative to identify the strengths and weaknesses of the non-farm sector in India to focus on, in order to alleviate poverty. The strengths and weaknesses of rural non-farm sector in India as highlighted by Mukherjee and Zhang (2005) have been discussed below.

Strengths

Institutional basis for rural non-farm sector: In India, the institutions underlying the development of the rural non-farm sector are very strong. These include secure property rights; a well-developed financial system with preferential access to credit for the sector; supporting institutions such as the KVIC, State Khadi Board, NHHDC, Small Industries Development Bank of India (SIDBI), State industrial corporations; policies and programs promoting linkages with agriculture, especially agro-industries; domestic marketing channels for rural non-farm production; as well as government support in export promotion. The institutional mechanisms for a rapid growth of the rural non-farm sector are already in place.

Decentralization process: Over the last two decades the State governments in India have been able to exercise far more independence in decision-making than in the pre-1980 period. Regional parties are an integral part in coalition governments at the Center. In turn, they have negotiated economic autonomy in the formation of state specific policies for development. Moreover, with the opening up of the economy in 1991, foreign direct investment (FDI) has come to play an important role in the overall policy environment. State governments are in competition with one another to attract higher FDI levels both in manufacturing and infrastructure. In some ways, it mirrors the path followed by China, although the volume of FDI coming to India is less than 10 percent of what is flowing into China. On the positive side, however, this creates an opportunity for higher levels of investment in the future.

Weaknesses

Infrastructure: The most significant bottleneck in generating higher levels of rural non-farm activity in India is the quantity, quality and reliability of infrastructure. For example, the World Bank Investment Climate Survey for India indicates that power outages were one of the most serious obstacles to the development of the non-farm sector (Economist, 2005; World Bank, 2005). Although corrective steps are now being taken, increased infrastructure remains the most important priority for the future. To achieve a sustained growth rate of 8-9 percent, the investment rate has to be stepped up from the current level of 24 percent to nearly 35 percent over the next decade, with investment directed at the rural sector (Planning Commission, 2000).

Regulatory restrictions on small-scale sector: Regulation of the small-scale sector constitutes an important aspect of non-farm development policy in India. In the initial stages, capital investment restrictions were imposed to protect the small-scale sector, especially in rural areas, from predation by large industry. Reservation of products for the sector was initiated to create a domestic market and quantitative restrictions imposed to protect them from competition from imports. At the end of the 1990s, however, these very policies have become detrimental to the dynamism of the small-scale sector, especially in the rural areas. Capital investment limits have discouraged economies of scale, and concessions offered to small industry have created adverse incentives against re-investment. Several official reports have recommended a substantial increase in the capital investment limit (from the present level of around $200,000) to make better use of technology and improve productivity (Planning Commission, 2000). However, no such policy announcement has been made as yet.

Quality of manpower: High levels of illiteracy in rural India have hampered the growth of the rural non-farm sector. Education has both intrinsic and instrumental value. Apart from having a positive correlation with wages, a minimum basic standard of education is necessary to apply for credit, to be aware of one's rights and responsibilities and to deal with instances of corruption and malpractice. Often, a lack of education is intrinsic to poverty, which seems to have been the case in India until recently.

In the rural areas, lack of education leads to labor being stagnant in agriculture, or moving to casual work occupations in the non-farm sector, and not to salaried employment with higher wages and benefits. Together with lack of technical skills, there is little incentive for rural firms to invest in technology, leading to low levels of labor productivity in the rural manufacturing sector compared to urban manufacturing (Chadha, 2003).

Forward and backward linkages: Absence of appropriate forward and backward integration greatly affects performance of non-farm activities in rural areas. Forward linkages of the RNF sector serve as inputs to other sectors. Also, in backward linkages the RNF sector demands the outputs of other sectors. Empirical studies indicate that forward linkages from RNF activities to agriculture (rurally produced agricultural inputs) are particularly important where traditional agricultural technologies are utilized, while in case of backward linkages between RNF activities and agriculture, especially the linkages between rural agricultural processing and the agriculture sector and between rural transport and rural marketing activities are quite significant for rural economic development. However, gaps in the integration of the production linkages brought about by poor infrastructure, low accessibility of market, support service weaknesses and intervention of middlemen have constrained the development of non-farm enterprises in India.

CONCLUSION

The Rural Non-farm sector is increasingly playing an important role in the development of rural areas in Asia and the Pacific region. Specifically, the role of agriculture in the region declines in importance in terms of its contribution to the economy, the Rural Non-farm sector will need to become more and more a major provider of employment and income to many rural folks. It should be noted, however, that Rural Non-farm Employment are not a substitute for employment in agriculture but rather as a supplementary measure. Agricultural development is still important and should be pursued as a necessary precondition. The promotion of Rural Non-farm Employment also should be undertaken within the broader context of rural development. The aggregate all-India figures on non-farm employment since 1993-94 can, however, be taken to suggest that demand-pull has played a greater role in the more recent periods. In the non-farm sector, manufacturing is the largest source of non-farm employment in rural India. Its proportion rose from 6.2 per cent in 1977-78 to 8.1 per cent in 2004-05. This kind of diversification does seem to tally with conventional expectations of diversification away from agriculture to more productive manufacturing, supporting the role played by demand-pull factors representing a degree of dynamism. The manufacturing sector employment is also crucially important both for male and female workers. The proportions were more than 7 per cent in 2004-05 for both the gender groups. A greater reliance on the non-farm sector would therefore provide a demand-pull to rural economy and also ensure welfare for rural workers. However, the sector has been contending with a number of factors like inadequate rural infrastructure, particularly roads, electricity and communication facilities, lack of sufficient skilled labour and adequate access to credit, information and training facilities, etc. So, formulation of possible approaches is more important with a view to promote rural-non-farm-sector for Rural Development.

References

Asian Productivity Organization (2004), "Non-farm Employment Opportunities in Rural Areas in Asia", Tokyo, p. 13.

Chadha, G.K. (2003), "Rural Non-farm Sector in the Indian Economy: Growth, Challenges and Future Direction". *Mimeo*. International Food Policy Research Institute, Washington D.C.

Economist (2005), "India's Electricity Reforms: A Power Shortage May Thwart India's Rush to Modernity", *Economist*, September 22: 83-84.

Mukherjee, A. and Zhang, X. (2005), "Rural Non-farm Development in China and India: The Role of Policies and Institutions".

Development Strategy and Governance Division, International Food Policy Research Institute, pp. 33-35.

National Sample Survey Organization (NSSO) (1981), "Survey on Employment and Unemployment": All India NSS Thirty Second Round, *Sarvekshna*, Vol 5, Nos. 1 and 2.

Planning Commission (2000), "Report of the Task Force on Employment Opportunities". Government of India: New Delhi.

Saith, A. (1992), "The Rural Non-farm Economy: Processes and Policies, International Labour Force", Geneva (*mimeo*), pp. 12-16.

Saxena, M. (2004), "Non-farm Employment Opportunities in Rural Areas in Asia", Tokyo, p. 90.

CHAPTER

29

Developing a Non-Farm Activity with Farm Orientation : The Case of Kewda Industry in Orissa

DHANA LAXMI PATNAIK AND BHAGABATA PATRO

INTRODUCTION

Emphasis on value addition through non-farm activity has gained importance in the development literature in the recent years. It is important to raise the level of living of the rural sector. Production in primary sector which is more labour intensive and did not give adequate return to labour power. The traditional argument of comparative cost advantage as applied to developing countries, "dictates the production of primary commodities and a pattern of trade that put these countries at a relative development disadvantage." (Thirlwal; 2003). One of the agendas in agriculture and employment-based development strategy is to promote a diversified, non-agricultural, labour-intensive rural development activity along with improvement in agriculture productivity. (Todaro and Smith; 2004)

THE PROBLEM AND SCOPE

The common name of Kewda (Pandanus fascicularis) is Fragrant Screw Pine, Umbrella tree, Screwpine and Screw tree. It is a species of Pancianus group found in Southern Asia. The Kewda or Screwpine is a densely branched shrub, delicious and unisexual, grows widely in the riversides and water logging places of seacoast of the Indian Peninsula on both sides and in the Andamans, Iran, Malaysia, Mauritius and Myanmar. Over 500 species are known of which the largest genus *Pandanus* has 36 species in India. The district of Ganjam of Orissa has its peculiar climatic conditions which is best suitable for a variety of plantation like Kewda, Lemon, Guava, etc. These crops play a significant role in the development of non-farm activity in this part of the state. The main concentration of Kewda plant is in three blocks viz Chatrapur, Rangeilunda and Chikiti of Ganjam district in Orissa. Kewda plantation are also found in the district of Balasore, Cuttack and Puri in the recent years. In Uttar Pradesh, Kewda oil is grown on a plantation scale in Badagun, Balia and Ghazipur districts About 36 species of Pandanus and about plants such as Kadam (Anthocephalus cadamba), Parijat (Nycanthes arbortris-tis), Bakul (Mimopsus elengi), Lotus (Nelumbo nucifera), Water Lily (Nymphae nouchalii), Golden Champa (Michelia Champaca) and Kewda (Pandaus odoratissimus) has been available in coastal districts of Odisha, Andhra Pradesh, Tamilnadu, and Gujarat and to some extent in Badayun, Balia, and Gahipur districts of Uttar Pradesh. The plants often form thickets along roadsides, borders of these are found in an area comprising of about 100 villages accounting for 85-90 percent of total production of Kewda in the country. This paper makes an attempt to understand the problems faced by these units and the ways for solving it through government and non-government organizations.

THE PROCESS OF PRODUCTION

In Ganjam district, the Kewda activity is an unorganized sector. Kewda is a hardy plant. It grows on sandy, saline wastelands and marshy land on the river and seashore bunds. There is no need of fertilizer or any special care for cultivation of the Kewda crop. However, the commercial cultivation of kewda would prefer a fertile and well-drained soil. Kewda grows naturally along the seacoast binds soil and checks soil erosion. The soil is mostly alluvial, black laterite, red and saline. Generally the silted soils are available near river banks. Soil salinity is found in blocks of Ganjam District such as Khallikote, Ganjam, Chatrapur, Rangeilunda and Chikiti. The tropical climate with heavy rains is favorable to the growth of Kewda crop and which does not require any special care. The Kewda is densely branched shrub growing up to 6 mts high with glaucous green leaves with 0.9-1.15 m long and 4 cm wide. During the monsoon season, these Kewda flowers are abundantly available and are harvested early in the morning as they lose their fragrance quickly after opening; thereafter, they are immediately subjected to hydro distillation in one of the countless small distilleries found in the region. The plant starts flowering after 4-5 years and yields about 35-40 flowers per year with an average age of 40-50 years. The flowering is maximum between 15 and 25 years. The plant is not generally affected by any pests or diseases of serious nature rather they under natural climate conditions. The female flowers do not have any characteristic smell but bear aggregate fruits comprising of numerous ovaries that develop into drupes. In the harvest season, every day men and women gather the flower spikes using long hooked poles that allow them to reach high up into the tree and snap them loose. These spikes are collected in manageable piles and transported to the distilleries. This area of India is still largely undeveloped. Though a number of distilleries producing high quality of Kewda products yet there is no proper infrastructure or road connectivity to reach these distilleries. The conservation, propagation and utilization of *Pandanus fascicularis* are mostly restricted to Ganjam district coastal area as compared to the other regions of the coastal line. The Kewda plantation is a time consuming activity as the shrub starts flowering after 4-5 years. The existing distillation units are adopting the age old Deg-Bhabka technology. There is no scientific and modern technology to improve the time and quality of the distillation process. Due to change in the market scenario, there is no certainty or guarantee in the business of Kewda industry and no assured income to the people whose main livelihood is Kewda business activity. There is no direct link between distilleries and traders and due to ignorance of the technology, better market acceptability and better realization of value by the farmers, distilleries, the commission agents of traders of Kannauj, Lucknow, Kanpur and neighboring parts of North India have maintained their superiority and monopoly on the Kewda trade. It is estimated that about 300,000 to 400,000 trees produces about 10,000,000 male inflorescences which are used for the preparation of attar, ruh and hydrosol mostly in the rainy season. About 139 distilleries installed with 911 degs (vessels) are working in these three blocks of Ganjam district and about 365 lakh flowers are processed during the monsoon season with a turnover of about 60 crores in a year, The distillation of Kewda flowers is carried out in two seasons which begins from the month of June and last up to September and during this period about 70 to 80 percent of flowers are produced and the rest 20 to 30 percent are produced during the remaining period of the year. The distilleries produce different grades of essences and after mixing the sandalwood which produces a product called *attar kewra* which is the common grade in perfumery; the pure essential oil is known as *ruh kewra*. The existing marketing channels of the Kewda industry contains various steps such as producers of raw materials (farmers), agents for procurement of flowers (pluckers/ Commission Agents), distillers and traders who are operating domestic and international market, the price of Keora flowers are fixed by the Keora flowers associations like Kia Phul Samiti in Keluapali Mandi and Chatrapur Block and Keora Union in Agraham Mandi.

IMPACT OF KEWDA INDUSTRY ON INCOME AND EMPLOYMENT

It has been identified that Kewda Trade is a potential area for creating employment and income generation opportunities on a sustainable basis with low level of financial investment. The inflorescence of the Kewda plant of Ganjam coast is reportedly of higher quality than from other regions of the country. Kewda flowers have a sweet, perfumed odor with a pleasant quality similar to rose flowers. The male flowers are very fragrant, which is used for the preparation of Kewda attar, Kewda water and Kewda oil and are used in other forms such as:

Medical Purpose

The Kewda oil used for treatment of antispasmodic disease which is reported to relieve headache and rheumatism. Apart from this the leaves are of value in the treatment of leprosy, small pox, scabies and leucoderma and diseases of the heart and brain and also the Anthers of male flowers are considered for curing blood diseases. It provides relief to the patient who suffers from insomnia. The Kewda flowers are considered acrid, bitter, aphrodisiac, demulcent and anodyne and are used for treatment of variety diseases such as ulcers, skin diseases, flatulence, colic, fever, diabetes, sterility also for cure of head and skin problems and wounds.

Other Purposes

Kewda attar and Kewda water, the most popular preparation of Kewda perfume, are mostly used in flavouring tobacco and food. The Kewda oil is used as main constituent for preparing Pan Masala, Ghutaka and other tobacco products. Apart from this, it is used for manufacturing of bouquet, lotions, cosmetics, soaps, hair oils and agarabati. Kewda water is used as food flavouring in sweets, syrups and soft drinks. The leaves are used for covering huts and used for paper making and for making mats, ropes, baskets and brushes. The leaves are used for covering huts and used for paper making and for making mats, ropes, baskets and brushes and Kewda attar and Kewda water used in flavouring tobacco and Pan

Masala, Ghutaka and other tobacco products and agarabati. The Kewda water is used as food flavouring in sweets, syrups and soft drinks. The exotic scents have played a very important role in Indian culture and have been used in indigenous medicines, perfumes, and cosmetics since ancient times. With the rise of aromatherapy as an important aid in re-establishing physical, mental and emotional equilibrium through the gentle agency of fragrance a new interest in the ways ancient cultures have used their natural aromatic resources has awakened. The Kewda (Pandanus odoratissimus) is very agreeable perfume. One of the flowers which dearer to the Indian heart is Kewda (Pandanus odoratissimus). spa dices (popularly called flowers) in the production of Kewda perfume. Kewda distillation was first started about 200 years ago in the suburbs of Berhampur and Chatrapur in the Ganjam districts of Odisha by some Muslims from erstwhile Punjab Province of undivided India. Thereafter some others—from Kannauj, Gajipur, Lucknow, and other parts of the north India—entered this industry. Even today the industry is monopolized by a small group of people from northern India. India possesses a great wealth of aromatic plants which, until recent times, have been little known outside the country, fields, banks of rivers and canals, and in coastal sands. It reveals from the available data, that during the year 1980, there were 50 distillation units in the district, at present there are 108 distillation units in the three blocks—Rangeilunda, Chatrapit and Chikiti. The distillation units are generally operated by the local and outside entrepreneurs. These units generally employ both skilled, semi-skilled labourers in the distillation related operations such as sorting, counting of flowers, carrying put distillation, wood cutting and controlling firing of the furnace. During the cropping season 3 mandays of labour for thousand of flowers distillation. 1,20,000 mandays of labours were engaged during the year 2008. The average wage paid to the skilled labourer increased from Rs. 100 to Rs. 120 per day. Semi-skilled labourers were paid Rs. 70 to Rs. 80. On the seasonal basis the distillation units were functioning on an average 120 days in a year and each distillation units is engaged with two persons which include owners and assistants help in counting flowers and deals financial matter, maintaining records and accounts themselves.

Price of Keora flower increased considerably from Rs. 1 in the year 1991 to Rs. 9.50 in the year 2007 and latter decreased to Rs. 7.00 in the year 2010. Keora industry generates employment opportunities to the local people who are engaged in collection, procurement, transport and distillation process. At present annual turnover of keora trade in the district is about Rs. 40 crores.

PROFILE OF GANJAM DISTRICT

Ganjam District is situated in the coastal region of the state surrounded on the north by Khurda district, on the east by Bay of Bengal, on the west by Kandhamal and Gajapati districts and on the south by Andhra Pradesh. The district has its peculiar climatic conditions which is best suitable for a variety of plantation like Kewda. The district is characterized by an equitable temperature all through the year, particularly in the coastal regions. The winter starts from December and lingers till February. It is followed by summer from March to May. The month of October and November constitutes the post-monsoon transition period. The average annual rainfall of the district is 1276.20 MM. The rainfall generally increases from the coast towards the interior hilly tracks of the district. The annual rainfall of the district starts from second week of June and ends early in October. On an average the district has 95 rainy days in a year but this of course varies from year to year. The cultivation of *P. odoratissimus* is almost limited to the state of Orissa and more precisely, the Ganjam district in Southern Orissa. During the monsoon season, the Kewda flowers are most abundantly available and generally they are harvested early in the morning, as after opening they lose their fragrance more quickly and subsequently hydro-distillated.

THE SIZE OF THE KEWDA INDUSTRY

In Ganjam district about 5000 hectares of waste land are under the Kewda cultivation. Major part of this is located within five kilometer distant from the sea coast. The detail coverage is given in the Table 1.

TABLE I

Coverage of Kewda Industry in Ganjam District

1.	Area covered by Kewda plantation	5,000 Ha
2.	From Sea-coast, the Kewda plantation area	
	Within 5 Kms	80-85%
	Within 5-10 Kms	15-20%
	More than 10 Kms ———	
3.	Total flower production in the district	375 lakhs (annually)
	(@ 7500 flowers/Ha)	
4.	Cost of flowers Rs. 28.75 crores	
	(@ Rs. 7 per flower)	

Table 2 gives the average final output of Kewda industry per annum. The total value of the output is around 30 crores per annum.

TABLE 2

Final Output of Kewda Industry

	Quantity (In kgs.)	*Rate (Rs.)*	*Value (Rs. in lakhs)*
Kewda Oil (Rooh)	346	2,70,000	865
Attar (Sandal wood oil base)	8150	8150	1630
Kewda water	1,55,000	300	465

Source : Technology Support Centre for Kewda Industry, Berhampur (Ganjam). Estimated data are approximate only.

DISTILLATION PROCESS

The distilleries mainly consist of a row of earthen pots that are filled with flowers and water, and a second, parallel row of water basins used to cool the distillate, which is collected in copper pots immersed into that water. Distillation takes place in a closed apparatus without any control of

TABLE 3

Block-wise Coverage of Kewda Industry

Sl. No.		Name of the Blocks				
		Ganjam	Chatrapur	Rangeliunda	Chikiti	Total
1.	Total geographical area (in hect.)	22,031	21,312	19,910	21,587	84,840
2.	Area spread over by Kewda plantation (in hect.)	5,907	12,368	14,827	4,365	37,467
3.	Canopy Kewda area (in hect.)	95	1,688	2,490	522	4,795
4.	Total No. of Gram Panchayats (pre-divided) (pre-divided)	10	17	21	13	61
5.	GPs covered with Kewda plantation	3	7	16	5	31
6.	Villages	35	67	73	25	200

Sources : Technology Support Centre for Kewda Industry, Berhampur (Ganjam), Orissa.

TABLE 4

Pattern of Land Utilization (Area Utilized in Percentage basis) of Kewda Spread Area

Sl. No.		Name of the Blocks			
		Ganjam	Chatrapur	Rangeliunda	Chikiti
1.	Kewda wild	2	11	14	9
	Kewda cultivated	0	3	3	3
	Total	2	14	17	12
2.	Area under habitant	6	6	17	7
3.	Area under other cultivations	73	66	60	62
4.	Cultivable fallows	9	6	7	9
5.	Waste land	10	8	6	10
	Total	100	100	100	100

Sources : Technology Support Centre for Kewda Industry, Berhampur (Ganjam), Orissa.

temperature or pressure, the steam flowing from the heated earthen pot to the cooled copper pot in bamboo pipes that are sealed with mud. In a year, the distillation of Keora flowers carried in two seasons which begins from the month of June and last up to September and during this period about 70 to 80 percent of flowers are produced and rest 20 to 30 powers are produced during the remaining period of the power. The distilleries produce different grades of essences and after mixing the sandalwood which produces a product called *attar Kewda* which is the common grade in perfumery; the pure essential oil is known as *ruh Kewda*. Kewda water is a by-product of hydro-distillation which best suited for culinary use. During the dry season, the flowers produced with lower quality which can be processed into Kewda water. The chief constituent of Kewda oil is methyl ether of beta-phenyl ethyl alcohol which gives the characteristic aroma of the flowers. This aromatic component is synthesized on a large scale in India and it is widely used to produce so-called Kewda

TABLE 5

Block-wise Kewda Production and Distillation

Sl. No.		*Name of the Blocks*				
		Ganjam	*Chatrapur*	*Rangeliunda*	*Chikiti*	*Total*
(1)	Flowers utilized - (in lakhs)	—	183	160	17	360
	Value (in lakhs) @ Rs. 5 per flower	—	915	800	85	1800
(2)	Distilation Charges (in lakhs) @ Rs. 360 (per 1000 flowers)	—	65.88	57.60	6.12	129.60
(3)	Sandal wood oil	—	6,662 kg	1,077 kg	84kg	7823 kg
	Cost of absorbent (in lakhs) @ Rs. 10,000 per Kg.	—	666.20	107.70	8.40	782.30
	Total Exp.					Rs. 2711.90

Sources : Technology Support Centre for Kewda Industry, Berhampur (Ganjam) Orissa.

products. There is difference between natural and synthetic products. Simple go/ms will easily detect this adulteration and identify complexity of the real oil. Due to market complications, the sincere and regular buyers of essential oils, attars and absolutes, use reputable go/ms services to assist in the identification of pure oils.

TABLE 6

Production of Kewra Perfumes and its Value

Sl. No.		Name of the Blocks				
		Ganjam	Chatrapur	Rangeliunda	Chikiti	Total
1.	Kewda concentrate (room in kg.) values in lakhs	—	130	172	30	332
	@ Rs. 2.25 lakhs/kg.	—	325	430	30	830
2.	Kewda Attar (with absorbent in kg.) values in lakhs	—	6,662	1,077	84	7,823
	@ Rs. 20,000/kg.	—	1,332.40	215.40	16.80	1,564.60
3.	Kewda water (in kg.) Values in lakhs	—	1, 44,318	4,682	—	1,49,000
	@ Rs. 300/kg.	—	432.954	14.046	—	447.000
	Total value in lakhs (Rs.)	—	2090.34	659.446	91.80	2841.60

Source : Technology support Centre for Kewda Industry, Berhampur (Ganjam) Orissa.

MARKETING OF KEWDA PRODUCTS

The existing marketing channels of the Keora industry contains various steps such as Producers of raw materials (farmers) agents for procurement of flowers (pluckers/ Commission Agents), distillers and traders who are operating domestic and international market. A five years ago, the traders used to pay the advance money to the agents for regular supply of the flower at pre-determined price. But, now, No such payments were made. Generally as per cost of production and demands, the price of Keora flowers are fixed by the Keora flowers associations. The commission agents or Middlemen are used to collect flowers through pluckers and supply the flowers to the micro-entrepreneurs (distilleries) to purify it by process of hydro-distillation using some chemicals and produce semi-finished products. Later these semi-finished products used to supply to the perfume and ghutaka traders at big cities like Lucknow, Mumbai and other North-east areas. In this process the distilleries get the margin profit of 15 to 20 percent. The cost of production and profit varies from season to season.

GOVERNMENT SUPPORT TO KEWDA INDUSTRY

In the year 1991, after the economic liberalization in the country, the growth of Kewda Industry was encouraged and in the year 1993, NABARD has identified Kewda industry as one of the important industrial growth sector in Ganjam district and has undertaken a pilot project named District Rural Industrialization Programme (DRIP). The Government of India, under the Mission for Millennium, 2000 of Small Industries Development Organization under the Ministry of Small Scale Industries, with the assistance of the state Government, set-up a Technology Support Service Centres for Kewda cluster at Berhampur (Orissa). The Quality Assurance Laboratory has been made functional and the Extension officer has been conducting Training/Awareness Programmes to the farmers and distilleries every month. The aim and object of service centre is to serve the farmers and the entrepreneur and the industry by way of providing, assistance to the farmers in adopting agronomical practices for Kewda essential oil bearing plants of higher yields and Technical support on post-harvest technology, storage, packaging, sampling and marketing and Provision for testing and Quality Control of essential oils, aromatic chemicals, raw-materials/products, etc. and consultancy services. A diagnostic study report has been prepared and plan of action has been earmarked for change of technology, increasing marketing network and formation of S.P.V. by the stakeholders. In the year 2006, recognizing the importance of Kewda activity, The Government of India, identified Kewda industry as Micro small and medium enterprises. As per PLP (potential linked credit plan) Report of 2008-09 of NABARD, The Kewda plantation carrying in about 10,000 hectare area under three blocks of Ganjam District, viz. Chatrapur, Rangeilunda and Chikiti and about 100 processing plants existed. And, there are 10 units registered with DIC with Credit flow given to the activity of Rs. 150 lakh which generated employment to 60 persons. About 90 Kewda processing units (distilleries) operating in the district are yet to register themselves with DIC. Bank credit loan of Rs. 30,000 each given to 3 units for Kewda oil extraction activity.

PROBLEMS OF THE KEWDA INDUSTRY

(1) In Ganjam district, The Kewda activity is an unorganized sector as majority of the distilleries are not registered either with District Industry Centre or any other Government organization.

(2) The farmers are getting income from the Kewda flowers without any investment and in addition to their original paddy cultivation activity. So, they never concentrate on Kewda production though it is a cash crop production.

(3) The Kewda plantation is a time consuming activity. The shrub starts flowering after 4-5 years. And, 5 year old plant bears about 4 flowers in a year and after 20 years, the fully matured flowers starts production 30 to 40 spadices in a year.

(4) In Ganjam district, The existing distillation units are adopting age old Deg-Bhabka technology, which is time consuming. There is no scientific and modern technology to improve the quick and quality of the distillation process.

(5) The Kewda oil is used as main constituent of Ghutaka and tobacco products which has been banned by the Government. Due to raise in the cost of Kewda oil, The ghutaka manufacturing companies though backdoor method using synthetic adulteration in their products, which is cheaper than Kewda oil and the Indian perfumeries are adopting

alternative material of western means by synthesizing various aromatic chemicals and were composing Champa, Bakul, Hina, etc. in their perfume products.

(6) Five years ago, the commission agents of the traders of tobacco, ghutaka, perfume manufacturers use to give advance money which financially helps the distilleries but now no body giving advance to these distilleries.

(7) Due to changing of market scenario, there is no certainty or guaranteed business to Kewda industry and no assured income to the people whose main livelihood is Kewda business activity.

(8) There is no unity among the farmers, distilleries and commission agents. The rate of flowers fixed arbitrarily by the local samities for their benefit and no common platform for the farmers, pluckers, commission agents and distilleries to put forth their grievances.

(9) In the recent years, Though, The Kewda cultivation propagated as cash crop because of number of benefits derived from it. Due to illiteracy and ignorance the farmers are not able to understand the importance of the Kewda crop and not interested to develop the crop production.

(10) There is no modern means of transportation for transport of Kewda flowers to long distances without losing their volatile aromatic constituents; hence the perfume traders had to go the raw materials.

(11) There is no direct link between distilleries and traders. The entire business has been monopolized by a small group of commission agents of the traders from northern India, who get the business to the local entrepreneurs.

MEASURES TO IMPROVE KEWDA INDUSTRY

State or Central Government should address the aforesaid problems and take sufficient measures to promote, develop and enhance the competitiveness of Kewda Industry of Ganjam district. More over, the Central Government has banned the manufacturing of consumption of Jarda, Ghutka and tobocco products, where the Kewda oil has been used. The State or Central Government should conduct regular awareness programmes relating to medicinal uses of Kewda oil and plants for cure of difference diseases, so it will increase the demand for Kewda oil. There should be a State level Board constituted by the State Government for supervising and development of kewda industry.

I. In Ganjam District, Kewda industry is age old activity and the existing units/distilleries are adopting traditionally old methods and equipment, where much attention is to be given on technological progress and modernisation of distilleries.

II. It is necessary to conduct research studies on kewda plant. Institutional finance and other facilities should be provided to the distilleries in terms of MSME act of 2006.

III. The Local entrepreneurs in Ganjam district of Odisha should be provided export license for marketing their own products directly to abroad.

IV. The distilleries are in need of working capital and Banks should come forward and provide credit loans to the distilleries and financial facilities to the farmers for commercial plantation of Kewda in coastal areas of Ganjam district.

V. The State Horticulture department should take an active role in extending technical support to the distilleries for higher yield of Kewda essence and extending support to farmers for commercial cultivation of Kewda and to increase its production and to improve the quality of Kewda perfume.

VI. A State level Kewda development board for the supervision and development of Kewda industry should be constituted by the Government of Orissa.

VII. The Government of India or State Government should create awareness to the general public of medicinally uses Kewda oil and leaves for treatment of different diseases, so that the demand for Kewda flower or Kewda product increases.

References

Annual Credit Plan—2008-09, Andhra Bank, Ganjam District.

Kewda Perfume Industry in India, by P.K. Dutta, H.O. Saxena and M. Brahmam.

Potential Linked Credit Plan—2008-09, NABARD, Ganjam District.

Information supplied by Shri V.V. Rama Rao, Extension Officer, Technology support centre for Kewda Industry, Berhampur, Ganjam District, Orissa.

Sahu, P. (1995), Perspectives of Keora Industries in Ganjam District (Odisha), *Odisha Economic Journal*, Vol. XXVII, Nos. 1 and 2.

Rao, M.S. (1998), A Study of Perfume Industry with special reference to Kewda in Ganjam District of Odisha", A study report of NABARD, Regional Office, Bhubaneswar.

Reddy, P.A. (2001), "Status of Keora industry in Ganjam".

CHAPTER

30

Performance of Handloom Co-operative Societies under Non-Farm Sector in Prakasam District—A Review

P. Purna Chandra Rao

INTRODUCTION

Indian hand woven fabrics have been known since time immemorial. Poets of the Mughal durbar likened our muslins to baft hawa (woven air), abe rawan (running water) and shabnam (morning dew). A tale runs that Emperor Aurangzeb had a fit of rage when he one day saw his daughter princess Zeb-un-Nissa clad in almost nothing. On being severely rebuked, the princess explained that she had not one but seven jamahs (dresses) on her body. Such was the fineness of the hand woven fabrics.

The handloom industry produces a staggering variety of clothes and handloom products, with each State having its own specialities, due to the different ethnic traditions as well as craftsmanship. Banaras and Kanchivaram Saries, Kashmiri Shawls and Moradabadi Carpets all are unique in themselves. Given the rich cultural heritage and uniqueness of Indian handlooms across various States, it could be further upgraded into a 'collector's item' through well planned promotional efforts by Handloom Export Promotion Council (HEPC). India also produces a range of home furnishings, household linen, curtain tapestry and yardage of interesting textures and varying thickness, which have been devised by using blended yarn. Muslims were forbidden the use of pure silk, and the half cotton half silk, fabrics known as mashru and himru were a response to this taboo. Seeing the wide and exciting range of handloom it is not surprising that the rich and beautiful products of the weavers of India have been called "exquisite poetry in colorful fabrics."

Gandhian Approach: Gandhiji had realized the dent made in the rural society and its economy by the machine economy, when he advocated Khadi and Village Industries and self-sufficiency of the village. He worked for the removal of the stigma of unsociability and low status of the artisan so that the artisan could get his due place and share in the rural society, its economy and their development. These and other ideas culminated in the "Sarvodayas" concepts had facilitated the development of rural economy. (Sankaran, A., 2007).

HANDLOOMS THE LARGEST COTTAGE INDUSTRY IN INDIA

Handlooms are an important craft product and comprise the largest cottage industry of the country. Millions of looms across the country are engaged in weaving cotton, silk and other natural fibers. There is hardly a village where weavers do not exist, each weaving unit the traditional beauty of India's own precious heritage. The sector has supported traditional skilled weavers from generation to generation. Handloom weaving is a full-time family profession, involving all members of the family. It provides both self-employment and casual employment. There is equal participation of men and women in this industry.

Handloom industry in this country comprises of three sections: 1. Cotton weaving, 2. Wool weaving, and 3. Silk weaving. About 56.62 meters of cloth is available per head in India. On an average 1.12 meters of handloom cloth is wearing by the common man, its value is Rs. 136 only but 17.92 meters of power loom cloth is wearing by the Indians, its value is Rs. 1135. On an average, Indian uses 21.4 meters of cloth per year of which urban people uses 11.91 meters whereas rural people uses 6.92 meters of cloth only. (Suresh, P., 2010). Mr. Dayanidhi Maran, The union Textile Minister announced first time that such a large amount of subsidy is being released in a single tranche, i.e. 2,546 crs. in three days under Technology Upgradation Fund Scheme (TUFS) to underpin its modernization efforts for meeting marketing challenges and enabling it to stay competitive in quality and price. Rs. 3,150 crs. have been provided in the budget for releasing to the industry under the scheme. In all, 12,514 applications have been cleared for this benefit, the maximum amount of subsidy goes to textile units in Tamil Nadu. In all, 3,286 applications were from the State involving a subsidy of Rs. 643 crs. followed by Maharastra, a subsidy of Rs. 637 crs., Punjab Rs. 352 crs., Gujarat Rs. 275 crs., from Andhra Pradesh, for 287 applications a subsidy of Rs. 132 crs. benefited. (*Business Line*, Aug. 7th 2009).

ATLAS—A GUIDE TO THE WEAVERS' COMMUNITY

To help unorganized handloom sector for survival in the tough time Handloom Export Promotion Council (HEPC) has come out with the idea of bringing out Handloom Atlas of India and circulate it through various external affairs ministry missions. The atlas is in four main international languages—English, German, French and Japanese along with an interactive CD. The atlas will help people to source the right kind of products from the right manufacturers from various clusters in India. The atlas covers topics like production of yarn, number of spinning mills, weavers service centers and list of textile research associations. One can also get details of handloom clusters, exports of Indian handloom items and number of spindles across the country. It also provides information about each state's products and specialization.

IMPORTANCE OF HANDLOOM SECTOR

The handloom sector plays an important role in the Indian economy since it is very large in scale and provides the largest employment opportunity, next to the agriculture sector. "Handloom industry occupies a place of importance in our country's economy chiefly by virtue of its employment potential, production and export orientation." (Rao, C.S., 1973). The major handloom weaving states in India are West Bengal, Tamil Nadu, Uttar Pradesh, Andhra Pradesh, Assam and Manipur. The world's installed capacity of handloom is 4.60 million. Out of this 85 per cent is installed in India. In India, there are more than 3.8 million handlooms in operation, with the livelihood of about 12.5 million people dependent on handloom weaving. It is one of the largest economic activities providing direct employment to over 65 lakh persons engaged in weaving and allied activities. This makes up 23 per cent of the total textile production in the country. Overall, this sector plays a vital role in local economies, especially in rural areas. The handloom sector holds an important position in India's economy mainly because of three reasons: 1. It provides great employment opportunities, 2. It earns foreign exchange, 3. It is responsible for one-fourth of the total textile production in the country.

ALL INDIA HANDLOOM BOARD

The board was set-up in 1952 for dealing with the various problems of development of Handloom industries. It renders assistance in organizing cooperatives in handloom, textile industries, establishing sales emporia and in providing grants and loans. It has set-up a central marketing organization which looks after the sales of the cloth in the home market as well as in the foreign markets. The all India Institute of handloom technology established at Salem started functioning from Sept. 1960. It conducts a diploma in handloom technology. Technical advice is given for design, quality improvement standardization and equipment modification.

ROLE OF COOPERATIVE SOCIETIES

Co-operative is a value-based socio-economic system that has emerged in the world to achieve social cohesion aiming at providing organizational strength to the weak to withstand the pressures and pulls of exploitative tendencies of market economy. In this context, cooperation as a system, on the one hand, inculcates in the people cooperative values and attitudes that strengthen the social base of the community and on the other, undertakes various activities for socio-economic development of the people belonging to those sections which have remained depressed, both economically and socially.

For the cooperative movement in India which has a vast rural network, the rural market is the main target for cooperative business. In order to maintain increasing market share in rural areas the cooperative sector has to explore many new fields and maintain existing status of the business smoothly in the rural market. In course of the increasing process of economic liberalization there is a probability of more competition to be faced by rural cooperatives. The market orientation should be pursued for the profitability and sustainability of the cooperatives without affecting the principles, values, ethos and ethics of cooperation in India. (Sanjib Kumar Hota, 2000). Cooperatives are based on the values of self-help, democracy, equality, equity and solidarity. Cooperative members believe in the ethical values of honesty, openness, social responsibility and caring for others.

In India, the second plan envisaged the establishment of the socialistic pattern of society. The building up of a cooperative sector as a part of the scheme for planned development is thus, one of the central aim of national policy wherein cooperatives are considered as powerful tools for bringing about rural transformation. According to the father of white flood revolution and pioneer of cooperative movement in India, Verghese Kurien, "what is harmful to the cooperatives is the spoon-feeding, we should note that the long-term assistance means long-term illness". This is quite a conspicuous point and it either directly or indirectly suggests the self-sufficiency concept (Thangirala, H.S.K., 2006).

Societies in Andhra Pradesh are organized as a two-tier structure with the primary societies in the village level called Primary Handloom Weavers' Cooperative Societies (PHWCS) and the APEX Society at the State Level called A.P. State Handloom Weavers' Cooperative Society Ltd., (APSHWCS), popularly called *APCO*. Since the handloom weavers were facing problems for getting inputs as well as for marketing their finished products, the creation of an APEX Society was thought of to address these twin problems by making timely availability of inputs and by providing effective marketing strategy. Further, the Government is aware that protection to the sector is required to be continued because of the pious obligation to protect the heritage of the State and support is required to be continued to offset the built-in cost handicap of the handlooms, the focus of all its policy has been to provide a level playing field so as to enable the weaver to stand up on his own and face the market in a level playing field.

APCO: The Andhra Pradesh State Handloom Weavers' Cooperative Society Ltd., (APCO) is the APEX Handloom Weavers' Cooperative Society of Andhra Pradesh. The Society was registered in the year 1976 with registered No. T.P.W.44 under the Andhra Pradesh Cooperative Societies Act 7 of 1964 with the following *main objectives, inter-alia,* among others are—

- To purchase raw materials and appliances, tools and machinery including spares as may be required for the industry and to sell the same to the member-societies.
- To purchase or receive for sale, the finished products of the member-societies by opening Sales Units, Go downs, Exhibitions, etc., within the State.
- To establish and run sales units within and outside the area of operation for Exhibition and sale of products of member-societies.
- To hold, own, establish and hire processing units to undertake and provide processing including dyeing, mercerizing, printing and furnishing, etc. For all

types of yarn and cloth, to the member societies and to other institutions and individuals.

- To advise and render assistance to member-societies in preparation of production programmes, design development, product diversification and such qualitative improvement as may be necessary in the products produced by the member-societies to be in tune with the consumer tastes and market demand.
- To provide training in improved methods of weaving and latest techniques for the weavers and also for the training of its own employees in sales techniques, accountancy and business administration, etc.

The National Cooperative Union of India (NCUI) estimates that primary handloom cooperatives are responsible for 58% of total annual handloom production.

MAJOR REASONS FOR THE POOR PERFORMANCE OF SOCIETIES

However, the failure of cooperatives to adapt to market shifts has perhaps been the biggest reason for declining handloom production and the loss of handloom-based livelihoods. The following are the *main reasons for the poor performance of cooperatives:*

- *Poor policy outlook and program design:* Cooperatives continue to be shielded from markets by a web of government programs and subsidies, and consequently, their ability to respond to market realities is low.
- *Lack of transperarency: Current* audited financial statements of most Weavers' Co-operative Societies (WCSs) are not available for public scrutiny. Financial information that is available is not in a standardized format. A review of financial statements of selected cooperatives show that, in many cases essential records of product inventory are not being maintained on a regular basis.
- *Poor Marketing:* Marketing refers not only to the

process of getting finished products to consumers, but also to range of activities designed to enhance customer-experience and perception of the product. This includes activities such as developing new designs and blends, improving product quality and ensuring that the product is available to customers through an appropriate outlet. There is minimal investment made at the WCS level on these activities, and they are addressed only as part of government schemes that promote a limited set of activities such as exhibitions and trainings.

- *Lack of timely access to credit:* WCSs face difficulty in procuring working capital credit on a timely manner. Credit delivery through primary handloom cooperatives has become defunct in most areas.
- *Master weavers' dominance*: The decline of cooperatives has been paralleled by the rise of master-weavers as intermediaries, coordinating production of weavers in their area and forgoing marketing links for the supply of woven cloth. They capture and Swallow majority of the profits of handloom production without making sufficient capacity-building investments into the sector.
- *Entrance of Design Firms*: Post-liberalization, anew category of intermediary firms have grown rapidly. These firms—Fabindia and Tulsi being two of the most prominent examples—have made capital investments in new product development, design innovation, product packing, branding and retail outlets, etc.

Varieties of Dyes Used

The process of resist dyeing, tie-dyeing and yarns tie-dyed to a pattern before weaving were the basic techniques of indigenous dyeing of village cloth. Shellac was used for reds, iron shavings and vinegar for blacks, turmeric for yellow and pomegranate rinds for green, extracted before the artificial synthesis of indigo and alizarin as dye stuffs, blues and reds were traditionally from the plants indigofera, anil and rubia tintorum (madder-root). These were the main sources for traditional Indian dyes. Even today, the Kalmkari cloth of Andhra Pradesh is printed with local vegetable dyes. The colors being shades of ochre, deep blue and a soft rose derived from local earths, indigo and madder roots printing: Andhra Pradesh has made a significant contribution to the history of hand-printed textiles in India. Printing is native to the land, its pigments being obtained from the flowers, leaves and barks of local trees and its chemicals obtained from clay, dung and river sands. A new technique has been developed in the northern sectors where warp threads are lined, measured and tied to the loom and then printed. The warp-printed material is a specialty of Haryana and Uttar Pradesh. The ideal seasons for block printing are the dry months. Excellence is achieved only if the block is freshly and perfectly chiseled. The designs are produced by artists and the designing is kept within the discipline imposed, the type of yarn, the dyes used and the weaving techniques, by the nakshabandhas (graph-paper designers).

HANDLOOM INDUSTRY IN A.P.

Andhra Pradesh has an old tradition of making hand woven clothes. The State is very much famous for its handlooms and textiles; its silk sarees are among the best produced in the country. Kalamkari, a special art of dyeing clothes with vegetable oils still exist in the state. These are called handlooms. The style of fabric and design vary from place to place. Sarees are the most exquisite among all fabrics. The most popular handlooms in the state are found at these following places: Chirala Textiles, Chirala Handlooms, Dharmavaram Sarees, Dharmavaram Handlooms, Eluru Carpets, Eluru Handlooms, Gadwal Sarees, Gadwal Handlooms, Ikat Weaving Handlooms, Kalamkari Fabrics, Machilipatnam Handlooms, Mangalagiri Sarees, Mangalagiri Handlooms, Uppada Sarees Handlooms, and Venkatagiri.

The once thriving handloom industry in Andhra Pradesh is going through tough and troubled times in the last few years. To mitigate their troubles, Government of Andhra Pradesh announced a series of measures in the wake of weavers suicides in Sircilla in Karimnagar district. Among the series of relief packages, a loan up to Rs. 50,000 at a nominal interest of 3 percent per annum. He also announced some other welfare policies including providing grains at subsidized rates, employment to 5,000 youth of the region.

HANDLOOM MARKETING (MARK)

The Office of the Development Commissioner for Handlooms, Government of India has been implementing a number of developmental schemes and programmes to protect the interest and welfare of the weavers. It is proposed to introduce the "Handloom Mark" which will provide a collective identity to the handloom products and can be used not only for popularizing the hand woven products but can also serve as a guarantee for the buyer that the product being purchased is genuinely hand woven. Besides, this would provide a distinctive name in identifying the product or the manufacturer. The Handloom Mark would, therefore, be a hallmark of powerful creative work that defines the product with clarity, distinguishing it from competition and connecting it with customers.

The creation of handloom mark was entrusted to the National Institute of Design, Ahmedabad. The form of the logo has been derived from the interlocking of the warp and the weft. These threads stand for the collaborative institutes giving their inputs and the weavers giving their skills. The interaction between them is leading to a close network. The warp and weft have been molded to form a three dimensional cube. The mark is in two forms. One for Domestic use: the word Handloom is written beneath the logo and the other for International marketing: same logo with the word Hand woven in India written beneath it. Application has been made for the registration of Certification Trade Mark under the Trade Mark Act, 1999 as well as for Copyright under the Copyright Act, 1957. Each label is coded on its backside for easy identification/classification. e.g. DF followed by coded number for fabric for domestic sale, DM followed by coded number, for made-ups and garments for domestic sale: EF followed by coded number for fabric for export and EM followed by coded number for made-ups and garments for export.

OBJECTIVES OF HANDLOOM MARKETING

- Promote handloom products in Domestic as well as International Market.
- Provide assurance to the consumers about the genuineness of the product origin.
- Improve international marketing linkages to the handloom weavers.
- Strengthen supply chain for Handloom products.
- Improve price realization of the Handloom products in Domestic as well as in Foreign.
- Improve the earnings of the handloom weaver community.
- Facilitate uninterrupted workflow throughout the year to the handloom weavers.

CONSTRAINTS OF THE INDUSTRY

However, in India majority of handloom weavers are facing manifold problems like poverty, low standard of life, hunger, poor health, malnutrition, illiteracy, diseases, poor housing and poor communication, handloom and general infrastructure facilities. Besides, growing indebtedness, increasing risk, price uncertainty, and low incomes are the other problems. All these trigger to suicides of weavers in the community. (Venkateswara Rao, N., 2010). Most of the weavers are leaving their profession and have diverted alternative works. The scarcity of yarn is the cause and consequence for crisis in industry. He suggested that the yarn should be supplied to the weavers at subsidy rates. (Kotaiah, P., 1992) The unorganized handloom sector suffers from basically three main problems such as—availability of timely and adequate credit; technology up-gradation; and marketing support. Apart from there is a high pressure of dumping from China and insensitive government policies.

A CASE STUDY

Prakasam District Profile in Brief

Prakasam district came into existence in 1972 in the state

of Andhra Pradesh. It occupies an area of 17626 sq. kms., with a density of 174 persons per sq. kms. whereas it is 277 for the state. This district has 1023 kms. of coastline spread over 10 mandals. There are 1093 villages existing in the district, of which 1058 are revenue villages and the remaining 35 are forest villages. Out of 1093 villages, 1002 villages are inhabited while the balance 91 un-inhabited villages. About 1043 are panchayats and 4 municipalities. The total population is 30.59 lakhs. Total Geographical Area: 43,55,720 Acres.

There are number of handloom Production centers in Prakasam District. There are about 20,000 weavers families and 25,000 handlooms engaged in the production of Jacquard sarees, semi Gadwal sarees, Lungies, Dress Material, Towels, Bed-Sheets, Kerchiefs, etc., produced in the counts of 40s, 80s, 100s, 2/120s Gas Cotton with silk and Zari, etc. All the weavers in this District are mostly traditional belonging to the weavers community. 16,802 Photo identity cards were issued to the weavers, of which 7,491 were issued to the Co-operative fold weavers and 9,311 issued to the outside Co-operative fold weavers. There are 47 Weavers' Co-operative Societies (45 cotton+2 Silk) in Prakasam District of which 36 (35 cotton+1 Silk) are working and 11 (10 cotton+1 Silk) Weavers Co-operative Societies are in Dormant stage. Out of the 36 working societies, 20 active societies have been selected for the study.

Some of the Welfare and Developmental schemes are being implemented to the Handloom weavers such as Mahatma Gandhi Bunkar Bhima Yojana, Housing under Indiramma, Cash credit, Credit cards, Health Insurance Scheme, Marketing Incentive under IHDS, Pavlavaddi, and Yarn subsidy, etc., Chirala is a prominent coastal town in the district with an average elevation of just 4 meters above mean sea level. Earlier known as "*Small Bombay*" because of the huge varieties of sarees that are produced there, Chirala was originally named Kshirapuri, which is translated to "Here the sea looks as white as milk". It also became the hub of the handloom industry, providing employment to at least 30,000 craftsmen. In 1959, a cooperative spinning mill was also started. Today, Chirala is a major commercial center between Ongole and Tenali, and lies on the main railway between Kolkata and Chennai.

Methodology

The study is descriptive and analytical in nature. It is mainly based on the secondary data.

Objectives

- To study the role of cooperatives in promoting the handloom industry.
- To study the performance of select WCSs in the Prakasam district.
- To study the significance of Handlooms in general

Scope

- The study is confined to 20 selective WCSs in the Prakasam district out of 36 working WCSs in the District.
- The study covers only handlooms not powerlooms.
- The study period is between 2004-05 to 2008-09.

The statement is showing the incentive sales of selective cooperative societies in the Prakasam district during 2004-05 to 2008-09. In case of the Vetapalem HWCS, 2008-09 is observed as the year of highest sales in tune with Rs. 12,41,416 (26.41%), whereas it is lowest in 2004-05 in tune with Rs. 6,52,604 (13.88%) during the study period. In case of the Sivaji Abyudaya HWCS, the sales trends reveal that during 2004-05, it is highest in tune with Rs. 12,64,244 (47.09%) whereas it is lowest in 2007-08 in tune with Rs. 97,560 (3.64%). In case of Abyudaya Ramnagar HWCS sales, during 2007-08, it is highest in tune with Rs. 5,09,531 (41.75%) whereas it is lowest in 2004-05 in tune with Rs. 1,02,349 (8.38%). In case of the Sitharama Raju HWCS, the highest and lowest sales percentages are found during 2008-09 and 2006-07 in tune with Rs. 2,26,150 (52.04%) and 4,30,650 (10.06%) respectively. In case of the Perala HWCS, the highest sales are observed during 2004-05 in tune with Rs. 3,45,932 (30.55%) whereas it is lowest in 2007-08 in tune with Rs. 1,07,029 (9.45%).

In case of the Chirala HWCS, the highest sales are recorded during 2005-06 in tune with Rs. 4,34,950 (25.70%)

Statement of Marketing Incentive Sales of Selective Active Co-operative Societies in the Prakasam District under IHDS

Sl. No.	Name of the Society	2004-05	2005-06	2006-07	2007-08	2008-09	Total
(1)	(2)	(3)	(4)	(5)	(6)	(7)	(8)
1.	The Vetapalem HWCS Ltd., Vetapalem	6,52,604 (13.88)	10,56,356 (22.48)	9,98,118 (21.24)	7,51,307 (15.99)	12,41,416 (26.41)	4699801.00 (100.00)
2.	The Sivaji Abyudaya HWCS Ltd., Pandillapalli	5,96,951 (47.09)	12,64,244 (22.23)	4,28,134 (15.95)	97,560 (3.64)	2,98,010 (11.09)	2684899 (100.00)
3.	The Abyudaya HWCS Ltd., Ramnagar	1,02,349 (8.38)	2,63,225 (21.56)	1,98,945 (16.30)	5,09,531 (41.75)	1,46,660 (12.01)	1220710 (100.00)
4.	The Sitharama Raju HWCS Ltd., Hastinapuram	5,44,383 (12.73)	4,73,184 (11.06)	4,30,650 (10.06)	6,03,647 (14.11)	22,26,150 (52.04)	4278014 (100.00)
5.	The Perala HWCs Ltd., Perala	3,45,932 (30.55)	2,43,098 (21.47)	2,28,374 (20.16)	1,07,029 (9.45)	2,08,045 (18.37)	1132478 (100.00)
6.	The Chirala HWCS Ltd., Chirala	3,93,379 (23.24)	4,34,950 (25.70)	2,78,954 (16.48)	2,60,257 (15.38)	3,24,989 (19.20)	1692529 (100.00)
7.	Sri Seetharama Zari HWCS Ltd., Perala	7,64,587 (12.95)	9,31,848 (15.79)	5,13,997 (8.71)	8,69,466 (14.73)	28,22,140 (47.82)	5902038 (100.00)
8.	Sri Kasivisweswara HWCS Ltd., Adivarapupeta	3,72,326 (10.25)	9,94,115 (27.38)	15,39,076 (42.39)	4,44,323 (12.24)	2,81,245 (7.74)	3631085 (100.00)
9.	Sri Srinivasa HWCS Ltd., Ankalammapeta	4,54,976 (13.20)	10,26,375 (29.78)	10,38,436 (30.13)	3,87,540 (11.24)	5,39,210 (15.65)	3446537 (100.00)

(*Contd.*)

(Contd.)

(1)	(2)	(3)	(4)	(5)	(6)	(7)	(8)
10.	Sri Ahobilam Laxmi Narasimha HWCS Ltd., Papaipalli	7,06,450 (13.98)	14,41,490 (28.53)	11,19,395 (22.16)	10,05,385 (19.90)	7,79,789 (15.43)	5052509 (100.00)
11.	Sri Srinivasa HWCS Ltd., Ammanabrolu	4,55,559 (20.75)	2,60,047 (11.85)	9,12,779 (41.57)	2,36,040 (10.75)	3,31,155 (15.08)	2195580 (100.00)
12.	Sri Venkateswara HWCS, Epurupalem	2,67,690 (11.14)	2,03,431 (8.46)	2,96,468 (12.34)	5,53,577 (23.04)	10,81,892 (45.02)	2403058
13.	Chimakurthi HWCS, Chimakurthi	1,66,630 (21.31)	3,61,215 (46.19)	2,54,154 (32.50)	NA	NA	781999 (100.00)
14.	Sri Rama Swamy HWCS, Ulchi	1,24,842 (7.62)	3,28,586 (20.04)	3,91,780 (23.90)	3,12,085 (19.04)	4,81,960 (29.40)	1639253 (100.00)
15.	Sri Veeranjaneya HWCS, Valaparla	1,54,337 (5.15)	7,68,581 (25.65)	11,22,745 (37.46)	505,563 (16.87)	4,45,780 (14.87)	2997006 (100.00)
16.	The Sivajyothi HWCS, Bapaiahnagar	2,13,825 (8.92)	7,07,460 (29.50)	4,46,400 (18.62)	4,92,000 (20.52)	5,38,000 (22.44)	2397685 (100.00)
17.	Sri Jagadamba Silk HWCS, Desaipeta	7,25,496 (10.84)	13,48,242 (20.15)	16,25,162 (24.28)	17,16,436 (25.65)	12,77,119 (19.08)	6692455 (100.00)
18.	The Thatiparthivaripallem HWCS, Thatiparthivaripallem	NA	1,28,793 (4.13)	1,71,916 (5.50)	4,84,150 (15.51)	23,37,255 (74.86)	3122114 (100.00)
19.	The Subhodaya HWCS, Amodagiripatnam	NA	NA	2,21.095 (33.53)	2,02,475 (30.70)	2,35,891 (35.77)	659461 (100.00)
20.	Sri Raja Rajeswari HWCS Ltd., Somavaripeta	NA	NA	1,08,045 (20.95)	1,66,810 (32.34)	2,40,940 (46.71)	515795 (100.00)

Source : A.D. Handlooms and Textiles, Prakasam, Ongole, NA = Not Available.

whereas it is lowest in 2007-08 in tune with Rs. 2,60,257 (15.38%). In case of Sri Seetharama Zari HWCS, 2008-09 and 2006-07 years are found highest and lowest sales in tune with Rs. 28,22,140 (47.82%) and 5,13,997 (8.71%) respectively. In case of Sri Kasivisweswara HWCS, 2006-07 and 2008-09 are found highest and lowest sales years in tune with Rs. 15,39,076 (42.39%) and 2,81,245 (7.74%) respectively. In case of Sri Srinivasa HWCS, Ankalammapeta, in tune with Rs. 10,38,436 (30.13%) is the highest sales found during 2006-07 and in tune with Rs. 3,87,540 (11.24%) is the lowest sales in the 2007-08 fiscal years. In case of Sri Ahobilam Laxmi Narasimha HWCS, during 2005-06 and 2004-05 are found the highest and lowest years in terms of sales in tune with Rs. 14,41,490 (28.53%) and in tune with Rs. 7,06,450 (13.98%) respectively. In case of Sri Srinivasa HWCS, Ammanabrolu, in tune with Rs. 9,12,779 (41.57%) and 2,36,040 (10.75%) are the highest and lowest sales found during 2006-07 and 2007-08 respectively. In case of Sri Venkateswara HWCS, 2008-09 is the record sales year in tune with Rs. 10,81,892 (45.02%), whereas it is lowest in 2005-06 in tune with Rs. 2,03,431 (8.46%) in the study period.

In case of Chimakurthi HWCS, as per the data, 2005-06 and 2004-05 are the highest and lowest sales years in tune with Rs. 3,61,215 (46.19%) and 1,66,630 (21.31%) respectively, among the available statistics during the study period. Under Sri Rama Swamy HWCS, during 2008-09 and 2004-05, the maximum and minimum sales are noticed in tune with Rs. 4,81,960 (29.40%) and 1,24,842 (7.62%) respectively. In case of Sri Veeranjaneya HWCS, 2006-07 and 2004-05 are the maximum and minimum sales noted in tune with Rs. 11,22,745 (37.46%) and 1,54,337 (5.15%) respectively. Under the Sivajyothi HWCS, 2005-06 and 2004-05 are the highest and lowest noticed sales years in tune with Rs. 7,07,460 (29.50%) and 2,13,825 (8.92%) respectively.

In case of Sri Jagadamba Silk HWCS, 2007-08 is recorded as the highest sales year followed by 2006-07 in tune with Rs. 17,16,436 (25.65%) and 16,25,162 (24.28%) respectively. The lowest sales are observed during 2004-05 in tune with Rs. 7,25,496 (10.84%). Under the Thatiparthivaripallem HWCS, Significantly the highest sales are noticed during 2008-09 in tune with Rs. 23,37,255 (74.86%), whereas the lowest sales are found in 2005-06 in tune with Rs. 1,28,793 (4.13%) followed by

2006-07 in tune with Rs. 1,71,916 (5.50%) .In case of the Subhodaya HWCS, among the available figures, it is observed that the sales trends are more or less similar in tune with Rs. 2,21,095 (33.53%), 2,02,475 (30.70%), 2,35,891(35.77%) during the consecutive years 2006-07, 2007-08 and 2008-09 respectively. In case of Sri Raja Rajeswari HWCS, among the available statistics, it is found that in tune with Rs. 2,40,940 (46.71%) is the highest sales during 2008-09, in contrast to the lowest sales that in tune with Rs. 1,08,045 (20.95%) recorded in 2006-07.

The above statement clearly shows about the sales trends of 20 active cooperative societies in the Prakasam district during 2004-05 to 2008-09. The calculated percentages of each society reflects that there are remarkable fluctuations in the sales turnover during the years. Due to the host of constrains, as mentioned below, encountered by the societies from time to time.

Some of the Measures adopted by HEPC for Promoting the Industry are as follows:

- Though, the industry in the past may have gone into declines from time to time, today that the industry is vibrant and confident of its survival due to its intrinsic strengths and opportunities.
- Realizing the need for a proactive approach to augment the export potential of the industry and keep it in a state of preparedness to meet the global challenges in the post-WTO world, the HEPC has formulated a number of strategies for increasing exports.
- The HEPC has identified five handloom production clusters to be developed as Handloom Export Zones. A scheme has been formulated and submitted to the Ministry of Textiles for sanction. The scheme aims at upgradation of infrastructure support and skill for handloom weavers, besides strengthening the production base by modernizing the looms.
- A school of weaving on the pattern of Gurukulam will not only nurture the talents of the 'disciples', but will surely create a congenial atmosphere to make

them proud of their profession. Through this concept, there will be a continuous transfer of skills from the master craftsman to the trainees. This will augment the available base of skilled workers, besides widening the production base for intricate weaves and reviving the rare weaving methods.

- The HEPC is seized of the fact that design adds value and contributes to the salability of any textile product. Therefore, the HEPC proposes to upgrade the existing CAD design studio into a high-class textile design studio with world-class packages.
- The HEPC has embarked upon a project called 'Handloom Expo Mart' to give a competitive edge to the exporters. The establishment of an international market will be the right strategy to attract buyers who are increasingly looking up to India as a major sourcing hub for handloom textiles.
- The Textile Ministry is taking steps to brand handloom products to prevent duplication by the power loom industry. The Ministry has also entrusted the textile committee to develop a handloom mark for the industry in the lines of wool mark and silk mark.
- The Centre has, over the years taken steps to protect the handloom industry. The Handlooms (Reservation of Articles for Production) Act, 1985 was enacted to reserve certain items for the industry.

CONCLUSION

As a major source of rural non-farm livelihoods, handlooms can play a critical role in promoting employment and raising rural, regional incomes. The handloom industry, as a whole, is getting a face-lift, by the policy-makers at the Centre and in the States, the various councils and associations, and the weaker handloom weavers. Indeed, the industry has the unending strength to survive. Develop database on the handloom supplies and weavers that will help in supporting the weavers through the existing schemes which are being implemented by the Govt. of India and States. As a result of

Government intervention through financial assistance and implementation of various developmental and welfare schemes, the handloom sector, to some extent, has been able to tide over these disadvantages. Moreover, the government needs to play a more proactive role by creating a policy regime that encourages greater public and private partnership (PPP) into the sector and promotes competitiveness of handloom products in mainstream markets over the power loom products, instead of mere measures and meager increases in budget allocations. Particularly, innovations and drastic steps are to be needed for the survival and success of the Industry at this juncture.

References

Adarsh Kumar, Increasing Competitiveness, *Yojana*, July 2007, pp. 31 and 33.

Thangirala, H.S.K. (2006), "Professionalism in Cooperatives", *Kurushetra*, Feb. 2006, p. 14.

Sankaran, A. and Jagadish Kumar, A., Performance of Khadi and Village Industries in India, *Southern Economist*, May 1st 2007, p. 35.

Sivayya, K.V. and V.B.M. Das, Indian Industrial Economy, S. Chand and Co. Ltd., N. Delhi, 2004, p. 122.

Business Line, Textile Units get Rs. 2,546 Crs. Subsidy under TUFS, Aug. 7th 2009, p. 15.

Jeyakodi, K., Strategy for Handloom Sector: An Evaluation, *Southern Economist*, June 15th, 2007, p. 15 .

Suresh, P., Separating Thread Relation, Eenadu, May 18th 2010, p. 4.

Maheswari, T., Handlooms in Crisis, Feb. 2010.

Venkateswara Rao, N., "The Handloom Weavers of Prakasam Dist., *Kisan World*, July 10, pp. 34 and 35.

Sanjib Kumar Hota, "Cooperatives in Rural Economy", *Kurukshetra*, August, 2000, pp. 2-6.

Kotaiah, P., Loom Tragedy, *Frontline*, 3rd July 1992, p. 19.

Rao, C.S., Employment in Handloom Industry, *Kurukshetra*, Vol. 22, No. 1, Oct. 1973.

The Hindu, July 7th 2009, pp. 1 and 14.

A Report on Handlooms and Textiles, Prakasam, Ongole.

http://www.fibre2fashion.com

http://www.thehindubusinessline.com

http://www.india-crafts.com/textile/indian_handlooms.

www.bharatonline.com/andhra-pradesh

http://www.fibre2fashion.com/news/textiles-policy-news/newsdetails.aspx?news_id=64652

http://andhraonline.in/Profile/Culture/art_craft.asp

http://www.apcofabrics.com/aboutus.html

http://www.merinews.com/article/hepcs-efforts-to-rescue-handloom-sector-in-crisis/15751671.shtml

CHAPTER

31

Non-Farm Sector and Rural Development, with Special Reference to Poultry Farming in Namakkal District : An Analysis

A. SHANMUGASUNDARAM AND R. VENKATACHALAM

INTRODUCTION

India has predominantly rural economy. Despite large scale migration of people from rural to urban areas, 73% of Indian population still lives in about six lakh villages. Development of the Indian Economy depends largely upon the development of rural areas. Rural areas continue to be the major places of residence for majority (72.22 percent) of Indians. A large proportion of them are engaged directly in agriculture.

Though the focus of traditional agricultural extension system is on improving the crop production and productivity, the income of marginal and small farmers cannot be improved. Hence, it is necessary to improve the non-farm sectors in the rural areas.

Poultry Farming is one of the important non-farm sector provides an effective source of subsidiary income to the rural sector. Further, it provides quality diet in the form of animal proteins which can help to solve the problem of malnutrition in the country. Amongst farm animals poultry is one of the quickest and most efficient converters of plant products into food of high biological value and broiler meat is the most widely accepted meat in India and cheaper than mutton or goat meat. The requirement of small space, low capital investment, quick returns on investment and well distributed turnover throughout the year, make poultry farming remunerative in both rural and urban areas. Poultry rearing can also be adopted to suit a wide range of climatic conditions and can be combined with other farm activities

The discussion made so far is significant in understanding the relative importance of agriculture sector in the development context. It is a well-recognised fact that agriculture alone is not capable of meeting the challenges of unemployment and poverty reduction of a country. A meaningful development in a developing country like India, where the majority of people live in villages, the emphasis must be on the development of the rural masses.

The development strategy of India must include every sector of the economy. When the primary sector is agriculture, diversification of agriculture, encouraging and promoting the farm-based allied activities are indispensable not only for reducing the unemployment and poverty but also for the overall economic development of the country and improving the quality of life of the rural masses.

Though India's rural population declined from 82.71 percent in 1951 to 72.22 percent in 2001, still over two-third of the population depend on agricultural sector for their livelihood.

Rural development has emerged as a strategy designed to improve the economic and social life of specific group of people known as rural masses. It is much wider as compared to agricultural development and concern with overall development of rural areas.

STATEMENT OF THE PROBLEM

The major components that play key role in sustainable social and economic development, particularly in the rural areas, are education, health, sanitation, safe drinking water, credit institutions, transfer of technology in agriculture, diversification of agriculture and diversified non-farm activities in rural areas. Of these, allied agricultural activities are more relevant as more than two-third of population live in rural areas. Further, out of nearly 25 percent agricultural contribution to the GDP in India one-fourth is through livestock products. The value of livestock output was about Rs. 1.55 lakh crores in 2001-02.

The egg production in 1980 was 12 billion, increased to 23 billion in 1990 and further increased to 42 billion in 2003-04.

The poultry sector, with total value of output exceeding Rs. 15,000 crores and providing direct and indirect employment to over three million people provided around 1.9 million tonnes of chicken meat and about 42 billion eggs.

In poultry farming, Tamil Nadu stands in number two place in the country. In Tamil Nadu, Namakkal district is the backbone of poultry. The egg production in Namakkal district accounts for 49 percent. Inspite of all these noteworthy significance of poultry industry, the industry faces many a problem like continuous rise in cost of feeds, uncertain market situation, wide variation in the price of eggs, media involvement, etc.

The price variation ranges from 30 paise to Rs. 2 an egg in 2006. There has been growing interest among the academicians, researchers and policy-makers to have first hand information on poultry industry.

Thus, the present study is being pursued considering the above discussed issues on poultry farming with the following objectives.

OBJECTIVES

1. To examine the cost-return of poultry farming of different sizes.
2. To study the quantum of employment promotion and income generation of poultry farming of different farm size.
3. To evaluate the sectoral linkages and multiplier effects of poultry industry in study area.
4. To find out the poultry of poultry farming and provides suggestions.

HYPOTHESES

1. Poultry farming provides considerable employment potential in the rural area of Namakkal district.
2. There is no significant variation in the cost of production of different farm sizes.

SELECTION OF THE STUDY AREA

Namakkal district is very famous poultry farming in Tamil Nadu. About half of the egg production in Tamil Nadu produced from Namakkal district. Namakkal has a surname of "Egg City". Hence, Namakkal district is purposively selected for the present study.

SELECTION OF SAMPLE DESIGN

The Namakkal district has 15 panchayat unions. The distribution of poultry farms is given in the Table 1.

Of the 15 panchayat unions 10 unions are considered for the present study. Further, the broiler farms of 40 are excluded. There are 812 layer farms in the selected 10 unions of the study area.

These farms are stratified into four groups on the basis of number of birds and the feed manufacturing infrastructure. The Group A has the bird strength upto 20,000, the Group B from 20,001 to 50,000 birds, the Group C from 50,001 to one lakh and the Group D has above one lakh birds.

TABLE I
Panchayat Union-wise Poultry Farms in Namakkal District

Sl. No.	*Name of Panchayat Union*	*Total Farms*	*Broiler Farms*	*No. of Layer Farms*
1.	Namakkal	184	4	180
2.	Mohanur	185	—	185
3.	Puduchatram	114	5	109
4.	Sendamangalam	19	1	18
5.	Erumaipatti	10	—	10
6.	Paramathi	33	3	30
7.	Kabilarmalai	7	1	6
8.	Rasipuram	55	—	55
9.	Vennanthur	67	7	60
10.	Namagiripettai	29	9	20
11.	Kollimalai	—	—	—
12.	Tiruchengode	153	8	145
13.	Elachipalayuam	2	2	
14.	Mallasamudram	—	—	
15.	Pallipalayam	—	—	—
	Total	858	40	818

Source : Animal Husbandry Department, Namakkal zone and Rasipuram zone.

Simple random samples of 20 percent from each group are selected and thus a total of 163 farms are taken for the study. Table 2 shows the selection of sample farms.

METHOD OF DATA COLLECTION

The primary data were collected with the help of Interview Schedule. The interview schedule was reformulated on the basis of pilot survey so as to cover all the objectives of the study.

TABLE 2
Selection of Sample Units

Group	Farm size	Total Number of Farms	Sample farms
A	Below 20000	512	102
B	20001-50000	183	37
C	50001 to one lakh	73	15
D	Above one lakh	44	9
	Total	812	163

STATSTICAL TOOLS

Coefficient of correlation has been applied to study the relationship between bird strength, i.e. the farm size and profit. Correlation analysis is also used to find out the relationship between investment and profit. The relationship between the size of the farm and profit is established through regression analysis. Regression line is fitted by the method of least square.

The t-test of significance was done to find out the significant difference between the family labour and hired labour for the different farm sizes and the significant difference between the farm size and output. Break-even analysis is used to find out the no-profit no-loss point. The limitations of these tools are applicable to this study also.

RESULTS AND CONCLUSION

The per bird cost structure of poultry farming shows that the total fixed cost is almost similar to Group C and Group B farms whereas it is comparatively higher (Rs. 105.65) for Group B and still higher (Rs. 112.70) for Group A.

The components of fixed and variable costs of poultry industry show that the cost of shed and implement is comparatively higher for Group A and it declines as the farm size increases

The study reveals that the economies of scale holds good only in the large farms.

In the present study, all the farm groups show a positive sign of earning profit. It could be noted that the net farm income is minimum for Group A. If the farm size is further less the net income may be negative.

Egg income is the maximum, 94 percent in the total income and the culled bird income is 4.9 percent. If the price of the egg fluctuates, the share of egg income in the total income will change accordingly.

The employment generation of the poultry farms of different sizes with respect to family labour and hired and causal labour shows that family labour is more (61.78%) in Group. Since majority of the farms use mechanical devices in the farming activities, the average employment generation is less.

It is observed from the investigation that poultry industry has strong backward and forward linkages. Encouraging the poultry industry further will promote rural development.

The cost-revenue ratio in Group A is 1:1.07 whereas it is as high as 1:1.18. This means that Group D earns 11 paise per bird more than that of Group A.

The break-even prices for groups B, C and D are Rs. 1.54, Rs. 1.48 and Rs. 1.45 respectively. Hence, the respective per egg profit for these groups are 6 paise, 12 paise and 15 paise.

The value of correlation coefficient of Group A (0.74), Group B (0.84), Group C (0.86) and Group D (0.81) show that there exists high order positive relationship between farm size and the profit. The values further reveal that profit increases with the increase in the bird strength of the farms.

The correlation values of farm size and fixed cost varies from 0.81 to 0.96, though they are highly positive. The correlation values of farm size and variable costs are almost similar to all the farms.

The correlation values between total cost and profit show that these two variables have a strong, positive association. The profit increases as the total cost increases

The calculated 't' values of all the farm groups individually and also all the groups combined for farm size and profit are more than the table value both at 5 percent and 1 percent levels and hence the correlation co-efficient are significant. It is found that majority of farmers (95%) are backward community and the remaining 5 percent are most backward community.

SUGGESTIONS

Majority of the farms, as they are large, use mechanical devices which replace labour to the maximum. If the size of the farm is small the employment generation will be more. Hence, the poultry sector needs to be given special attention to encourage the small farms considering the fact that it provides mass rural employment so as to prevent migration of people from rural areas.

The key factor for the success in the poultry farming is to reduce feed costs. There is an urgent need to set-up cold storage facilities in sufficient numbers. Further, the government should set-up egg pre-processing units, processing liquid eggs into powder form or the government should encourage the setting up of such units in the private sector by providing incentives and subsidies.

REFERENCES

Banarjee, G.C., "Poultry", Oxford and I.B.H. Publishing Company, New Delhi, 1976.

Bhat, G.A., "Problems of Marketing Poultry and Poultry Products", Poultry Guide, 1990.

Imas, "Indian Poultry Looks-up", Indian Poultry Year Book, 10th Edn., Babha Barkha Nath Printers, New Delhi, 1994.

Indian Poultry, Year book, 10th Edn., Babha Barkha Nath Printers, New Delhi, 1994.

Panda, "Poultry Production", Indian Council of Agricultural Research", 1998.

Rajannair, "Marketing", Sultan Chand and Sons, 2000.

Reddy, R.R., "Price Spread in Tamil Nadu Egg Marketing", 1989.

Mahapatra, S.B. *et. al.*, "Sources of Supply, Consumption, Price Spread and Market Margins of Eggs at Bhubaneswar", Indian JANIM, Production Management, 1991.

CHAPTER

32

Determinants of Rural Non-Farm Sector in Andhra Pradesh : A Discriminant Analysis

M. KOTESWARA RAO

I. INTRODUCTION

The discussion on rural development policy mostly focused on tenurial, institutional, technical, infrastructural and economic aspects of agricultural development. In India like other developing countries, labor absorption in agriculture and in urban and service sectors have not been enough to absorb the growing labour-force. The diversification of rural economy appears to be an important element of development strategy. Rural Non-Farm Sectors (RNFs) plays a positive role in the removal of poverty, generation of employment and decentralization of urbanization. It is a significant source of income to small and landless farmers during the slack season. It also facilitates structural transformation. Being a long-term strategy, non-farm avenues of employment and earnings are inevitable for rural households. The rural non-agriculture activity is essentially a residual category. The growth of non-farm sector has become an important field of research in the developing countries. In recent years many studies on non-farm activities have come up and the understanding of non-farm sector is gradually growing.

2. TRENDS IN RNFE IN ANDHRA PRADESH

The State of Andhra Pradesh exhibits regional diversities in terms of agro-climatic factors, resource endowments, historical and cultural influences which will have a bearing on both farm and non-farm sectors. There is significant difference in the relative shares of farm and non-farm employment at all levels. But Andhra Pradesh maintained a higher level of RNFE than all India level both in 1971 and 1981. But there is a substantial change in 2001. It had increased by about 150 per cent between 1971 and 2001. However, the growth rate is much higher at all India level than in Andhra Pradesh. The shares of males and females in both at national and state level. The differences in shares of males and females in RNFE continuously increasing both at All India and Andhra Pradesh. The relative shares of males and females in RNFE in Andhra Pradesh are higher than All India level during the four census periods. RNFE and changes over time at All India level and Andhra Pradesh. During the four census periods (1971, 1981, 1991 and 2001) the share of males in RNFE is greater than females both at All India and Andhra Pradesh. The relative shares of males and females in RNFE in Andhra Pradesh are higher than All India level during the four census periods. Though the percentage of male participation is more or less equal in 1991, Andhra Pradesh had an edge over All India level. In case of female participation there has been a continuous decline at state level by 1991 but increased at 2001. But it is slightly increased by 1991 compared to 1971 at national level. A significant rise can be observed by 2001. Considerable variations are observed in the extent of non-farm employment across regions and districts. Coastal Andhra recorded the highest percentage of non-farm employment followed by

Telangana and Rayalaseema regions. The districts which have experienced fast urbanization have recorded a greater decline in the rural non-farm unemployment. Rayalaseema records the lowest percentage of non-agricultural workers. Within Coastal Andhra except East Godavari district other affluent districts in which deltas are located i.e., West Godavari, Krishna, Guntur, Nellore recorded lower percentages of non-agricultural workers. In Rayalaseema, Chittoor and Anantapur record the lowest percentage of non-agricultural workers in 1981 but by 1991 Chittoor had shown fast growth rate and maintained the same in 2001 also. In Telangana region alone, half of the districts, i.e. five out of 10, recorded a lower percentage of rural non-agricultural employment which may be due to migration to urban areas as a result of decline in traditional industries. Apart from growing mechanization, reduced size of holding and other resource constraints, the Telangana region has been subjected to serious distress conditions in agriculture limiting the scope for greater absorption of labor. Caste hierarchy and agrarian movements have played an important role in the region.

The trends of growth of RNFE among different districts during 1971-2001 show interesting results. It is found that some are recorded with continuous rise, while the others have erratic growth. However, the RNFE rates of growth are examined at regional level for the state of Andhra Pradesh. Geographically the state of Andhra Pradesh is divided into three regions, namely, Coastal Andhra, Rayalaseema and Telangana. Among these three regions, three districts are identified which are having continuous increase in the rate of growth of employment in non-farm sector. These districts are West Godavari in Coastal Andhra, Adilabad in Telangana and Chittoor in Rayalaseema region.

3. NEED FOR THE STUDY

Studies on RNFs for the last few decades have examined various aspects largely based on secondary data at the national and state levels. Some studies in India suggest that the growth of agricultural production is likely to stimulate the growth and development of RNFE (Bhalla, 1990, 1993; Dev, 1990; Papola, 1987, 1992; Shukla, 1992; Unni, 1991, 1994). The studies have also identified and tested many factors responsible for the emergence and growth of the RNFE. It is sometimes observed that the growth of RNF is due to various factors including agricultural growth (Eapen, 1994, 1995; Shukla, V., 1989, 1992; Singh, 1994; Unni, 1991). However, majority of the studies are not only based on secondary data but the analysis is mostly limited to the macro-level aggregates only. Further, not much attention has been paid to estimate the composition of RNFE as well as its determinants at micro-level. Hence, there is every need to take up studies on RNFE which reflect its composition and determinants at the micro-level with disaggregated data to analyze the emerging trends in greater detail.

4. OBJECTIVES OF THE STUDY

An attempt is made to analyze the determinants of RNFE in the three survey districts viz. West Godavari, Chittoor and Adilabad. In the literature on RNFE researchers have identified a number of determinants. The number of determinants considered varied widely depending upon the nature of the study—national, regional or micro-level. Among the variables identified as determinants of RNFE in the earlier literature as well as availability of data at Mandal levels, 11 variables are considered in the present study to explore the determinants of RNFE in the three sample districts of Andhra Pradesh. The results of the discriminant function used for each of the variable considered are examined in detail in the following pages.

5. METHODOLOGY

Specification of Variables

The variables considered for the purpose of discriminant analysis among the three sample districts are examined briefly. These variables are examined under three main heads: namely, (i) Agro-characteristics, (ii) demographic characteristics and (iii) the infrastructural characteristics. They are:

(a) Percentage of Area Under Food Crops

The percentage of area under food crops is considered as a vital determinant of RNFE in the study is, because industrial structure of the districts is centered round agro-based industries.

(b) Percentage of Irrigated Area to Total Cropped Area

In view of the significance of irrigation, percentage of gross irrigated area to gross cropped area is considered as a discriminating variable. It also serves as proxy for stable and increased incomes of agricultural population.

(c) Number of Tractors used

The use of tractors facilitates saving of time and labour cost, which also enables the peasant to reduce other incidental costs like transportation costs, etc. This reflects the level of mechanization in agriculture and indirectly indicates the level of agricultural development of an area.

(d) Cropped Area Per Agricultural Worker

The extent of cropped area is an indicator of the possible income of agricultural population in a district. Hence, the cropped area per agricultural worker is considered as a better indicator of employment.

(e) Cropping Intensity

Cropping intensity is considered as one of the variables in view its significance in relation to RNFE. It indicates the level of development of agricultural sector in the district. It also influences incomes and employment in the rural economy.

(f) Urban Proximity

Urbanization influences RNFE in two ways viz., creating jobs and demand for skilled and unskilled labour force of rural areas. It is considered as a good proxy for infrastructural development. Since urban centers influence the nearby rural economy. Urbanization was found to be critical determinant of rural manufacturing activity.

(g) Density of Population

Density of population is not only an important indicator of the demographic pressure but also attainment of efficient scales of production in any sector of the economy. The pressure of population will stimulate the economy in different channels like increase in population, sectoral shift in employment, leading to economics of scale, unemployment, etc.

(h) Level of Literacy

The role of literacy in economic development is well recognized to capture variations in RNFE. Studies in the area of research show the positive relation between RNFE and literacy. Hence, this variable is considered as one of the determinants of RNFE in the study area.

(i) Road Length

The promotion of RNFE significantly based on the expansion of infrastructural facilities of the area. In this context, the availability of roads is considered as one of the causes for expansion of RNFE, since it facilitates exports of raw materials, product goods, agro-produce and imports of inputs to agriculture, consumer items and machines and tools. Hence, the ratio of road mileage (in kms) to geographical area is considered.

(j) Number of Bank Branches

The number of bank branches per 10,000 populations is taken to analyze the character of mandals regarding total RNFE and its composition.

Statistical Technique

To identify the characteristics of the mandals which are grouped as the highest level group and the lowest level group based on the share of RNFE in rural workers and share of major industry group of employment in RNFE, the technique discriminant analysis is used.

The discriminant analysis, a multi-variate statistical technique is used in the analysis of determinants. By considering the standard coefficients discriminant variables are linear function are as follows:

$$D = K_1x_1 + K_2x_2 + K_3x_3 + \dots\dots\dots\dots K_nx_n$$

where K_1, K_2, K_3 are standardized coefficients.

x_1, x_2, x_3, are variables selected to identify the determinants.

The Wilks Lamda value has a Chi-square transformation for testing the significance of the discriminant function is computed.

6. DETERMINANTS OF RNFE IN WEST GODAVARI DISTRICT

The variables selected to identify the determinants of RNFE across mandals:

1. Percentage of Food Cropped area to Gross Cropped Area
2. Percentage of Canal irrigated area to GCA
3. Percentage of Tube well irrigated area in GCA
4. Ratio of Road length to Geographical area
5. Number of Tractors per thousand hectares of Cropped Area
6. Cropped area per agricultural worker (in hectares)
7. Cropping intensity
8. Urban Proximity
9. Density of Population
10. Percentage of literates to total population
11. Number of Bank branches per 10,000 population

The mandals with the highest percentage of RNFE (Cluster I) and those with lowest (Cluster II) used in the discriminant function are shown in Table 1. As shown by discriminant variable there are marked variations in agro-economic characteristics between the two groups of mandals. Cluster I is marked by high proportion of cropping intensity, density of population, literacy levels, ratio of road length, number of bank branches, compared to Cluster II. On the other hand, Cluster II showed higher percentage of food cropped area, tube-well irrigated area, number of tractors than Cluster I.

TABLE I

Distribution of Mandals According to Share of RNFE in Total Rural Employment in West Godavari District

CLUSTER I *(Highest percentage Group)*		*CLUSTER II* *(Lower percentage Group)*	
J.R. Gudem	32.41	Achanta	22.47
Tanuku	31.98	Nidamarru	21.54
Palacole	31.52	Pedavegi	19.75
Eluru	27.54	K. Kota	19.54
Chintalpudi	26.54	Kovvuru	19.54
Ganapavaram	25.65	Polavaram	19.41
Pentapadu	24.68	Nallazerla	18.94
Attilli	24.58	Denduluru	18.71
Mogaltur	23.98	Devarapalli	18.51
Tallapudi	23.98	Pedapadu	18.47
Palacoderu	23.78	Peravali	18.45
Nidadavole	23.51	Undi	17.95
Poduru	22.95	J. Milli	16.64
Narasapur	22.84	T.P. Gudem	16.54
Gopalapuram	22.55	Penugonda	16.11

The results of discriminant analysis are shown in Table 2. As could be seen from the table that, the means of variable in Cluster I are greater than those of Cluster II in respect of other sources of irrigation, cropping intensity, density of population, literacy, ratio of road length, number of bank branches. In Cluster II the variables such as food cropped area, tube-wells, number of tractors, cropping intensity were marked high proportion than Cluster I. The differences in means are higher in density of population and food cropped area. With respect to coefficients, food cropped area, cropping intensity are higher in both Clusters and these are higher in Cluster II than Cluster I.

The distribution of mandals in West Godavari District according to probability attached discriminant function is shown in Table 3. The probability of arbitrary classification of variables of Cluster I represents 73.33 per cent of correctness

TABLE 2

Means and Coefficents of Discriminant Function of RNFE in Total Rural Employment in West Godavari District

	Means		*Coefficients*	
	Cluster I	*Cluster II*	*Cluster I*	*Cluster II*
1. Percentage of Food Cropped Area to Gca	69.1020	71.0229	-3.260	-3.258
2. Percentage of Canal Irrigated Area to Gca	81.8387	81.7495	.231	.223
3. Tube-well Irrigated Area	12.3756	30.4576	-.487	-.458
4. Ratio of Road Length to Geographical Area	14.4920	14.3705	-1.964	-1.978
5. Number of Tractors Per Thousand Hectares of Land	4.3873	4.6191	-4.364	-4.235
6. Cropped Area Per Agricultural Worker	.6860	.7928	105.102	106.452
7. Cropping Intensity	150.9609	149.9964	1.364	1.362
8. Urban Proximity	496.7328	497.3476	47.829	47.872
9. Density of Population	491.2407	475.4224	.475	.475
10. Percentage of Literates to Total Population	50.3267	50.2652	-344.567	-345.599
11. Number of Bank Branches Per 10,000 Population	.7581	.7407		

and it is about only 60 per cent in case of Cluster II. It is found that four mandals such as Pentapadu, Attili, Palacoderu and Nidadavole have exhibited the characteristics of Cluster II. It could be due to low level of urban proximity, literacy levels and density of population in these mandals. On the other side, Pedavdgi, Devarapalli, Pedapadu, Undi, T.P. Gudem, Penugonda have shown the characteristics of Cluster II. This may be due to food cropped area, tube-wells, literacy and road length in the mandal. The distribution is shown in the following equation by using standardized coefficients.

TABLE 3

Distribution of Mandals According to Probability Attached by Distriminant Function in West Godavari District

Distribution	*Probability*	
	Cluster I	*Cluster II*
Cluster I	J.R. Gudem	Pentapadu
	Tanuku	Attilli
	Palacole	Palacoderu
	Eluru	Nidadavole
	Chintalpudi	
	Ganapavaram	
	Mogaltur	
	Tallapudi	
	Poduru	
	Narasapur	
	Gopalapuram	
Cluster II	Pedavegi	Achanta
	Devarapalli	Nidamarru
	Pedapadu	K. Kota
	Undi	Kovvuru
	T.P. Gudem	Polavaram
	Penugonda	Nallazerla
		Denduluru
		Peravali
		J. Milli

$D = 0.002x_1 - 0.009x_2 + 0.034x_3 - 0.016x_4 + 0.015x_5 + 1.580x_6 - 0.002x_7 - 0.051x_8 + 0.001x_9 + 0.704s_{10} - 1.207x_{11}$.

These results indicate that x_6 and x_{11} have the greater discriminating power among 11 variables chosen for this analysis, of which x_2, x_4, x_7, x_8, x_{11} are negative and remaining are positive. It can be said by observing the above equation that the variables such as, cropping intensity and number of bank branches are discriminating the distribution.

7. DETERMINANTS OF RNFE IN CHITTOR DISTRICT

The mandals with the highest percentage of RNFE (Cluster I) and those with lowest (Cluster II) used in the discriminant function are shown in Table 4. There are marked variations in agro-economic characteristics between the two groups of mandals as shown by discriminant variables. Cluster I is marked by cropped area per agricultural worker, urban proximity, density of population, literacy levels, ratio of road length, number of bank branches, cropped area per agricultural worker compared to Cluster II. On the other hand, Cluster II showed higher percentage of food cropped area, tube well irrigated area, other sources of irrigation than Cluster I.

The results of discriminant analysis are shown in Table 5. As could be seen from the table that, the means of variable in

TABLE 4

Distribution of Mandals According to Share of RNFE in Total Rural Employment in Chittoor District

CLUSTER I *(Highest percentage Group)*		*CLUSTER II* *(Lower percentage Group)*	
Thirupati (Rural)	43.82	Varadiahpalem	18.41
Thirupati (Urban)	42.58	Karivetinagaram	18.03
Narayanavaram	41.61	Penumer	17.87
Nagari	32.85	Vadamalpeta	17.84
Gurramkonda	32.02	Pulicherla	17.71
Renigunta	31.77	Puttur	17.42
Chandragiri	28.74	B. Kothakota	17.33
Kurabalakota	27.63	Tnavanampalle	17.29
Madanapalli	25.8	Kalikiri	16.5
Chittoor	25.12	Chinnagottigallu	16.28
Rompicherla	22.27	Yadamarri	16.08
Piler	22.23	Irala	15.22
Vayalpadu	19.51	Gangavaram	14.95
Sahtyavedu	19.08	G.D. Nallore	14.74
Yerpadu	18.83	Nagalapuram	16.11

TABLE 5
Means and Coefficents of Discriminant Function of RNFE in Total Rural Employment in Chittoor District

	Means		*Coefficients*	
	Cluster I	*Cluster II*	*Cluster I*	*Cluster II*
1. Percentage of Food Cropped Area to Gca	63.4030	74.0394	-0.0380	0.0470
2. Percentage of Canal Irrigated Area to Gca	0.0100	0.0107	7624.872	7889.778
3. Percentage of Tube well Irrigated Area to Gca	7.8195	11.7784	1.054	1.265
4. Percentage Ratio of Road Length to Geographical Area	40.6764	38.1787	-0.0986	-0.0810
5. Number of Tractors Per Thousand Hectors of Ropped Area	10.0637	12.4896	-2.288	-2.400
6. Cropped Area Per Agricultural Worker in Hectares	2.9971	2.3831	9.745	11.060
7. Cropping Intensity	.2977	.3368	189.270	219.841
8. Urban Proximity	136.0572	8.8619	0.0740	.102
9. Density of Population	217.0743	116.4752	-.155	-.210
10. Percentage of Literates to Total Population	62.4457	59.6079	3.693	3.582
11. Ratio of Road Length to Geographical Area	54.9067	50.8467	-.172	-.175
12 Number of Bank Branches Per 10,000 Thousand Population	.8220	.7233	-21.777	-25.530251

TABLE 6

Distribution of Mandals According to Probability Attached by Distriminant Function in Chittoor District

Distribution	*Probability*	
	Cluster I	*Cluster II*
Cluster I	Thirupathi (Rural)	Kurabalakota
	Thirupathi (Urban)	Rompicherla
	Narayanavaram	Yerpadu
	Nagari	
	Gurramkonda	
	Renigunta	
	Chandragiri	
	Madanapalli	
	Chittoor	
	Piler	
	Vayalpadu	
	Sahtyavedu	
Cluster II	B. Kothakota	Varadiahpalem
	Kalikiri	Karivetinagaram
		Penumer
		Vadamalpeta
		Pulicherla
		Puttur
		Tnavanampalle
		Chinnagottigallu
		Yadamarri
		Irala
		Gangavaram
		G.D. Nallore
		Nagalapuram

Cluster I are greater than those of Cluster II in respect of other sources of irrigation, cropped area per agricultural worker, urban proximity, density of population, literacy, ratio of road length, number of bank branches. In Cluster II the variables such as food cropped area, tube-wells, number of tractors, cropping intensity were marked high proportion than Cluster I. The differences in means are higher in urban proximity, density of population and food cropped area. With respect to coefficients, cropping intensity, cropped area per agricultural worker is higher in both clusters and these are higher in Cluster II than Cluster I.

The distribution of mandals according to probability attached by discriminant function is shown in Table 6. The probability of arbitrary classification of variables of cluster I represents 80 per cent of correctness and it is about 87 per cent in case of Cluster II. It is found that three mandals such as Kurbalkota, Rompicheral and Yerpadu have exhibited the characteristics of Cluster II. It might be due to low level of food cropped area, urban proximity, and density of population in these mandals. On the contrary, two mandals B. Kothakota and Kilikiri have shown the characteristics of Cluster II. This may be due to tube-wells, literacy and road length in the mandal. The distribution is shown in the following equation by using standardized coefficients.

$$D = 0.918x_1 + 0.245x_2 + 704x_3 + 0.486x_4 - 0.347x_5 + 1.346x_6 + 1.483x_7 + 2.539x_8 - 3.425x_9 - 0.315x_{10} - 0.018x_{11} - .0572x_{12}$$

These results indicate that x_6, x_7, x_8 and x_9 have the greater discriminating power among 12 variables chosen for this analysis, of which x_6, x_7, x_8 are positive and x_9 became negative. It can be said by observing the above equation that the variables such as, cropped area per agricultural worker, cropping intensity, urban proximity, density of population are discriminating the distribution.

8. DETERMINANTS OF MAJOR INDUSTRY GROUPS OF RNFE IN ADILABAD DISTRICT

The distribution of mandals by share of rural non-farm employment in Adilabad district both with the highest levels (Cluster I) and the lowest (Cluster II) is presented in Table 7.

TABLE 7

Distribution of Mandals According to Share of RNFE in Total Rural Employment

CLUSTER I (Highest percentage Group)		CLUSTER II (Lower percentage Group)	
Sarangapur	63.99	Bela	29.93
Leheswara	53.11	Siripur (U)	29.36
Nirmal	50.66	Dandepally	28.68
Boath	46.87	Dahagaon	26.58
Neredigonda	45.1	Tanoor	25.85
Kubeer	44.34	Talamadugu	24.07
Bhainsa	41.32	Wankadi	23.65
Mamda	41.05	Laxmana-Chan	22.73
Khanapur	40.81	Vemanapalle	22.09
Kaddam	40.61	Lakshettipet	20.18
Dilwarpam	35.75	Krethala	19.96
Mudhole	32.21	Kuntala	19.96
Tamsi	30.68	Bajjur	19.75
Jannarm	30.61	Icoda	18.74
Bheemini	30.04	Kotapalle	18.23

Many mandals of Cluster II had lower proportion in respect of area under food crops; tube-well, number of tractors, cropping intensity, urban proximity, density of population and literacy than Cluster I. It is observed in Cluster I cropped area per agricultural worker, number of bank branches marked with lower proportion, compared to Cluster II.

The results of discriminant analysis are shown in Table 8. As seen from the table the means of variable in Cluster I are greater than those of Cluster II in respect of cropped area per agricultural worker, canal irrigated area, tube-well irrigated area, number of tractors, density of population, levels of literacy and ratio of road length. In Cluster II the variables such as number of bank branches, cropping intensity and urban proximity were marked high than Cluster I. The differences in means are higher in cropped area per agricultural worker,

TABLE 8

Means and Coefficents of Discriminant Function of RNFE in Total Rural Employment

	Means		*Coefficients*	
	Cluster I	*Cluster II*	*Cluster I*	*Cluster II*
1. Gross Cropped Area Per Agricultural Worker	52.2900	46.8967	.510	.512
2. Number of Bank Branches Per 10,000 Population	.9120	.9433	13.907	11.952
3. Canal Irrigated Area	11.8433	7.4340	-1.703	-1.540
4. Density	139.5333	117.2667	-.113	-0.0844
5. Percentage of Food Cropped Area to Gca	67.4487	66.7573	1.767	1.595
6. Cropping Intensity	76.8567	79.9773	.427	.390
7. Percentage of Literated to Total Literates	20.8660	17.6480	3.999	3.629
8. Road Length	22.5287	18.6780	.250	.199
9. Percentage of RNF Workers to Total Workers	41.8100	23.3173	2.007	1.463
10. Number of Tractors	.3140	.2420	7.815	.832
11. Tube-well Irrigated Area	7.6800	1.6853	-.431	-.387
12. Urban Proximity	24.3347	38.1347	-.139	-.108

density of population and urban proximity. With respect to coefficients number of bank branches, number of tractors is higher in both Clusters.

The distributions of mandals according to probability attached by discriminant function are shown in Table 9. The probability of arbitrary classification of variables of Cluster I represents 94 per cent of correctness and it is about 100.00 per cent in case of Cluster II. It is found that only one mandal, i.e. Tamsi exhibited the characteristics of Cluster II. It could be due to low level of urban proximity, density of population, levels of literacy in the mandal. The distribution is shown in the following equation by using standardized coefficients.

TABLE 9

Distribution of Mandals According to Probability Attached Discriminant Function in West Godavari District

Distribution	*Probability*	
	Cluster I	*Cluster II*
Cluster I	J.R. Gudem	
Cluster I	Sarangapur	Tamsi
	Leheswara	
	Nirmal	
	Boath	
	Neredigonda	
	Kubeer	
	Bhainsa	
	Mamda	
	Khanapur	
	Kaddam	
	Dilwarpam	
	Mudhole	
	Jannarm	
	Bheemini	
Cluster II		Bela
		Siripur (U)
		Dandepally
		Dahagaon
		Tanoor
		Talamadugu
		Wankadi
		Laxmana-Chan
		Vemanapalle
		Lakshettipet
		Krethala
		Kuntala
		Bajjur
		Icoda
		Kotapalle

$$D=0.008x_1+0.566x_2+0.003x_3+0.004x_4+0.037x_5-0.006x_6+0.201x_7+0.002x_8-2.192x_9+0.105x_{10}-0.006x_{11}$$

These results indicate that x_2, x_9 have the greater discriminating power while x_6, x_9, x_{11} are become negative. Wilks Lamda is 66.50.

However, apart from the analysis of broad determinants of RNFE among the three districts, an attempt is also made to analyse the determinants in various sub-sectors of the economy at mandal level. The activities considered for the analysis of determinants of RNFE are manufacturing, construction, trade and commerce, transport and other services. The result of discriminant function of RNFE and the probability attached by discriminant function are explaining mixed pattern. Hence to obtain a comprehensive vision of growth of RNFE, it requires a detailed enquiry at the micro-level, particularly at village level.

References

Bhalla, Sheila, 1990, 'Employment in Indian Agriculture : Retrospect and Prospect', *Social Scientist*, Nos. 192 and 193.

Bhalla, Sheila, 1993, 'Trends of Some Propositing About the Dynamics of Changes in the Rural Workforce Structure', *Indian Journal of Labour Economics*, Vol. 36, No. 3, July-September.

Chadha, G.K., 1992, *Non-Farm Sector in Rural China : Trends and Variation in Labour Productivity*, ILO/ARTEP Working Paper, New Delhi, May.

Chadha, G.K., 1992, 'Non-Farm Employment for Rural Households : Evidence and Prognosis', *Indian Journal of Labour Economics*, Vol. 38, No. 3, July-September.

Dev, S. Mahendra, 1990, 'Non-Agricultural Employment in Rural India : Evidence at a Disaggregate Level', *Economic and Political Weekly*, Vol. XXV, No. 28, July 14.

Eapen, Mridul, 1994, 'Rural Non-Agricultural Employment in Kerala : Some Emerging Tendencies', *Economic and Political Weekly*, May 21.

Eapen, Mridul, 1995, 'Rural Non-Agricultural Employment in Kerala : Inter-District Variation', *Economic and Political Weekly*, Vol. XXX, No. 12, March 25.

Haggblade, S., P. Hazell and J. Brown, 1989, 'Farm-Non-Farm Linkages in Rural Sub-Saharan Africa', *World Development*, Vol. 17, Nos. 8 and 13.

ILO, 1983, *Prevention of Employment and Income for Rural Poor Including Rural Women through Non-Farm Activities,* International Labour Organisation, Geneva.

Lanjouw, Jean O. and Peter Lanjouw, 1995, *Rural Non-Farm Employment : A Study,* Policy Research Working Paper 1463, World Bank, Washington.

Papola, T.S., 1992, 'Rural Non-Farm Employment : An Assessment of Recent Trend', *Indian Journal of Labour Economics,* Vol. 35, No. 3.

Papola, T.S. and V.N. Mishra, 1980, 'Some Aspects of Rural Industrialisation', *Economic and Political Weekly,* Vol. 15, Nos. 41-43.

Shukla, Tara, 1989, 'Rural Non-Agricultural Employment', *Economic and Political Weekly,* Vol. XXIV, Nos. 35 and 36, September 2-9

Shukla, Vibhooti, 1989, 'Rural Non-Farm Activity : A Regional Model and its Empirical Application to Maharashtra', *Economic and Political Weekly,* Vol. XXVI, No. 45.

Shukla, Vibhooti, 1992, 'Rural Non-Farm Employment in India : Issues and Policy', *Economic and Political Weekly,* Vol. XXVII, No. 28, July 11.

Todaro, M.P., 1989, 'A Model of Labour Migration in Urban Employment in Less Developed Countries', *American Economic Review,* Vol. 59, No. 1, March.

Unni, Jeemol, 1989, 'Changes in Women's Employment in Rural Areas 1961-83, *Economic and Political Weekly,* Vol. 24, No. 17, April 29.

Unni, Jeemol, 1991, 'Regional Variation in Rural Non-Agricultural Employment : An Exploratory Analysis' in *Economic and Political Weekly,* Vol. XXVI, No. 3, January 19.

Unni, Jeemol, 1994, 'Inter-regional Variations in Non-Agricultural Employment in Rural India : An Exploratory Analysis in P. Visaria and R. Basant (eds.), *Non-Agricultural Employment in India : Trends and Prospects,* Sage Publications, New Delhi.

Vaidyanathan, A., 1983, 'Labour Use in Rural India', *Economic and Political Weekly,* Vol. 21, No. 52.

Vaidyanathan, A., 1986, 'Labour Use in Rural India : A study of Spatial and Temporal Variation', *Economic and Political Weekly,* Vol. 21, No. 52, December 27.

Visaria, Pravin, 1995, 'Rural Non-Farm Employment in India : Trends and Issues for Research', *Indian Journal of Agricultural Economics,* Vol. 50, No. 3, Conference Number, July-September.

Visaria, Pravin and Rakesh Basant, (Eds.), 1994, *Non-Agricultural Employment in India : Trends and Prospective,* Sage Publication India Pvt. Ltd.

CHAPTER

33

Rural Non-Farm Sector in the North-Eastern States : Challenges and Perspective

APURBA SAIKIA

In the present days the role of rural non-farm sector in the alround development of the rural economy is recognized all over the world. The result of various studies reveal that this sector plays an important role in providing employment and earning to a substantial section of the rural population. It must be admitted that the rural economy characterized by heavy demographic pressure on land, small and fragmented holdings, iniquitous land distribution structure need strong support from non-farm sector to solve the problems of rural unemployment and under unemployment. This paper makes an attempt to throw light on various aspects of rural non-farm sector in the north-eastern states specially its present role and future prospect. The paper also discusses the composition and growth of rural non-farm sector in the north-eastern states and its place in the overall livelihood diversification of rural household. The paper has been prepared based on information and data found in the report of economic census 2005, Economic survey, Govt. of India, Planning Commission documents, various periodical publications, books and journals.

INTRODUCTION

The North-eastern Region accounts for 7.9% (2,62,500 sq. km.) of the national territory of India and about 3.7% of the national population. In 2001 the population of the region was 3.9 crores as against India's total population 102.7 crores. Largely considered as a land locked region of India the region is constituted of eight states viz. Arunachal Pradesh, Assam, Manipur Meghalaya, Mizoram, Nagaland, Tripura and Sikkim. Decadal growth rate of population in most of the states of the region has been higher than all India average. In the decade 1991-2001 the decadal growth rate of population of Arunachal Pradesh, Mizoram, Sikkim and Nagaland was 26.21%, 29.18% 32.98% and 64.41% respectively as against all India average 21.34%. The region is predominantly rural with over 83% of the population living in rural area. The per capital income of north-eastern states was Rs. 12918 in 2001-02 as compared to all India average of Rs. 17823. Despite being rich in natural resources with fertile land, rich forest and mineral deposit development of the region has been slow and problem like poverty, unemployment, etc. have emerged as the biggest threat to the region.

THE RURAL NON-FARM SECTOR : CONCEPTUAL ISSUES

The rural non-farm sector comprises of all those non agricultural activity which generate income to the rural household either through waged work or in self-employment. Since it is defined negatively as non-agriculture. It incorporates a wide range of activities including household and non-household manufacturing, processing, repairing, construction mining and quarrying, community, social and personal service, health and social works, post and telecommunication, petty trading, etc. There are two alternative approaches to define rural non-farm activities. The first is the locative approach in which the primary criterion is that an activity to be called as rural non-farm activity when it is performed in a location falling within a designated rural area. The second is the linkage approach under which the activity of an industrial enterprise generating significant development linkage leading to rural development is called as non-farm activity. The rural non-farm sector provides supplementary employment to small marginal farmer reduce income inequalities and rural-urban migration.

RURAL NON-FARM SECTOR IN THE NORTH-EASTERN STATES

For a long time the people of the north-eastern states depended heavily on agriculture as the only source of livelihood. However in the last few decades the situation has changed. These states have shown significant economic diversification with expansion of the non-farm sector. The growth as well as present scenario of the rural non-farm sector can be understood from the growth of rural non-farm labour force in these states. It is shown in the Table 1.

This Table 1 shows that the percentage of non-farm rural labour force in the north-eastern states has increased at a rate much higher than national average during the period 1993-94 to 1999-00. The percentage of rural non-farm labour force in the NE states is shown in Figure 1.

The growth of non-farm sector in the north-eastern states can be shown in terms of rural non-farm own account establishment and number of workers therein. It is presented in the Table 2.

This Table 2 shows that the NE states had around 5% rural non-farm own account establishment and 8,33,573 persons were working in such farm in 2005. One important aspect which should get special mention is that as against all India employment rate 1.30 person per establishment employment rate in the rural non-farm own account establishment in the NE state were 1.38. It is a matter of pride that NE state Nagaland recorded highest rate of employment in the country.

TABLE I

(*figures in percentage*)

States	*1993-94*			*1999-2000*		
	RM	*RF*	*RP*	*RM*	*RF*	*RP*
Arunachal Pradesh	20.93	3.84	13.61	24.38	4.89	16.58
Assam	22.36	1.65	21.41	35.35	20.60	32.32
Manipur	33.49	39.56	35.88	22.05	30.43	24.70
Maghalaya	17.46	9.48	14.01	14.03	12.72	13.45
Mizoram	13.65	6.73	11.22	15.99	12.53	14.49
Nagaland	32.13	10.81	25.55	29.47	8.07	20.34
Sikkim	43.02	32.97	40.86	43.16	29.93	39.25
Tripura	51.56	45.48	50.52	54.12	50.93	54.29
All NE	25.32	18.82	23.76	34.86	19.97	31.34
All India	25.90	13.8	21.60	28.60	13.70	23.70

Source : (NSS 50th and 55th rounds, extracted from household level data on CD-Rom).
RM—Rural Male, RF—Rural Female, RP—Rural Person.

FIG. I

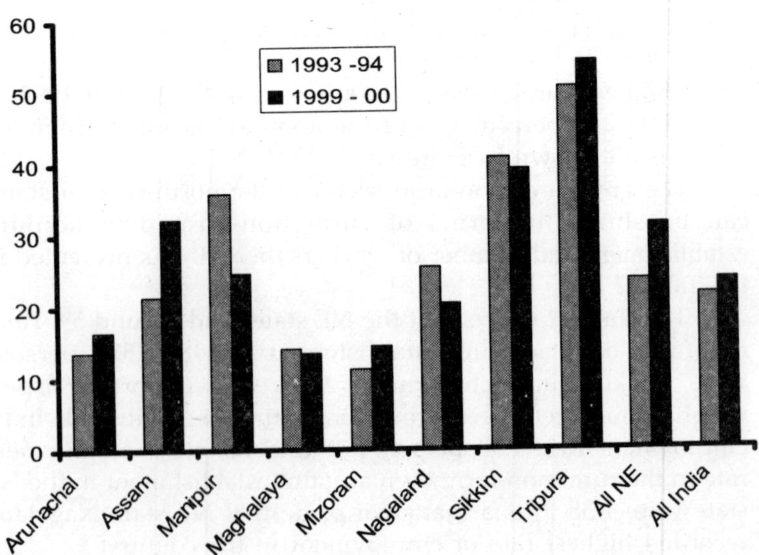

TABLE 2

Sl. No.	NE States	Rural		Number of persons working			
		Number	Percentage in All India	Total	Female	Child	Employment Rate
(1)	(2)	(3)	(4)	(5)	(6)	(7)	(8)
1.	Sikkim	6350	0.05	9404	3303	170	1.44
2.	Arunachal Pradesh	9186	0.07	13121	3362	376	1.43
3.	Nagaland	10,823	0.08	17384	5842	212	1.61
4.	Manipur	41211	0.41	63730	25820	1983	1.55
5.	Mizoram	6929	0.07	8080	3461	9	1.20
6.	Tripura	97760	0.84	125463	16427	4368	1.28
7.	Meghalaya	22119	0.19	27964	10396	792	1.26
8.	Assam	445820	3.36	568427	37092	6919	1.28
	Total in the NE Sates	639998	5.07	833573	105703	14829	1.38
	All India	13,262173		17302128	3216655	326967	1.30

Source : Economic Census, 2005.

TABLE 3

Sl. No.	NE States	Rural NFE with hired workers		Average number of persons working in RNFE			
		Number	Percentage in All India	Total	Female	Children	Hired worker
(1)	(2)	(3)	(4)	(5)	(6)	(7)	(8)
1.	Sikkim	4935	0.08	5.70	1.18	0.04	5.39
2.	Arunachal Pradesh	9271	0.14	5.54	1.27	0.05	5.18
3.	Nagaland	8154	0.12	6.77	2.22	0.04	5.74
4.	Manipur	10,619	0.16	4.37	1.18	0.11	3.54
5.	Mizoram	6888	0.10	2.89	0.85	0.00	2.78
6.	Tripura	31386	0.48	3.76	0.76	0.05	3.37
7.	Meghalaya	30064	0.46	3.40	0.94	0.06	2.89
8.	Assam	196053	2.99	4.05	0.70	0.09	3.24
9.	NE States	297361	4.33				
	All India	6564894		3.75	0.91	0.11	3.22

Source : Economic Census 2005.

The present picture of non-farm enlistment in the NE states can also be shown in terms of rural non-farm establishments with hired worker and average number of hired workers working in such establishment. It is shown in the Table 3.

This table shows that the NE States had less than 5% non-farm establishment with hired workers in 2005. In terms of average number of working Persons NE states like Sikkim, Arunachal Pradesh, Manipur, Nagaland and Assam showed much better result than the all India average.

Although available data present before us an increasing trend of rural non-farm sector in the north-eastern states looking at the ground realities of this region one must say that only through substantial development of the rural non-farm sector the NE states can get rid of its serious problem.

IMPORTANCE OF RURAL NON-FARM SECTOR IN THE NORTH-EASTERN STATES

The rural non-farm sector has to play an important role in the North-Eastern States mainly due to the following reasons:

(a) Decadal growth rate of population in the north-eastern states has been higher than national average. In the growth rate of state domestic product these states have been lagging behind rest of the country or the national average. Because of these two reasons the problem of unemployment and poverty have emerged as the biggest threat to the region.

(b) As against the national average of 70% more than 80% of the population in the NE states lives in the rural areas. It shows the importance of rapid development of rural non-farm sector in the NE region.

(c) Development of rural non-farm sector will contribute in maintaining pace and tranquility in the region.

(d) A planned strategy of development of rural non-farm sector may prevent rural people of the region migrating to urban industrial and commercial centers.

(e) When the economic base of the rural economy extends beyond agriculture, economic gaps are found to get narrower along with beneficial effects in many of the sectors associated with the life and aspiration of the people.

(f) As the region is facing the problem of dearth of capital, labour-intensive rural industry will be of immense value.

(g) With the development of rural non-farm sector the rural income distribution in the NE states will be much less unequal.

(h) Agricultural growth has strong backward and forward linkages with industry and other non-agricultural activity. A healthy growth of the rural non-farm sector will contribute substantially towards a more productive and diversified agricultural growth in the NE region.

(i) Most of the NE states are known for rich natural resources. Growth of non-farm sector will help in the utilization of local talent and slack local resource of this region which cannot be easily shifted or utilized elsewhere.

(j) With the growth of rural non-farm sector women of the region will get much better opportunities to engage themselves in productive work.

(k) Farm activities in the NE region suffer from the problem of low productivity, low cropping intensity, inadequate irrigation facilities, frequent damage by natural calamities, insufficient infrastructural facilities, etc. Because of this constraint the people of the region have to depend on the non-farm sector.

CHALLENGES OF NON-FARM SECTOR IN THE NORTH-EASTERN STATES

North-eastern states have to face a number of challenges in the development of rural non-farm sector. Some of the challenges are discussed below:

(a) *Inadequate infrastructural facilities*: In road, transport, communication and energy supply the region lag far behind rest of the country. The per capital energy consumtion in the region was 20.4 KWH against all India figure of 75.2 KWH in 1999-2000. The NER has only 4% of the total network of Indian railway. During monsoon both rail and road link are disrupted. The high cost of transportation due to poor quality of services poses a big problem in carrying out the non-farm activities.

(b) Geographical isolation and difficult terrain reduces mobility among the people of NE region. In the hilly areas production and marketing of non-farm products become difficult.

(c) *Inefficient Governance*: Misutilisation and low utilisation of fund are two biggest challenges in the way of development of rural non-farm sector. Following Table shows utilization of fund released by the central government up to Dec. 2009.

(d) Limited access to credit.

(e) Low credit deposit ratio. The credit-deposit ratio during 2001 in the region was 28.31% as against the national average 57.20%.

(f) Lack of entreprenial zeal among the native people of the NER is a major cause of slow and tardy industrial growth in the region.

(g) Lack of adequate skilled labour in rural areas.

(h) Lack of adequate access of information/lack of awareness about non-farm employment opportunities.

(i) Inadequate supply of raw material for sustaining rural industries.

(j) Unfavorable law and order situation.

(k) Inadequate incentive and government support.

Government Measures for the Development of Rural Non-farm Sector in the NER

Both the central and respective state Governments of the region are taking a number of measures for the development of rural non-farm sector. Some of the measures are as follows:

TABLE 4

(In %)

Name of the Govt. Scheme	Name of NE States								
	Assam	Manipur	Meghalaya	Mizoram	Tripura	A.P.	Sikkim	Nagaland	Overall
(1)	(2)	(3)	(4)	(5)	(6)	(7)	(8)	(9)	(10)
SGSY	69.89	8.33	50.05	49.68	92.24	20.18	35.73	72.38	66.65
IAY	74.96	38.85	37.02	47.25	65.41	75.45	47.70	73.29	70.50
NRDWP	31.09	0	46.18	31.72	38.62	36.39	0	59.25	34.85
TSC	42.07	61.19	60	73.75	11.47	21.7	100	74.6	43.3

NRDWP = National Rural Drinking Water Programme.
TSC = Total Sanitation Campaign.

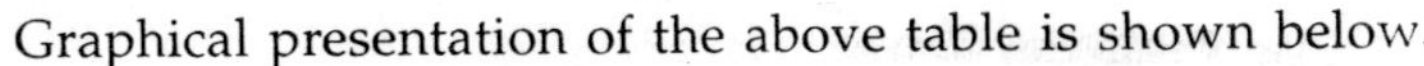

Graphical presentation of the above table is shown below.

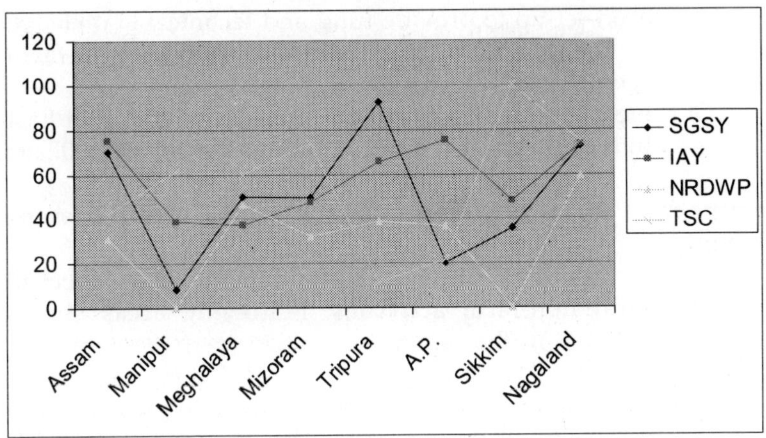

Source : Annual Report, 2009-10, Ministry of Rural Development, Govt. of India.

1. *Provision of fund* : To remove fund constraint 10% of Gross Budgetary Support (Plan) is reserved for the North-Eastern States. During 2008-09 and 2009-10 the allocation of fund to the region under Rural Department is as under.

		2008-09	*2009-10*
A.	Department of Rural Development	Rs. 2460	Rs. 2225
B.	Department of Land Resources	Rs. 166.51	Rs. 234.30
C.	Department of Drinking Water Supply	Rs. 859.99	Rs. 930.00
	Total	Rs. 3486.50	Rs. 3389.30

Source : Annual Report, 2009-10, Rural Development Ministry.

A. Significant part of the fund has been utilized for the development of non-farm activities.

B. *Development of institutional network* : An institutional mechanism has been developed for the rural non-farm sector of the region. Institutions like KVIC state Khadi Board, NHHDC, SIDBI, NABARD, State

Investment Corporation, Indian Institute of Entrepreneurship, SIRD, NIRD, NEDFI, etc. have been set-up to provide fund and technical knowledge to the people interested to undertake non-farm activities.

C. Incentives and concession offered under North-East Industrial and Investment Promotion Policy 2007.

Prospect of Rural Non-Farm Sector in the North-Eastern Region

All the north-eastern States have tremendous prospect to increase rural non-farm activities. Following areas shows prospect of RNFS in the region:

(a) Development of rural non-farm activities through expansion of border trade. The NER shares only 2% of its border with the mainland of the country and 98% is connected with the neighboring countries like China, Bangladesh, Bhutan and Myanmar. Because of its locational advantage the region could emerge as a business transit point for the neighbouring countries. This will significantly help in the expansion of rural non-farm sector of the region.

(b) Prospect of undertaking rural non-farm activities based on rich natural resources specially in the tourism sector.

(c) Low competition in the market.

(d) New opportunities under India's look East policy. Taking the new opportunities the region can develop eco-tourism, power generating projects and attract tourism and domestic investment due to its proximity to South-West China and SE Asia.

(e) Keeping in mind the increasing flow of fund from the centre one can expect that rural non-farm activities will develop in the coming days.

Policy Prescription

Rural non-farm sector should receive special focus from all concern to play its role in the overall development of the region. Some of such areas are mentioned below:

(a) Presently growth of rural non-farm sector seems to be restricted to a few well communicated rural location of the region. So, a policy of balance rural infrastructural development will bring about substantial development in the rural non-farm activities in the region.

(b) To ensure supply of quality manpower for the rural non-farm sector education should be made easily accessible and affordable to all.

(c) Take measures to develop the level of awareness of government incentive and programme relating to rural non-farm sector.

(d) Make banking facilities available in all parts of the region and increase credit-deposit ratio.

(e) Ensure proper implementation of various poverty alleviation/employment generation programmes to put a check on the rapidly increasing unemployment.

(f) Develop North-East as a common market. The NER is land locked politically and isolated from mainland with the creation of a common market the region will be able to take full advantages of India's Look East Policy.

(g) Take immediate measures to reopen the Stillwell road to develop trade link of the region with China, Myanmar and other South-East Asian Nations.

CONCLUSION

North-Eastern Region particularly rural area are lagging far behind the rest of the country. This is one of the reasons of increasing violence and youth unrest in the north east. Every year a large number of people of the region are migrating to different parts of the country in search of livelihood. But this is not possible for all and it should not be allowed to continue in the interest of the country. It is only through the development of rural non-farm sector the region can get rid of a number of problems and march ahead with peace and prosperity.

References

Rural Non-Farm Employment in India. Macro-Trends, Micro-Evidence and Policy Options, Brajesh Jha.

Rural Non-Farm Sector Employment in India, Distress Driven or Growth Driven. Vinoj Abraham, Centre for Development Studies, Trivandrum, Kerala.

Is Rural Non-farm Sector the last Report for Employment in India? Manoj Kumar, Indira Gandhi Institute of Development Research.

Role of the Non-Farm Sector in Transforming Rural India : A Study with Special Reference to Uttar Pradesh. By Nirankar Srivastava.

Rural Transformation in India : The Role of Non-Farm Sector. Edited Rohini Nayyar and Alakh N. Sharma, Institute for Human Development.

The Rural Non-farm Economy : Processes and Policies, Volume 1, Ashwani Saith.

Rural Non-Farm Employment in Arunachal Pradesh, Growth, Composition and Determinants. By Deepak K.R. Mishra.

Indian Economy and North-East Economy. Gautam Purkayastha, Publisher : Bani Mandir, Guahati.

Report of Economic Census, 1998 and 2005.

Report of Economic Survey, Government of India, Its Various Report.

Indian Economy, Mishra and Puri, Himalayan Publication House, Mumbai.

The Indian Economy, Dutta, R. and S.K.M. Sundaram, S. Chand and Co., New Delhi.

Annual Report, 2009-10, Ministry of Rural Development, Govt. of India.

PART IV

NON-FARM SECTOR : INFLUENCES ON INDUSTRIES, SPECIFIC GROWTH

CHAPTER

34

Rural Non-Farm Sector in India : Prospects of the Food Processing Industry for Development

G. RAJALAKSHMY AND MOHANA BANDKAR

INTRODUCTION

The growth and performance of the rural sector holds the key to India's development as a whole. There are about six lakh villages in India where nearly 70% of the population lives. It is true that the agricultural sector is the prominent sector which is important for its contribution to the gross domestic product and employment generation. However, in recent years, the non-farm employment and output growth have been significantly higher than that of the farm sector employment and output. This is indicative of the inability and saturation of the farm sector due to various problems faced by this sector. Though the non-farm sector is showing a higher growth trajectory, to maintain and accelerate the growth momentum of this sector it is necessary to nurture the farm sector also. The linkages and complementarities of the farm and non-farm sectors call for a balanced approach. The food processing industry, which has emerged as an important non-farm sub-sector has a tremendous potential to transform the rural sector. The food processing industry has a huge potential for unleashing the unexplored resources and skill in rural India. Realizing the potential growth prospects and the role it can play in transforming rural India, the Government has initiated a number of steps to promote and encourage this 'Sunrise Industry'. It is necessary to study the constraints and discuss plausible solutions in the light of the fact that this industry has become very relevant and important sub-sector of the emerging non-farm sector in rural India.

THE RURAL NON-FARM SECTOR IN INDIA

The non-farm activities contribute to 35-50% of incomes in rural areas in the developing countries. Increasingly the non-farm sector has become an integral source of livelihood opportunity in rural areas. In fact, the agricultural households are relying more on non-farm earnings to diversify risk, moderate seasonal income savings and finance agricultural input purchases. A high percentage of marginal and small farmers as well as landless agricultural workers depend on the non-farm sector for their livelihood and survival. The stagnation of the farm sector and growing disguised unemployment has triggered-off the shift to the non-farm sector over the years. This is true of poor agrarian countries which are reeling under the burden of growing number of marginal and unemployed people and experiencing poor agricultural growth. The capital requirements of many undertakings in the rural non-farm sector are minimal. It is essential to examine the key factors which affect the growth and equity of the rural non-farm sector which have bearing on the policy-making. Empirical evidence shows that the non-farm income accounted for about 35% of the rural income in African countries and 50% in Asia and Latin America.

TABLE I

Non-Farm Shares of Rural Incomes (%)

Region	*Total Non-Farm Earnings*	*Local Non-farm business $ and employment*	*Transfer remittances*
Africa	34	28	06
Asia	51	40	11
Latin America	47	41	06

Source : Haggblade, Hazell, Reardon (2007).

The figures pertaining to income shares amply bring out the economic importance of part time and seasonal non-farming activities.

The rural non-farm sector includes a heterogeneous collection of trading, agro-processing, manufacturing, and commercial and service activities. The scale of individual rural non-farm business varies enormously. It may be as small as part time self-employment in household-based cottage industries or large agro processing and warehousing facilities operated by large multinational firms. The employment in this sector is seasonal and non-farm activity depends on the availability of agricultural raw materials. Hence, most of the activities in the rural non-farm sector revolves around and are directly affected by the performance of the agricultural sector. In spite of efforts to promote rural industries, manufacturing accounts for only 20-25% of rural non-farm employment. Trade, transport, construction and other services account for 75-80% of rural non-farm employment. In all the above mentioned countries small retailers, cottage industries, basic farm equipment repair services and input supply firms are based in rural areas. Remittances account for a large share of rural income as in the case of the economies of Southern Africa.

Historical evidence suggests that in regions where agriculture has grown robustly, the rural non-farm economy has also enjoyed rapid growth. Empirical evidence shows that the growth linkages are such that each dollar of additional value added in agriculture generates $ 0.60 to $ 10.80 of additional rural non-farm income in Asia. A study conducted

TABLE 2

Non-Farm Income Shares in Rural Areas in India (1993-94)

Quintile	*Cultivation*	*Agricultural Wage Labour*	*Non-farm Labour*	*Non-farm Self-Employment*	*Non-farm Regular Employment*	*Total Non-farm Sources*	*Other*	*Real Per Capital Income*
(1)	(2)	(3)	(4)	(5)	(6)	(7)	(8)	(9)
Lowest	38.2	28.2	15.8	11.4	4.4	31.6	2.0	1146
Q2	38.0	21.3	14.7	16.8	7.0	38.5	2.3	2113
Q3	45.2	13.4	10.1	16.3	11.7	38.1	3.2	3141
Q4	50.1	7.5	6.1	14.6	18.6	39.3	3.2	4712
Highest	64.5	2.1	2.0	7.9	21.1	30.9	2.5	11226
Total	54.9	8.0	5.9	11.5	17.1	34.4	2.7	4468

Source : NCAER, 1993-94.

on the basis of data collected by the National Council of Applied Economic Research in 1993-94 based on rural data from 3200 households belonging to 1765 villages across all parts of India shows that non-farm incomes account for a significant proportion of household income in rural India with considerable variations across quintiles and across major Indian states. It showed that education, wealth, caste, village-level agricultural conditions, population densities and other regional factors affect influence the access to non-farm occupation. The direct contribution of the non-farm sector to poverty reduction is possibly muted since the poor lack assets.

TABLE 3

Distribution of Rural Workforce in Non-Farm Activities in India

(Percentages)

Year	*Persons*	*Males*	*Females*
1972-73	N.A.	16.7	10.3
1977-78	16.6	19.3	11.8
1983	18.5	22.2	12.5
1987-88	21.7	25.4	15.3
1993-94	21.6	26.0	13.8
1999-2000	23.7	28.6	14.6
2004-05	27.6	33.5	16.7

Source : NSSO 2005.

THE PROBLEMS AND PROSPECTS OF FOOD PROCESSING INDUSTRY

The food processing industry has emerged as a very important and strategically important sector for transforming rural India. This non-farm sub-sector has strong synergies with the inclusive growth mandate of the Government and also provide a platform to significantly transform the face of rural India (Chandrasekhar, 2010), Food processing as such is a large sector that covers areas like agriculture, horticulture, plantation, animal husbandry and fisheries. The food processing in India is one of the biggest industries. It ranks fifth in terms of production, consumption, export and expected growth. Though India is one of the major producers of food globally, it accounts for only 1.7% of the world trade in this sector.

TABLE 4
Level of Processing Across Segments

Segment	*India*	*Other Countries*
Fruits and vegetables	2.2%	USA-65%
		Phillipines-78%
		China-23%
Marine	22%	—
Poultry	6%	60%-70% in developed countries
Buffalo meat	20%	—
Milk	35%	60%-75% in developed countries

Source : Ministry of Food Processing Industries, Annual Report, 2003-04.

The table brings out the level of processing activity in both the categories of countries. The food processing industry in India is a highly fragmented industry. A large number of

TABLE 5
Food Processing Units in Organised Sector

(*Numbers*)

Flour Mills	516
Fish Processing units	568 (+482 cold storage units)
Fruit and vegetable processing units	5293
Meat processing units	171
Sweetened and aerated water units	656
Milk products units	266
Sugar mills	429
Solvent extract units	725
Rice mills	139208
Modernised rice mills	35088

Source : Ministry of Food Processing Industries, Annual Report, 2003-04

players in this industry are small sized and mostly in the unorganized sector. However, though the organized sector is small. It is growing fast. The unorganized sector accounts for 70% of the output in volume terms and 50% of the output in value terms (Ministry of Food Processing 2003-04).

DIAGRAM 1

The Structure of the Food Processing Industry (2004-05)

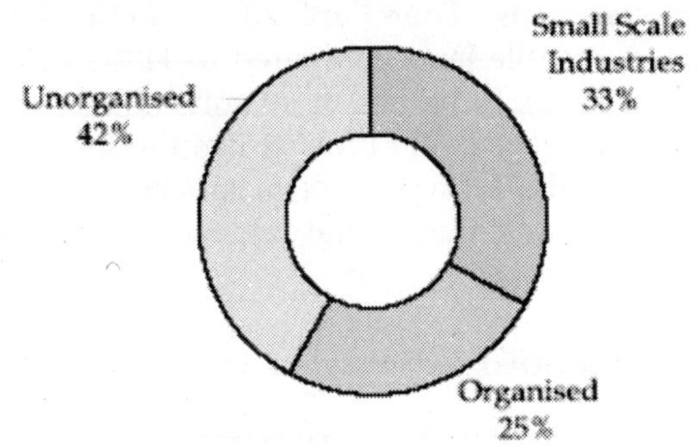

Source: FAIDA/Ministry of Food Processing Industries

DIAGRAM 2

The Share of Food Processing in Manufacturing (2004-05)

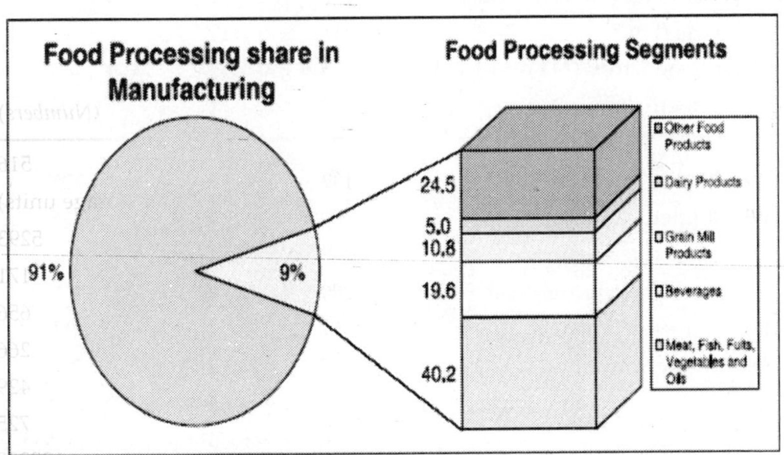

Source : D and B Research.

India's Strengths

The food processing industry is termed as the 'Sunrise Industry' and it has attracted the attention of the policy-makers in recent years due to its potentials. The industry is poised for growth due to the various advantages like the availability of raw materials and cheap labor. India has a huge supply advantage due to diverse agro-climatic conditions and wide ranging raw material base. India has all climates suitable for growing various crops. There are 20 agro-climatic regions. India has 52% cultivable land compared to 11% world average. (Chandrasekhar, 2010). India has more sunshine hours which is ideal for round the year cultivation. It has the largest livestock population and is the largest milk producer. It is the largest producer of cereals and the second largest fruit and vegetable

TABLE 6

India's Competitive Advantages in Food Processing

India's Competitive Advantages in Food Processing			
	India	*Global Rank*	*Share in Global Production*
Arable Land (million hectares)	161	2	—
Irrigated Land (million hectares)	55	1	—
Cost Line (km.)	8041	19	—
Major Food Crops (MT)	35	3	4%
Fruits (metric tonnes)	47	1	10%
Vegetables (metric tonnes)	82	2	10%
Rice/Paddy (metric tonnes)	132	2	22%
Wheat (metric tonnes)	65	2	12%
Milk (metric tonnes)	88	1	16%
Sugarcane (metric tonnes)	289	2	21%
Pulses (metric tonnes)	12	1	21%
Tea (metric tonnes)	0.88	1	28%
Edible Oilseed (metric tonnes)	25	3	7%
Cattle (million)	226	1	20%

Source : Cygnus report, India Food Processing Sector, 2005.

producer. India is among the top five producers worldwide of rice, wheat, groundnuts, tea, coffee, tobacco, spices, sugar and oilseeds. Besides, India has a huge consumer market for food. The industry growth drivers are urbanization, rise in disposable incomes and aspirations which have contributed to a significant increase in demand for functional foods. The emergence of nuclear families with working women has increased the demand for functional foods. It is necessary to tap the potentials favorably so that India can emerge as a preferred destination for food processing globally.

The food processing industry is employment intensive. It is estimated that for every Rs. 1 crore invested, it creates 18 jobs directly and 64 indirectly in the organized sector and 20 jobs in the unorganized sector across the supply chain. (Processing for Future Report of the Ministry of Food Processing, 2010) Most of the new jobs would benefit people who constitute the bottom of the pyramid. Thus, an investment of 1 lakh crore would create as many as 1 crore jobs in the rural economy by 2015. This is the only sector that gives economic sustainability to the farmers by providing opportunities for demand driven clusterised farming, which further give rise to ample employment opportunities from the landless labor up to highly skilled individuals including CEOs. (Ministry of Food Processing Report 2010)

Even the Government of India is allowing 100% Foreign Direct Investment in the food processing industries. The initiatives include :

(a) Full repatriation of profits and capital
(b) Immediate approvals of foreign investments up to 100%
(c) Import duty to be zero for 100% export-oriented units. Reduction in the customs duty for packaging machines
(d) Income tax rebates
(e) Financial aid for setting up Agro Food Parks
(f) Full duty exemptions on all imports for units in the export processing zone.

The Problem Areas

1. The supply chain is notoriously long. The logistics costs are high because of the limitations of infrastructure like roads, warehouses, carriers, cold chain, etc.
2. Indian consumers are by and large price conscious. The demand for processed food is income elastic as well as price elastic. The inclination of Indian consumers is for buying fresh food. Hence fresh food choice is a barrier to processed food even today
3. Fragmented processing capacities, poor scale economies, poor adoption of pre and post harvest technologies, on-farm and off-farm losses of fresh produce and the low level of processing are other features of the sector.
4. Though the peak rates of custom duty has been brought down, still a number of agricultural commodities bear high rates of duty. It is also necessary to remove non-tariff barriers and other technical hassles.
5. As per estimates, nearly 20%-25% of the production is lost during the various stages of cultivation. Agricultural productivity is also low due to factors like poor quality of seeds, planting materials, and sub-standard technology. There is an urgent need for strengthening backward linkages with the farmers with the help of contract farming and other means.

GOVERNMENT INITIATIVES AND FUTURE TRENDS AND PROSPECTS

It is widely accepted that food processing industry can do to the rural economy what the IT industry has done to the urban India. The major trends in the Indian food processing industry reveal an expanding product variety and improved packaging. A very welcome trend is that the regional brands are becoming popular and improvements are visible in food retail environment. Mergers and acquisitions are in process which will enable effective consolidation of capacities and modernization. The Government of India allows for 100% FDI in the food processing sector. In addition to the 10 mega food parks already being set-up, 5 more are planned to be set-up. As a part of the farm to market initiative, arrangements are being made by the Government to make available External Commercial Borrowings. A number of overseas businesses are considering India as an investment destination. Investment opportunities are available not only in food production but also in infrastructure development, marketing, inspection and testing services, technology development etc. Strategic measures like infrastructure development, provision of logistics solution, cold chain and transport including refrigeration, setting up food parks and food retail and giving a boost to contract farming can enable India to emerge as a sourcing hub.

The measures announced in the Union Budget 2010-11 also reflect the serious intentions of removing the drawbacks of this sector. The key areas of importance stressed were as follows :

1. A strong supply chain for perishable farm products to reach consumption and processing centres promptly.
2. Infrastructure and technology to convert such produce into value added products.
3. Infusion of technology to augment agricultural production.
4. The Government is intending to initiate similar steps in sectors like apiary, horticulture, dairy, poultry, meat and marine and aqua culture.

The Ministry of Food Processing has formulated a vision document, according to which the level of perishable is to be enhanced from 6% to 15%, the level of value addition from 20% to 35% and to increase India's share in Global Trade from 1.5% to 3% by the year 2015. The Ministry is inviting potential investors to invest in Mega Food Parks. The cost of a Mega Food is Rs. 150 crores, out of which Rs. 50 crores is the Government grant-in-aid. It is planned to give several IT and customs benefits to give a boost to this sector. Efforts are on to train personnel and new talent through the National Institute of Food Technology and management so as to fulfill the training needs of the industry.

Thus, it is clear that the Government has understood and acknowledged the potentials of this sector. Though the intentions seem to be genuine, it is equally important to implement the programmes. The infrastructural bottlenecks have to be removed fast and the industry needs to be given a favorable environment to experiment and expand its business. It is necessary to ensure efficient and affordable technological know-how for all the processes from harvesting, handling and transportation to processing, packaging, storage and distribution. It is essential to give importance to research and development in the food processing sector. It is also important to ensure standardization of processed food, proper packaging, value addition, improvement in shelf life testing for contamination, improved storage, transport and distribution. Another very important step to be taken is to implement contract farming which will be beneficial to both the farmers and the food processing industry. In view of the fact that contract farming can provide a viable solution to the problems in supply chain, it is necessary to expedite the consolidation of marginal and tiny farms .The farmers need not worry about the risks associated with cultivation since they will be renting out their farms on contract. The farmers are assured fixed returns and they will get salaries if they choose to work on the farms. The quality of produce will be better which will facilitate better processing. Therefore, it is felt that what is needed is industrialization of agriculture. Thus, pragmatic approach and progressive implementation of ideas will enable the growth of the farm and non-farm sector in India

CONCLUSION

It is true that agriculture alone cannot provide livelihood opportunities for the teeming millions in the rural areas and it is essential to give an impetus to the growth of the non-farm sector. In fact, both the sectors are complementary. Recent trends point to the fact that the non-farm sector's growth has overtaken the farm sector growth rate. The food processing industry, which is a part of the non-farm sector has emerged as a very important sector providing employment and generating income for the rural sector in India. To compete in the international market and to thrive in the domestic market it is necessary for this industry to tackle problems of lack of infrastructure, weak backward and forward linkages, wastes along the value chain, poor hygiene, safety standards and sub-standard technology. The seasonal nature of production and highly perishable nature of raw materials lead to weak supply chain. The processing units face the problem of not being able to reach optimum size and achieve economies of scale. Lack of consistent quality is also hindering the small scale units to build brand equity for themselves in the international and domestic markets. Most of these problems were recognized and steps are being initiated to address them. A very practical step to give a boost to this sector would be to encourage contract farming. In fact, this will prove to be the most viable and practical solution to many problems. Since there are a large number of small and marginal farmers, farming has become uneconomical and unmanageable for many. If the farms can be consolidated and rented out on contract, it will help the farmers as well as the processing industry. A number of schemes were proposed in the Union Budget, for 2010-11, which are intended to take care of the weaknesses of the food processing industry. The underlying issues of low agricultural productivity, poor infrastructure and out-dated technology have to be tackled on an urgent basis, if India's potential as a major food processing hub is to become a reality. It is equally important to push forward the much needed reforms to empower this sector to emerge as a major industry, generating income and employment for the rural population.

References

Basant, R. and B.L. Kumar (1989), 'Rural Non-Agricultural Activities in India : A Review of Available Evidence', *Social Scientist*, Vol. 17, Nos. 1, 2

Bhalla, S. (1997), The Rise and Fall of Workforce Diversification Process in Rural India in G.K. Chadha and A.N. Sharma (eds.); Growth, Employment and Poverty, Vikas.

Bhaumik, S.K. (2002), Employment Diversification in Rural India: A State Level Analysis, *The Indian Journal of Labor Economics*, Vol. 45, No. 4.

Bharadwaj, Anil (2009), 'Race Towards Summit', *Modern Food Processing*, Vol. 4, pp. 44-45.

Chandrasekhar. G. (2010) 'Food Processing Sector Offering Huge Opportunities', *Business Line*, The Hindu Group of Publications, June 4.

'Processing For Future'—A Special Report of Food Processing Industries-Government of India, *The Times of India*, March 26, pp. 18-21.

National Sample Survey Organization (1987) : Report on the Third Quinquennial Survey on Employment and Unemployment Survey Results—All India, Department of Statistics, Government of India.

CHAPTER

35

Prospects and Problems of Non-Farm Sector : A Case Study of Coir Industry in India

NAGARAJA, G.

INTRODUCTION

Coir industry is one of the most traditional cottage and agro-based industry with high employment potential and export prospects. Historically, the coir industry started and flourished in Kerala which has a long coast line, lakes, lagoons and backwaters providing natural conditions required for retting. However, with the expansion of coconut cultivation, coir industry has also picked up in Tamil Nadu, Karnataka, Andhra Pradesh, Orissa, West Bengal, Assam, Tripura, Pondichery and Union Territories of Lakshdweep and Andaman and Nicobar Islands.

The only raw material required for the production of coir is coconut husk. The Fibre Extracted from the Coconut husk is known as coir or coco-fibre. Coconut grows ideally in humid tropics regions in India. The west coast belt regions accounts for more than 80 percent of the area under this crop.

The Specific Objectives of the study are:

- To study the growth and development of Coir Industry in India
- To study the external market performance of Coir Industry in India
- To review the problems faced by the Indian Coir Industry during the period
- To study the concrete measures for raising the economic status of Coir Industry in India.

As far as the data is concerned Secondary Data is employed for the present study. The Secondary data has been collected from annual report of coir board, magazines, etc., for the period of fifteen years from 1995 to 2009. The study have concerned to production of coir and coir products, the market structure export performance and the constrains of this industry.

RESULT AND DISCUSSIONS

Production of Coconut

The Coir Industry consists of two segments; namely, white fibre and brown fibre. White or retted fibre is extracted from husk of matured coconut after a process known as Retting which is more suited for spinning coir yarn for further producing into finished products like door mats, mattings, carpets, geo-textiles, etc.

Brown fibre is extracted from coconut husk soaked in water for number of days by mechanical means. It is utilize for internal consumption for shifting upholstery manufacture of curled coir for rubberisation in the rubberized coir manufactory industry, a part of the export too. Brown fibre is also used for producing ropes and finished products though on limited scale only.

Development of coir industry depends on a largest extent of the availability of basic raw materials, which is derived from the coconut. Therefore, coconut is a great relevance and important in the coir sector, the trend of area, production, productivity, in India from the period from 1995-96 to 2007-08 is furnished in Table 1.

TABLE I

Area, Production and Productivity of Coconut in India During 1995-96 to 2007-08

Year	*Area ('000 ha)*	*Production (Million nuts)*	*Productivity (Nuts per ha)*
1995-96	1830.9	12952.3	7074
1996-97	1890.8	13061.0	6908
1997-98	1861.0	12717.3	6834
1998-99	1754.5	12535.9	7145
1999-2000	1768.1	12129.9	6860
2000-01	1823.91	12129.0	6951
2001-02	1932.3	12678.9	6709
2002-03	1921.8	12535.0	6523
2003-04	1933.7	12178.2	6298
2004-05	1935.0	12832.9	6632
2005-06	1964.8	14811.1	7608
2006-07	1939.9	15840.0	8165
2007-08 (P)	1954.8	15920.0	8265

Source : Coconut Development Board, India.

Above Table 1 shows the coconut Production, Area under Cultivation and Productivity. It is clear that the cultivation area under coconut has increased over the years. In 1995-96 it was 1830.9 thousand hectares, which increased to 1954.8 thousand hectares in 2007-08. The production of coconut increased from 12,952.3 million nuts to 15,920 million nuts. The Productivity of coconut showed a fluctuating trend, but steadily increased from 6298 nuts per hectare in 2003-04 to 8265 nuts per hectare in 2007-08.

The coconut production was slow down from 1998-99 to 2003-04 due to some disease in coconut trees which are the

TABLE 2

State-wise Coconut Cultivation Area and Production

States /Union Territories	2006-07			2007-08		
	Area '000 hectares	Production Million nuts	Productivity Nuts/hec.	Area '000 hectares	Production Million nuts	Productivity Nuts/hec.
(1)	(2)	(3)	(4)	(5)	(6)	(7)
Andhra Pradesh	105.0	1326.0	12629	101.32	1119.26	11047
Assam	19.0	153.0	8053	19.00	136.00	7158
Goa	25.5	126.7	4969	25.50	127.60	5004
Gujarat	16.4	138.3	8433	16.40	138.30	8433
Karnataka	401.0	1625.0	4052	405.00	1635.00	4037
Kerala	870.9	6054.0	6951	818.80	5641.00	6889
Maharashtra	21.0	175.1	8338	21.00	175.10	8339
Nagaland	0.9	0.2	222	0.90	0.20	222
Orissa	51.0	275.8	5408	51.00	275.80	5408
Tamil Nadu	374.6	5429.9	14495	383.37	4968.20	12959
Tripura	3.3	7.0	2121	5.80	11.40	1966
West Bengal	25.1	359.1	14307	28.60	355.50	12430
A & N Islands	21.4	89.0	4159	21.60	80.60	3730
Lakshadweep	2.7	53.0	19630	2.70	53.00	1930
Pondicherry	2.1	27.9	13286	2.20	26.60	12091
All India	1939.9	15840.0	8165	1903.19	14743.56	7747

Source : Directorate of Economic and Stastics, Ministry of Agriculture, Govt. of India.

responsible for the decrease in production, the objective of farming approach is to achieve desired growth rate in the production and productivity of coconut at farm level. The major thrust shall be on soil and water conservation, recycling of used waters and development of farming system conductive to steady increase in the output and income from the managed agro-eco system.

Table 2 shows state-wise area, production and productivity of coconut in India. In 2006-07 and 2007-08, Kerala occupied the first position with coconut production of 5641.00 million nuts from an area of 818.8 thousand hectares. Tamil Nadu ranked second followed by Karnataka and Andhra Pradesh.

According to Coconut Development Board the annual production of coconut 2008-09 is estimated at 15.920 million nuts of which 40 percent would be used as tender coconuts leaving behind 60 percent for industrial utilization.

PRODUCTION OF COIR AND COIR PRODUCTS IN INDIA

Production of coir products in India is estimated on the basis of actual exports, trends of internal consumption assessed from the movement of coir products by rail/streams/roads and estimated consumption in producing centers. The production of coir and coir products in India for the period of 2002-03 to 2008-09 furnished in Table 3.

Under the plan programmes while pursuing measured for advancement of fibre coir sector coordinated action and concentrated efforts were made with the involvement of the Industries Directorate of the government of the concerned states union territories and entrepreneurs whereas white coir production has remained stagnant due to non-availability of green husk in the required quantum.

Marketing of Coir Goods

The most important player in coir industry with reference to International trade is the exporters. There are two types of exporters in the coir industry, viz. : (a) Permanent Exporters, and (b) Provisional registration holders. In the export market

TABLE 3

Production of Coir and Coir Products

(Quantity in Metric Tonnes)

Item	2002-03	2003-04	2004-05	2005-06	2006-07	2007-08	2008-09*
(1)	*(2)*	*(3)*	*(4)*	*(5)*	*(6)*	*(7)*	*(8)*
Coir Fibre	3,53,700	3,64,000	3,85,000	4,10,000	4,30,000	4,37,800	4,91,000
Coir Yarn	2,26,800	2,32,500	2,45,500	2,70,000	2,88,000	2,90,000	2,92,900
Coir Products	75,750	77,900	98,000	1,20,000	1,70,000	1,72,000	1,73,550
Coir Rope	50,000	50,000	50,000	50,000	50,000	52,000	52,470
Curled Coir	28,000	29,500	36,500	38,000	48,000	49,000	49,540
Rubberised Coir	50,250	51,000	60,000	62,000	68,000	70,000	7,840

*Provision

Source : Annual Report, 2008-09, Coir Board.

according to Indian export destinatives about 67 countries all over the world imports coir and coir products. Major Indian products are coir fibre, coir yarn mat and mattings, coir rugs, mourzouk, rubberized coir, coir geotextiles, coir poles, cocologs and coir pith. The major importing coir and coir products from India are USA, European countries, Canada, Australia, Russia, etc.

India's export of coir and coir products in various segments in 2008-09 is represented in Table 4. Total exports decreased by 1.12 percent from Rs. 894.7 crore in 2007-08 to Rs. 636.36 crore in 2008-09. The segment "Handloom Mats" dominates coir exports even though there is a marginal decline of 3.14 percent in 2008-09.

Other segments registering a significant growth during the period 2007-08 and 2008-09 are: Coir fibre (95.31%), Rubberised coir (37.85%), Tufted mat (13.50%), Coir pith (32.54%), geo-textiles (10.13%) and Handloom matting (23.11%). On the other, the segments showing a steep decline between 2007-08 and 2008-09 are: Powerloom mat tings (3.64%), Coir Yarn (27.78%) Hand loom mattings (8.66%), Coir rugs (49.68%), Power loom mats (23.97%) and coir other sorts (67.10%).

India's Exports—Country-wise

Country-wise export trends, as may be seen from Table 5, show that USA continues to be the largest market for Indian coir and coir products in terms of rupee value. Exports to USA during 2008-09 registered marginal decline of 1.96 percent. Countries to which exports from the coir industry increased during the year were: Belgium (54.86%), Australia (10.43%), Germany (25.86%) and UK (5.75%). On the other, the countries showing a declining trend during the period 2008-09 comprised: Canada (-0.11%), Spain (19%), Italy (4.41%), Netherlands (2.82%), Spain (18.07%) and France (3.56%).

A major change that has happened in the Indian Coir Export is the change in Minimum export price policy with process of globalisation and liberalization of the Indian economy, the Coir Board Introduced MEP in 1996 to satisfy the requirements of Indian coir exports. The introduces of MEP ensured better export realization to the exporters. The exporter took up the opportunity of excess production capacity bargain

TABLE 4

Segment-wise India's Exports of Coir and Coir Products During 2004-05 and 2008-09

Segment	*2004-05*	*2005-06*	*2006-07*	*2007-08*	*2008-09*	*Percentage increase in 2008-09 over 2007-08*
(1)	*(2)*	*(3)*	*(4)*	*(5)*	*(6)*	*(7)*
Handloom mat	251.29	266.99	273.71	543.00	235.37	(-) 3.14
Tufted mat	94.99	116.06	182.13	199.11	225.98	13.50
Coir pith	30.42	38.73	53.82	63.84	84.62	32.54
Coir Yarn	33.58	30.19	31.62	26.66	19.26	(-) 27.78
Handloom matting	23.34	19.13	23.55	18.79	17.16	(-) 8.68
Geo Textile	10.50	11.41	13.35	14.44	15.91	10.13
Coir Fibre	1.86	1.96	10.76	12.24	23.90	95.31
Rubberised Coir	3.41	3.77	6.98	8.52	1.17	37.85
Coir rugs and carpets	10.03	7.30	3.29	1.34	6.76	(-) 49.68
Curled coir	0.12	—	2.08	1.52	2.23	47.21
Powerloom mat	9.31	10.27	1.69	5.26	4.00	(-) 23.97
Total (Incl others)	473.40	508.45	605.17	894.7	636.36	(-) 1.12

Source : Coir Board, Kochi.

TABLE 5

India's Exports of Coir and Coir Products to Major Countries

Segment	*2004-05*	*2005-06*	*2006-07*	*2007-08*	*2008-09*	*Percentage increase in 2008-09 over 2007-08*
(1)	*(2)*	*(3)*	*(4)*	*(5)*	*(6)*	*(7)*
USA	186.25	204.70	221.98	200.50	196.60	(-) 1.96
UK	49.42	48.46	55.39	49.50	52.35	5.75
Germany	30.11	38.15	45.92	42.00	52.87	25.96
Netherlands	36.57	35.40	41.95	49.54	48.14	(-) 2.82
Italy	24.19	21.30	26.29	28.16	26.92	(-) 4.41
Spain	18.49	19.47	25.39	24.23	19.85	(-) 18.07
Canada	10.10	12.91	19.66	20.49	20.47	(-) 0.11
France	19.62	17.93	17.15	19.69	18.99	(-) 3.56
Australia	10.73	9.63	15.10	17.28	19.08	10.43
Belgium	9.03	10.52	10.04	9.78	15.14	54.86
Total (incl others)	473.40	508.45	605.17	461.17	470.41	2.00

Source : Coir Board, Kochi.

with the manufactures to reduce the price under the new policy embellishment has been taken out of the purview of MEP and also certain products like power loom products and geo-textiles. The Coir products are in demand because of its special attributes, viz. : (1) fitness of purpose, (2) price, (3) craftsmanship, (4) quality, and (5) attractiveness.

PROBLEMS FACED BY THE COIR INDUSTRY

The Coir Industry has stuck to traditional methods of production for historical and sociological reasons. The industry cannot go on far into the future, with its present method of production, mechanization of defibering and spinning sectors would displace labour in large numbers. Inadequacy of raw material is one reason for the decrease in employment and the high cost raw material in the cost of production.

Primary coir is the large employer unit but mostly under employment, part time employment and secondary employment, with below part sustenance wages conditions of work to a large mass of unorganized casual workers. The entrepreneurial initiative in coir is essentially individualistic with a property orientation, even there are some corporate facades. The role of professionalism is some of what restrained by the very definition of trade structure large business houses of a country have chosen to keep away from coir and again expatriate interest from consuming countries and multinational have been kept away.

The organization of overall industry is fragile. The weakest link in its economic strength is a discarded waste husk becoming the primary raw material, lack of integration with the effective farming lobby of growers of coconuts, absence of large corporate and professionalism, absence of even remote nexus with product sector. These factors are inescapable realities not readily conducive for exponential forward in growth in volume value, product profile and quality of coir industry.

The growth of coir in quantity or quality terms in its traditional settings is not optimistically promising while the economic disabilities of the commodities and the weakest of trade and explicit coir areas, the social issues of the coir workers are compounding the complex problems.

When we take into account the market, the domestic market within the country largely by private traders. The internal consumption coir yarn in India is about 85 percent and only 40 percent of indigenous production of coir door mat, mattings, etc., is estimated to be consumed within the country.

Because of the emergence of plastic fibre, the substitute for fibre, the demand for coir fibre declining day-by-day. In spite of positive growth of coir products from India a general lowering trend has been noticed in the international market in the demand for coir products *vis-à-vis* other competing products the reasons for this trend could be attributed to the ever increasing shift competition synthetic floor coverings and cheapes natural substitutes like gross products by aggressive marketing efforts. The marketing efforts of coir exporters is more or less confined to the age-old convention pattern and have not been very successful to create a niche for this product in the target market. It is as fact that the lack of overseas market intelligence especially for favorite designs colours, etc. and insufficient marketing efforts are serious impediments to increase coir exports. The traditional coir item such as coir mats, mattings, rugs carpets, etc., have not shown appreciable growth exports.

The high tariff rate for coir existing in various importing countries is one of the main hurdles for promoting coir and coir products. The tariff regime prevailing now varies between 2 to 30 percent in various countries according to the information collected by the coir board. Apart from the tariff, very often several non-tariff barriers are impeding the growth of export market for coir products. They include allegations of child labours, restrictive banking facilities import quota restriction, antidumping measures, etc.

SUGGESTIONS

Though India has evolved several modern methods both in terms of product and marketing there are several potentials which we have not been able to tap. There are several countries where coir has not reached and several countries where coir products are available. The majority of the people do not know anything about coir. Coir and Coir-related products are definitely finding new applications and markets. However, to succeed in there efforts, the industry has to be given emphasis to the following factors:

- Upgradation of the skill workers though training for increasing productivity and income.
- Promotion of research and development in process improvement, product diversification modernization elimination of drudgery and pollution and extensions of fruits and R and D to the concerned sections in the coir industry.
- Improving the quality of coir yarn and coir products promoting new coir products like geo-textiles, rubberized coir, needed belt, coir pith, etc.
- Developing new pattern of designs.
- Expansion of domestic market through market development scheme. Particularly in domestic exhibitions and fairs by Coir Board, publicity in T.V. and news papers, etc.
- Increase in export through market development missions sponsoring delegations participation in trade fairs and exhibitions abroad.
- Congenial work atmosphere.

CONCLUSION

The liberalization policy of the government is certainly a welcome thing and should be continued in the interest of the trade. It should be carried out more R and D work in the spinning areas of remote Coastal belts in improve to method of production and also living conditions of coir workers. The raw material availability of coir industry. The coconut is available in abundance in India and further processing activities will generate employment to village artisans engaged in this sector.

REFERENCES

Coir Board, A Ten Year Plan for Development of the Coir Industry, 1996.

Coir Board, 2002-03, Fortyninth Annual Report of the Coir Board, Kochi, Kerala.

Coir Board, 2004-05, Annual Report of the Coir Board, Kochi, Kerala.

Coir Board, 2008-09, Annual Report of the Coir Board, Kochi, Kerala.

Coir Board, 2005, Unpublished Database, Kochi, Kerala.

Directorate of Coir Development (2003-04), Trivandrum, Kerala.

Kerala State Industrial Development Corporation Ltd. (KSIDC), 2004, Sector Study.

Report on Coir Industry, An Executive Summary, A.C. Rajan and M.T. Binil Kumar.

Kumaraswamy Pillai, M. (2005), *Technology Packages* for *Development of Coir.*

Clusters in India, paper presented in National Seminar on the Cluster/ Consortium-based Development of Coir Industry, Ashoka Hotel, New Delhi, 28 September.

Moir, B. (2002), Foreword of the Proceedings of the International Coir Convention, Colombo, Sri Lanka, 13 and 14 June 2002, published as CFC Technical Paper No. 20.

Michael, E. Porter (1998), "Clusters and the New Economics of Competition", *Harvard Business Review,* November-December.

Nair, G.K. (2005), Centre envisages regeneration scheme for coir industry, Kochi, *Business Line,* Oct. 7.

Parliamentary Standing Committee on industry on Coir Board—A Performance Review pertaining to the Ministry of Agro and Rural Industries, Parliament of India.

Rajya Sabha, 108th Report on action taken by the Government on the Recommendations contained in the 91st Report, December, 2002.

website:http//rajyasabha.nic.in; e.mail:rsc-ind@sansad.nic.in)

State Planning Board, 2004, *Economic Review,* Government of Kerala, Trivandrum, Kerala.

CHAPTER

36

Role of Non-Farm Sector in Rural Development (Special Reference to Small Scale Industry)

R.B. BHANDWALKAR AND R.R. GAWHALE

In India about 60 percent of the total workforce is engaged in the farm sector, contributing about 25 percent of the national Income. This reflects on the existence of widespread poverty in this sector. During the central planning era, rural development in transition economies was frequently associated with the agricultural economy. The Indian economy grew at an impressive rate in the last decade and demographic pressure also slowed. The incidence of unemployment towards the end of the 1990s was more than seven percent employment in rural sector which is associated mostly with agriculture has stagnated so that employment in the non-farm sector becomes an important option. In recent years the focus has begun to shift to also include the non-farm rural economy defind as being all those activities associated with wage work or self-employment in activities that are not directly derived crop and live stock production but located in rural areas. In India economic opportunities in the non-farm sector have also increased.

Rural non-farm sector holds the key to faster economic development of the country. It has potential and promise for generating employment and increased income in the rural areas. It is well known that agricultural growth stimulates non-farm activities through backward and forward linkage. The developmental factors like modernization of agricultural and its commercialization increased demand for non-agricultural goods and services, growing literacy, urbanization, have tried to pull the labour force away from the farm sector to more lucrative non-farm activities. The role of the rural non-farm sector there are at least four arguments in favor of promotion of the rural non-farm sector: (1) It provides employment for a growing labourforce. (2) It contributes to growth or development. (3) It slows rural-urban migration and helps control urban congestion and pollution. (4) It promotes an equitable distribution of income and contributes to the alleviation of poverty.

Growth of the rural non-farm sector especially industry in correlated not only with growth of the farm sector but also with growth of the urban industrial sector. There are segments of the rural non-farm sector especially those that produce tradable commodities, including small scale industries, that are in competition with urban activities or industries. Therefore, it is up to the more labour-intensive rural non-farm sector to absorb excess labour promote economic growth and diversity income sources. The percentage of rural workers employed in the non-farm sector averages 20-50 percent but varies from one country to another.

Demand for the goods and services produced by the rural non-farm sector is derived from several sources. The response of the non-farm sector to the demand for its goods and services depends on the availability of labour, access to capital or credit availability of infrastructure, and access to technology including production technology and marketing techniques. As agricultural production increases it generates demand for inputs such as seed, water and fertilizer and for farm implements produced by the non-farm sector. The need to process food and agricultural raw material also stimulates rural non-farm activities.

The role of small scale industry in non-farm sector is very important the small scale industrial sector which plays a pivotal role in the Indian economy in terms of employment and growth has recorded a high rate of growth. Since independence in spite of stiff competition from the large sector and not so encouraging support from the Government. This is evidenced by the number of registered units which went up from 16,000 in 1950 to 36,000 units in 1961 and to 33.7 lakh units in 2000-2001 and 91.46 lakh units in the unregistered sector thus recording a total number of 105.21 lakh units has increased from 79.6 lakhs in 1994-95 to 133.68 lakhs in 2007-08 indicating an annual average growth rate of 4.1 pereent but their production (at 1993-94 prices) increased from Rs. 1,09,116 crores in 1994-95 to 2,77,668 cores in 2005-06, i.e. an annual average growth of 8.8 percent. As a consequence of the increase in small scale industry sector more especially in the unregistered sector employment increased from 191.4 lakhs in 1994-95 to 322.3 lakhs in 2007-08 recording an average growth rate of 4.07 percent per annum. So far as exports by the small scale industry sector are concerned they increased from Rs. 29,068 crores in 1994-95 to Rs. 1,77,600 crores in 2006-07 recording a growth rate of 16.26 percent annum.

Obviously the growth rate of the small scale sector has been faster both in terms of output and employment. The rapid growth of the small scale industries has a great relevance in our national economic policies. The growth of the small sector improves the production of the non-durable consumer goods of mass consumption. The capacity utilization in the small sector as a whole was of the order of 53 percent. There were, however many units having high capacity utilisation, e.g. industries utilising 60 to 80 percent of the capacity included leather goods, readymade garments oils, woolen knitwear, etc. Industries like plastic products had very low capacity utilization 29 percent. The obvious conclusion is that the growth of small scale industries in terms of number and output is comparatively much higher in reserved items than in unreserved items. The policy of reservation has therefore positively helped the growth of this sector. After the announcement of industrial policy of 1991 the Government has been deserving more and more items of the small scale industry sector.

The expansion of roads, transport and communications infrastructure leads to specialization and division of labour by rural households. It promotes the development of a trade marketing and distribution network, including sub-contracting arrangements linking farm and non-farm sectors to local towns and big cities. Primany and secondary education promotes the growth of the rural non-farm sector. Economy-wide policies such as trade, exchange rate and general regulatory policies and sector-specific policies such as credit or technical assistance can stimulate the non-farm sector.

Import liberalization is also likely to improve the relative position of small-scale enterprises since the large-scale inducting are more likely to be adversely affected by import-competing products for example China's items, etc. A number of questions must be addressed regarding the evolving nature and future role of the non-farm sector in developing countries.

References

Nurul Islam, The Non-farm Sector and Rural Development.

Dutt, R. and S.K.M. Sundaram (2008), The Indian Economy, S. Chand and Co., New Delhi.

Mishra and Puri, Indian Economy, Himalayan Publishing House, Mumbai.

Govt. of India, *Economic Survey*, Relevant Issues.

CHAPTER

37

Tourism : A Key Factor to Solve the Problems of Unemployment and Poverty in Bihar

R.K. CHAUDHARY AND AJAY KUMAR RAO

I. INTRODUCTION

Bihar, the birth place of loard Mahavira, Gyan Place of Buddha, Tapo-Bhumi of Maharshi Valmiki, Karm-Bhumi of Rashtrapita Mahatma Gandhi and a State of Indian Union famous for a lot of other important historical places is now become the home of 83.0 million Indian.[1] After the bifurcation of the state from the southern plateau of Jharkhand, it features only a large stretch of plains with neither any well funded industry nor any major town except the state capital of Patna. This state is characterized as the least urban population (10% of total population of the state is urban as per 2001 census), Lowest literacy rate (47.00%), the highest poverty ratio (nearly 30% in 2006-07) and lowest per capita income at current prices in 2005-06 (Rupees 7875) among all the major states of union.[2] On the basis of above said characteristics, Bihar is characterized as the most backward state of India.

The GDP growth rate during 1993-94 to 2003-04 of the state was 5.03% in compression to 6.03%[3] for the country as a whole. At the same time the population of Bihar has increased by 23.4% during 1980 and 28.4% during the 1990, while the population of the nation as a whole has raised by 23.9% during 1980 and declined by 21.3% in 1990, this is why the population density of Bihar stands at a phenomenally high level of 880 persons per square km as against 234 persons for a country as a whole according to census 2001 at the same time, Bihar's per capita income, which was approximately 60% of all India average during early 1960, declined up to 40% in 1993-94, 31% in 2003-04 and 30.03% in 2006-07.

II. STRUCTURAL COMPOSITION OF GSDP OF BIHAR

In the above said economic constraints the other social performances of this state has become dismal that is reflected by it's structural composition of gross state domestic product GSDP of the state of Bihar and the India as a whole Table 1 highlights.

TABLE I

Structural Composition of Income (State of Bihar and the Nation as a Whole) during 2003-04 to 2006-07

Category	*Gross State Domestic Product at 1999-2000 prices in*			
	2003-04		*2006-07*	
	Bihar	*India*	*Bihar*	*India*
Primary Sector	40.00	24.03	35.00	20.55
Secondary Sector	10.20	24.54	11.20	24.71
Tertiary Sector	50.80	51.43	53.80	54.74
Total	100.00	100.00	100.00	100.00

Source : Economic Servey of India, 2007-08, Govt. of India, Ministry of Finance, Economic Division.

III. POVERTY AND UNEMPLOYMENT SCENARIO IN BIHAR

The Composition of GSDP of Bihar reflects that Bihar is still after the six decades of independence recognized as a predominantly an agricultural economy. And it is an open truth that the concentration of excess poverty is the basic characteristics of an agricultural economy. This is why the major population of poor reside in Bihar. The percentage residing in Bihar share of poor was 14.31% in 1983 (16.58% in rural and 6.25% in urban areas) increased up to 15.40% in 1993-94 (18.48% in rural and 5.57% in urban) and 16.36% (19.48% in rural and 7.33% in urban) in 1999-2000. After the bifurcation of Jharkhand has made the matters worse for the present Bihar. Apart from very high level of poverty, the state has also showed a regional concentration of poverty. Majority of poverty sections of this state has concentration in North Bihar nearly 57% of poor population of the state are concentrated in North Bihar region, the region lacking behind the industrialization. To have an exact idea about trend and pattern of poverty in Bihar it would be worth to go through the bar-diagrams of 1 and 2. As it has widely being accepted that poverty is the outcome of unemployment. Hence, it would be pertinent to throw a flash of high on the unemployment ratio in Bihar. A clean cut idea is reflected by diagram 2.

No doubt a lot of programs as well as schemes are undertaken by the central and state Governments to eradicate poverty by reducing unemployment in this state. It is an open truth that all the programs launched so far are either wage-employment or self-employment and the procedure of these implementations are designed in such a way so that a direct attack on poverty would be insured. No doubt, the poverty and it's intensity has substantially reduced but still it appears as a major problem before the Government and an acute obstacle before the path of development of the state. The declining trend can be seen in Diagram 3.

DIAGRAM 1

Proverty Ratio in India and Bihar

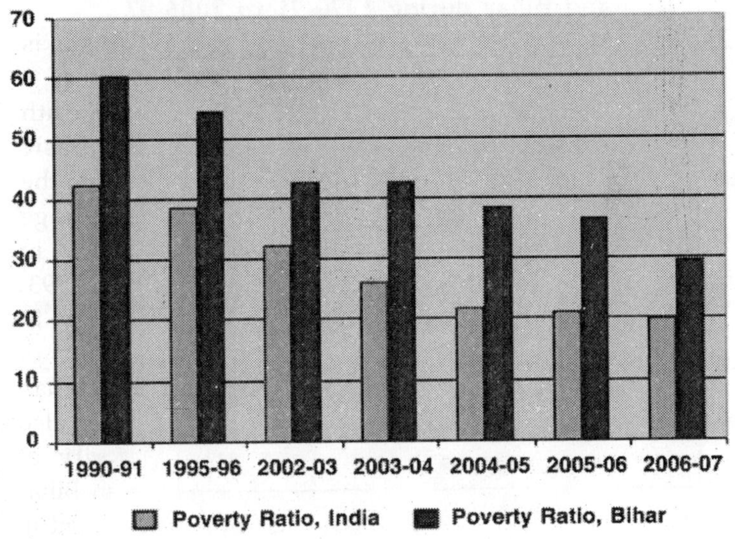

DIAGRAM 2

Un-employed in Bihar

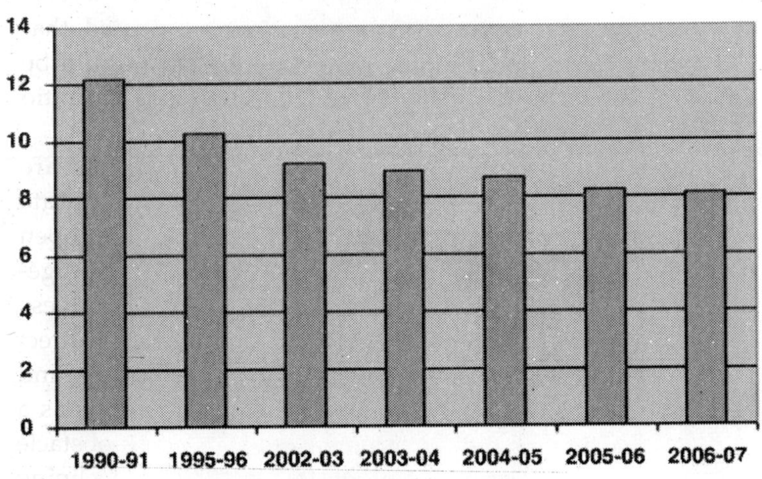

TABLE 3

Trend and Pattern of Poverty Ratio in India and Bihar during 1990-91 to 2006-07

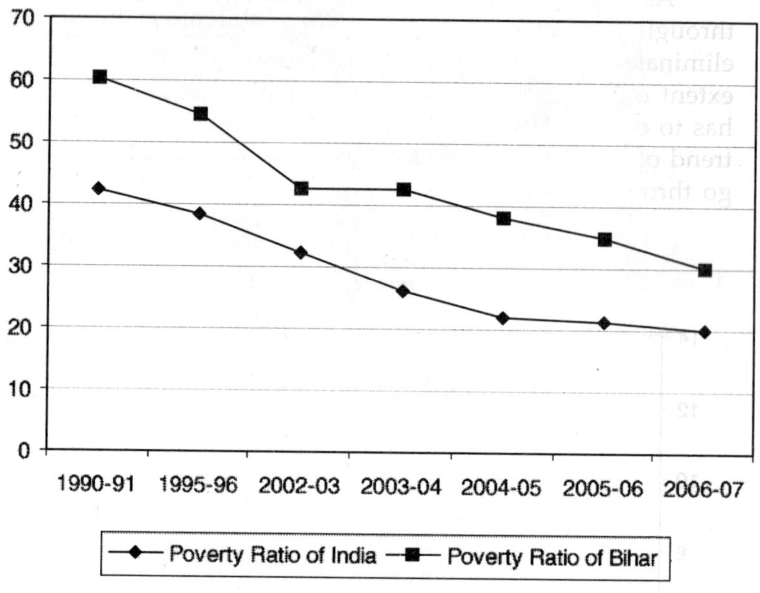

TABLE 2

Poverty Ratio and Employment Scenario in India and Bihar during 1990-91 to 2006-07 Period

Year	*India*		*Bihar*	
	Poverty Ratio	*Unemployment Ratio**	*Poverty Ratio*	*Unemployment Ratio*
1990-91	42.32		60.28	12.17
1995-96	38.47		54.42	10.26
2002-03	32.18		42.60	09.22
2003-04	26.03		42.60	08.93
2004-05	21.80		15.00	08.67
2005-06	21.13		15.00	08.26
2006-07	19.87		15.00	08.12

*Employment Ratio refers to active or working population.

IV. TOURISM AS A KEY FACTOR TO REDUCE POVERTY AND UNEMPLOYMENT IN BIHAR

As we know, poverty is the outcome of unemployment through a lot of provisions are made during planning era to eliminate or to reduce unemployment ratio and up to some extent our Government has succeeded to reduce it but still it has to cover a large distance in this regard. To understand the trend of reduction in unemployment ratio it would be better to go through diagram 4.

DIAGRAM 4

Unemployed Ratio Bihar

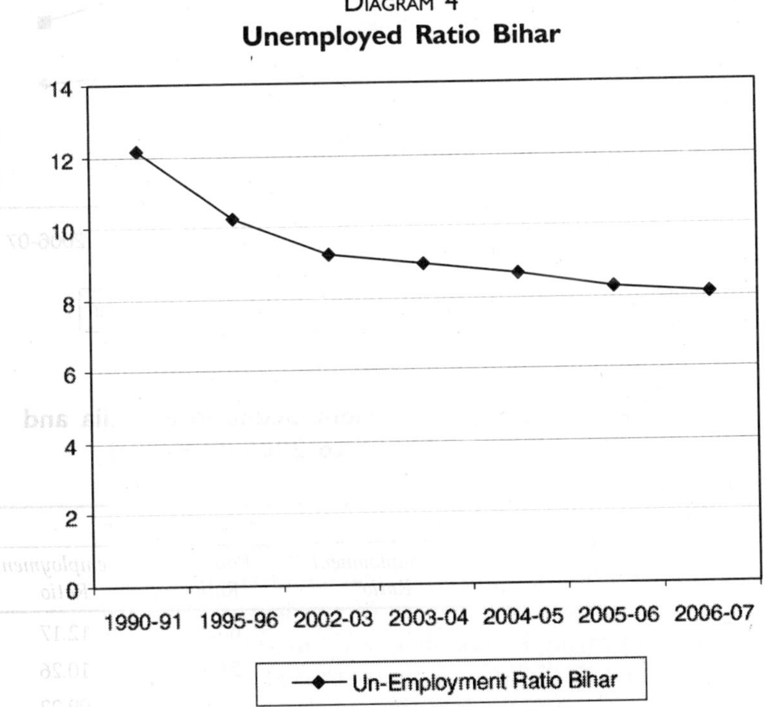

As we know, tourism has been associated with Indian traditions in the form of parikarmayataras and pilgrimages. Tourism is not only enables the human beings to enjoy the beauty of life through releasing the tension of fast life of modern world but accelerate the process of employment and

Income generation. Today tourism revolution is sweeping the globe, a involution promising much and delivering a great deal. It has emerged as a most lucrative business of the world exchange, fielding tax revenues, promoting growth of ancillary industries, generating income and employment and in the development of industrially backward regions through it's various linkage effects, Department of Tourism has adopted the definition as per recommendation of the UN conference of international Tourism, in Rome (1963). The richest cultural heritage of Indian civilization is most probably influenced by it's tourist business, because it is not only a financial transaction but also associated with the appreciation of cultural heritage. For our country in general and Bihar in particular possess thousand years of old civilization reflects it's glory through the history and heritage of large cities, architechctural beauty, religious monuments, including caves and their sculptures and paintings.

V. TOURISM ATTRACTIONS IN BIHAR

Bihar is enrich with monuments that bear the memory of lord Buddha and Mahabir Jain. It has ruins of year old enlightened educational place like Nalanda which is world famous and established in 5th century AD in spite of these places. Bihar tourism is enriched by so many historical places having greater interest of tourism. Besides, such enrich tourism places, due to inefficient law and order situation inadequate infrastructural facilities and acute dearth of world class hotels, restaurants as well as transport and communication Bihar has failed to exploit its tourist places in attracting tourist and mobilizing the tourist revenue in accelerating the process of development through enhancing the foreign earnings. As it has estimated that the visit of a single tourist in the state induced multiple impact on the process of income and employment generation. But the magnitude of per tourist earnings is much more depends upon the socio-economic status of the state in general and the process of development in it (in the state). As we see the earnings of per tourist in Bihar is quite very low (that is Rs. 5,510 only) in comparison to Kerala (Rs. 20,665)[4] and Goa (Rs. 38,780).[5] This least earnings in Bihar reflects about the cheep rate of goods and services in Bihar which in turn highlights the inferior quality of goods and services available in this state. Availability of superior quality of goods and services in a state is decided by the level of employment and generation in spite of it's bright and luminous tourist places, due to lack of political will ignorant attitude of administration, mass illiteracy, huge unemployment and excessive out-migration of labour. Bihar has failed to capture the momentum of growth along with the other states of Indian Union. The poverty of Bihar can easily be seen with the add of following picture. Having seen this facts one can easily understand the story behind the poverty of Bihar. No doubt a lot of efforts have been undertaken to solve the problems of poverty and unemployment and positive impact of those programs have also started to come. But it requires some greater efforts to reach the target in near future.

As it has reflected by the findings of economic survey of Bihar 2009-10, Bihar registered an annual growth 11.35% during 2004-05 to 2008-09. Per capita income of the state has increased from Rs. 7443 in 2004-05 to Rs. 13959 in 2008-09. The

TABLE 3

Year-wise No. of Foreign Tourists Visited in India and Bihar and Foreign Exchange Received during 1990-91 to 2006-07

Year	*India*		*Bihar*	
	No. of Visitors	*Earnings Rs. in Crore*	*No. of Visitors*	*Earnings Rs. in Crore*
1990-91	1025400	244	6270	3.45
1995-96	132100	8640	52840	29.11
2002-03	1523400	15991	42650	23.50
2003-04	2354100	23054	104160	57.39
2004-05	2848400	29858	133730	73.69
2005-06	3038600	34871	148820	81.99
2006-07	3227200	41127	174270	96.03

Source : For India all data are taken from Table.

TABLE 4

Foreign Exchange Earnings from Tourism in India (1991 to 2009#)

Year	*No. of Visitors*		*Earning (Rs. in Crore)*		*Proverty Ratio*		*Un-employeed Ratio*
	India	*Bihar*	*India*	*Bihar*	*India*	*Bihar*	*Bihar*
(1)	*(2)*	*(3)*	*(4)*	*(5)*	*(6)*	*(7)*	*(8)*
1990-91	1025400	6270	2444	3.45	42.32	60.28	12.17
1995-96	1132100	52846	4640	29.11	38.47	54.42	10.26
2002-03	1523400	42650	15991	23.5	32.18	42.6	9.22
2003-04	2354100	104160	23054	57.39	26.06	42.6	8.93
2004-05	2848400	133730	29858	73.069	21.82	38.23	8.67
2005-06	3038600	148820	34871	81.99	21.13	36.46	8.26
2006-07	3227200	174270	41127	96.03	19.87	29.49	8.12
2007-08	387698	346532	46932	108.36	NA	NA	NA

Source : Economic Survey of India, 2009-10, Publication Division, Ministry of Information and Broadcasting, Govt. of India, New Delhi.

construction boom has translated in to mushrooming growth of apartment and tale density has risen from 5.34% in 2006 to 22.18% in 2009. Internet connections in the rural Bihar are up to 4.99 lac in 2009-10 compared to 0.43 lac in 2008-09. Spared of road connectivity has grown at a faster rate which in town accelerates the internal as well as external connectivity of the people of Bihar.

The maintained law and order situation in Bihar has not only raised the faith of people in police administration but attracts tourist a lot. This is reflected by the increase in number of tourists 3.46 lacs in 2008 against 0.61 lac in 2003 as showing the data of Tables 3 and 4.

No. of foreign tourists visited tourist places in India and Bihar can be seen with the aid of Diagram 5.

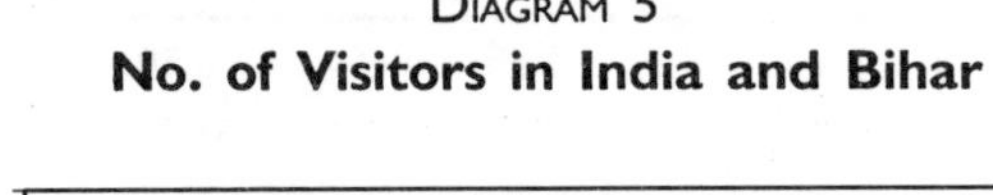

DIAGRAM 5

No. of Visitors in India and Bihar

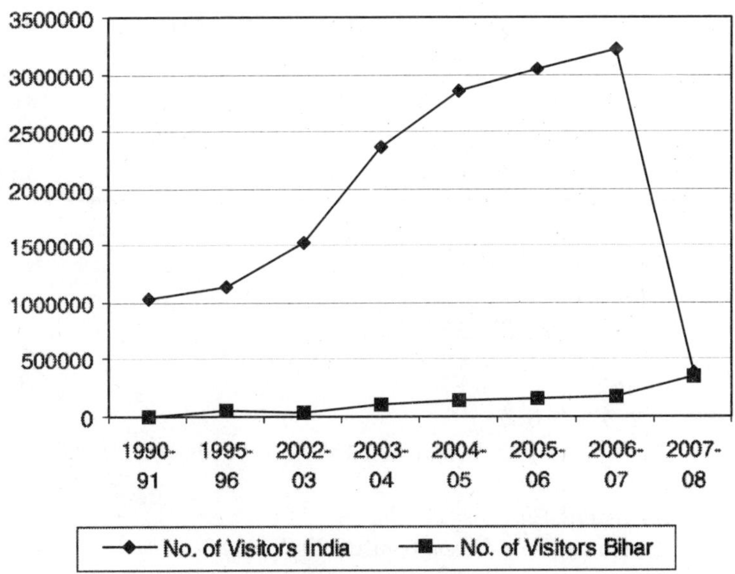

To have clear cut picture we would like to through a flash of light on poverty ratio, unemployment rate and foreign earnings in India and Bihar.

VI. ANALYSIS OF DATA

Year-wise Distribution : Foreign tourist, tourist revenue, poverty ratio and unemployment rates in Bihar during 1990-91 to 2006-07 is given in Table 5 which gives a basis for R-analysis to test the hypothesis.

H : Tourists earning is positively related with poverty and unemployment.

Ho : There is not any correlation between tourists earnings and poverty and unemployment.

TABLE 5

Year	*Year level*	*ear*	*pvr*	*une*
1990-91	1	3.45	60.28	12.17
1995-96	2	29.11	54.42	10.26
2002-03	3	23.50	42.60	9.22
2003-04	4	57.39	40.80	8.93
2004-05	5	73.69	38.70	8.67
2005-06	6	81.99	37.90	8.26
2006-07	7	96.03	36.20	8.12

>summary (lm(pvr~ear+une))
Call:
Lm(formula=pvr~ear+une)

Residuals:
1 2 3 4 5 6 7
-1.6746 4.3674 -1.1961 -0.8456 -1.1788 0.6081 -0.0803

Coefficients:
Estimate Std. Error t-value Pr(>|t|)
(Intercept) -11.97025 17.63909 -0.679 0.5346
ear -0.01145 0.06455 -0.177 0.8678
une 6.07760 1.55384 3.911 0.0174 *

Signif. codes: 0 '***' 0.001 '**' 0.01 '*' 0.05 '.' 0.1 ' ' 1

Residual standard error: 2.539 on 4 degrees of freedom

Multiple R-squared: 0.9495, Adjusted R-squared: 0.9243

F-statistic: 37.61 on 2 and 4 DF, p-value: 0.002550

Findings

If we are interested to reduce the poverty by 0.01145% we must have to increase the earnings of tourism by one Rupee, i.e. development of tourist places requires huge investment in the field of infrastructural facilities, road and transport (all road, rail and airways), telecommunication, Hotel Management, law and order situation and maintenance of Good Governance etc.

The value of R^2 is 0.9495 and adjusted R^2 = 0.9243 which concludes that there is very-very strong but negative correlation between earnings form tourism and poverty and unemployment. P-value is very-very low, so Ho can be rejected and H, is accepted.

VII. CONCLUSION

To conclude we may say that :

- Along with all poverty alleviation as well as employment generation programs such as NREGA, SGSY, PMRY, SJGSY, IAY, etc. Government must take notice to develop all tourist places in Bihar. All these development programs are tagged with poverty alleviation.
- An equal and proportionate distribution of funds must be allocated for acquiring balanced development.
- Availability of all necessary infrastructural facilities concerned with tourist development must be provided at an adequate basis and proper maintenance could also be ensured.
- Beautification of tourist places situated in Rural Bihar acts as an pull factor to raise employment opportunities in providing all necessary services to the foreign as well as national tourists on the one

hand, at the same time it have its multiple effect on accelerating the prcess of rural development.

On the whole, we may say that development of tourist places have direct and positive correlation with the problems of unemployment and poverty, i.e. Tourism Development has both pull and push effect on employment generation which in turn accelerating the process of Rural Development.

Notes

1. India Year Book, 2009, Publication Division, Ministry of Information and broadcasting, Govt. of India, New Delhi.
2. Economic Servey of India, 2008-09, Govt. of India, New Delhi.
3. *Op. cit.*
4. Information Bureau, Govt. of Kerala, 2008-09.
5. Information Bureau, Govt. of Goa, 2008-09.

References

Mondrika Dutta, "Tourism in Bihar, Study Material, Global Meet for Resergent Bihar, 19-21, January, 2007.

India Year Book, 2009.

Economic Survey of India, 2009.

Interpreting Mathematical Economics and Econometrics, Byron D. Eastman.

CHAPTER

38

Indian Handicraft Industry

MAHENDRA RANAWAT AND T.V. RAMAN

INTRODUCTION

India is one of the important suppliers of handicrafts to the world market. The Indian handicrafts industry is highly labor-intensive cottage-based industry and decentralized, being spread all over the country in rural and urban areas. Numerous artisans are engaged in crafts work on part-time basis. The industry provides employment to over six million artisans (including those in carpet trade), which include a large number of women and people belonging to the weaker sections of the society.

In addition to the high potential for employment, the sector is economically important from the point of low capital investment, high ratio of value addition, and high potential for export and foreign exchange earnings for the country. The export earnings from Indian handicrafts industry for the period 2009-10 amounted to Rs. 8718.94 cr.

Although exports of handicrafts appear to be sizeable, India's share in world imports is miniscule. It is a sector that is still not completely explored from the point of view of hidden potential areas. India, a country with 26 states and 18 languages and more than 1500 dialects offers an enormous range of handicrafts from each of the states. Major centers in Uttar Pradesh are Moradabad also known as the *"Peetalnagari" (City of Brass)*, Saharanpur for its *wooden articles*, Ferozabad for *Glass*. The North Western state of Rajasthan has to offer the famous *Jaipuri quilts*, Bagru and Sanganer *printed textiles* and *wooden and wrought iron furniture* from Jodhpur. The coastal state of Gujarat comes with *embroidered articles* from Kutch. Narsapur in Andhra Pradesh is famous for its *Lace and Lace goods*. But this is only a small part of the total product range. India offers much more.

Handicrafts are classified into two categories:

1. Articles of everyday use
2. Decorative items

The craftsmen use different media to express their originality. The diversity of the handicrafts is expressed on textiles, metals—precious and semi-precious, wood, precious and semi-precious stones, ceramic and glass.

CRAFT CONCENTRATION AREAS

A wide range of handicrafts are produced all over Indian artmetalware/EPNS ware, wood carvings and other wooden art wares, imitation jewellery, hand printed textiles, shawls as art wares, embroidered goods, lace and lace goods, toys, dolls, crafts made of leather, lacquer ware, marble crafts, etc. Although it is difficult to limit a specific place for the particular craft, the following places are listed for their particular crafts.

Artmetalware : Moradabad, Sambhal, Aligarh, Jodhpur, Jaipur, Delhi, Rewari, Thanjavur, Madras, Mandap, Beedar, Kerala, Jagadhari, Jaselmer

Wooden Art wares	:	Saharanpur, Nagina, Hoshiarpur, Srinagar, Amritsar, Jaipur, Jodhpur, Jagdalpur, Bangalore, Mysore, Chennapatna, Madras, Kerala and Behrampur (WB)
Handprinted Textiles	:	Amroha, Jodhpur, Jaipur, Farrukhabad, Sagru and Sanganer
Embroidered goods	:	Kutch (Gujarat), Jaisalmer, Baroda, Lucknow, Jodhpur, Agra, Amritsar, Kullu, Dharmshala/Chamba and Srinagar
Marble and Soft Stone Crafts	:	Agra, Madras, Baster, Jodhpur
Papier Mache Crafts	:	Kashmir, Jaipur
Terracotta	:	Agra, Madras, Baster, Jodhpur
Zari and Zari Goods	:	Rajasthan, Madras, Baster
Imitation Jewellery	:	Delhi, Moradabad, Sambhal, Jaipur, Kohima (Tribal)
Artistic Leather Goods	:	Indore, Kolhapur, Shantiniketan (WB)

Source : Export Promotion Council for Handicrafts.

COUNTRY-WISE EXPORTS OF HANDICRAFTS

The major buyers for handicrafts (other than carpets) are as under:

Art Metal wares	:	U.S.A., Germany, U.K. and Italy
Wood wares	:	U.S.A., U.K., Germany and France
Hand Printed and Textiles	:	U.S.A., U.K., Germany and Canada
Embroidered and Crochetted Goods	:	U.S.A., Saudi Arabia, U.K., Germany
Shawls as Art wares	:	Saudi Arabia, U.S.A., Japan and U.K.

Zari and Zari goods	:	U.K., U.S.A., Japan and Saudi Arabia
Imitation Jewellery	:	U.S.A., U.K., Saudi Arabia and Germany
Miscellaneous Handicrafts	:	U.S.A., Germany, U.K. and France

Source : Export Promotion Council for Handicrafts.

In the changing world scenario, craft products exported to various countries form a part of lifestyle products in international market. The impact is due to the changing consumer taste and trends. In view of this it is high time that the Indian handicraft industry went into the details of changing designs, patterns, product development, requisite change in production facilities for a variety of materials, production techniques, related expertise to achieve a leadership position in the fast growing competitiveness with other countries.

The 6 million craft persons who are the backbones of Indian Handicraft Industry as provided with inherent skill, technique, traditional craftsmanship but that is quite sufficient for primary platform. However, in changing world market these craft persons need an institutional support, at their places, i.e. craft pockets for value addition and for the edge with other competitors like China, Korea, Thailand, etc.

TRIBAL HANDICRAFTS

India has the largest concentration of tribal population in the world. The tribal are the children of nature and their lifestyle is conditioned by the eco-system. India due to its diverse ecosystems has a wide variety of tribal population. Tribes' people constitute 8.14% of the total population of the country, numbering 84.51 million (2001 Census). There are 697 tribes notified by the Central Government under Article 342 of the Indian Constitution with certain tribes being notified in more than one State. More than half the Scheduled Tribe population is concentrated in the States of Madhya Pradesh, Chhattisgarh, Maharashtra, Orissa, Jharkhand and Gujarat whereas in Haryana, Punjab, Delhi, Pondicherry and

Chandigarh no community has been notified as a Scheduled Tribe.

The tribal handicrafts are specialized skills which are passed on from one generation to another and these handicrafts are means of livelihood of the artisans. However, in absence of any organized activity in this sector and the products not being adequately remunerative, there is a possible likelihood of the artisans taking up alternate livelihood options (which may involve migration as well). In such a case this age-old activity will die its own death. At this stage it is very imperative to understand the problems faced by this sector and suggest the strategies for development of tribal handicraft based on which certain policy level interventions need to be taken by the government to sustain the traditional tribal handicrafts.

HANDICRAFTS OF CHHATTISGARH

Cane and Bamboo

Chhattisgarh also has a rich tradition of bamboo and cane crafts. Items like, mats, baskets, hunting and fishing tools, agricultural implements, etc. are made from bamboo. The tribal belt of Chhattisgarh and Bastar are main centers for production of items made from bamboo.

Bamboo thickets are common sight in the State and tribal of Chhattisgarh have been putting their craftsmanship to work. Craftsmanship of Chhattisgarh tribal can be seen from varying articles of craft produce they make out of bamboo. Articles for daily as well as decorative use are produced by these artisans. Some of them will known Bamboo produce include agricultural implements, fishing traps, hunting tools and baskets.

Bell Metal (DHOKRA)

The bell metal or Dhokra Craft is one of the earliest known methods of metal casting. Metal pieces akin to Dhokra objects and figurines have been discovered at Harrappa and Mohenjodaro, leading to the belief that this craft dates back to prehistoric times. Today this craft is practiced extensively in Chhattisgarh, in the areas of Lalitpur, Raigarh, Sarguja and most importantly in Bastar.

Dhokra essentially refers to the casting of bell metal or brass using the 'lost-wax' technique. In Chhattisgarh; it is the Ghadwa community that is associated with this craft. Interestingly, in local etymology, Ghadwa means 'to shape or create'. A variety of products are crated by the Ghadwas for local use such as effigies of local deities, vessels, and jewellery. These handicrafts include items like elephant, deer, horses, etc. of different sizes. Bell metal craft is of vital significant for it is basically a tribal craft. The bell metal craft of Bastar area popularly known as "Dhokra craft" is mainly found in Jagdalpur and Kondagao and Narayanpur blocks. The Dhokla craft is claimed to be age-old craft being transmitted from generation to generation.

PROBLEMS AND CHALLENGES IN HANDICRAFT BUSINESS

Handicraft market is fragmented, it is very difficult to predict demand.

- Supply is one of the major constraints in this business as products are hand carved.
- Exact duplication of product is very difficult or not possible.
- Lead time is very high (from order to delivery).
- Transportation is one of the constraints as products are delicate.
- Availability of quality wood.
- Availability of workforce.

Objectives

- The main objective is to know the scope of handicrafts of Bastar (Chhattisgarh) in terms of demand, margin, quality and commercial viability.
- Secondary objective to develop market for handicrafts of Bastar (Chhattisgarh) to start business in this field.

Scope of the Work

The scope of work of the proposed study is as follows:

- To study the scope of handicrafts of Bastar (Chhattisgarh).
- To study existing status of tribal handicrafts of Bastar (Chhattisgarh).
- To asses the support provided by the governmental and non-governmental institution to artisans.
- To study existing status of handicraft market of Delhi.
- To study commercial viability of teak wood handicrafts.

Export Data Analysis

Handicrafts export in the year 2009-10 stood at Rs. 8718.94 Cr. which is 6.5% higher than corresponding previous year which stood around Rs. 8183.12 Cr. Handicraft sector is one of the most affected sector by recession. This sector was showing increasing trend from 2001-02 to 2006-07 and was growing at cumulative average growth rate of 17.54% but in the middle of 2007 world economy went in the recession export of handicrafts in 2007-08 fell from Rs. 17288 Cr. to Rs. 14012 Cr. decrease of Rs. 3276 Cr. or 18%. But worst was yet to come from 2007-08 to 2008-09, there was steep decline of 41%. But as the economy started recovering in middle of 2009 handicraft sector started to revive and export increased by 6.5% in year 2009-10. (see Diagram 1)

From the Diagram 2 it is clear that USA is major export destination of Indian handicrafts followed by UK and Germany. USA and Europe are major handicraft trading partners of India, approximate 60% of all export goes to USA and Europe. USA and Europe were the worst affected from the recession in year 2008-09 that is why there was decline in export by 41% from Rs. 14,012 cr. to Rs. 8188 cr.

The major export markets for woodenware are Europe (Germany, France, UK, Italy, etc.) and USA. Total export of woodenware in 2009-10 stood around Rs. 717.24 cr. and 50% of which went to USA and Europe.

Diagram I
Exports of Handicrafts

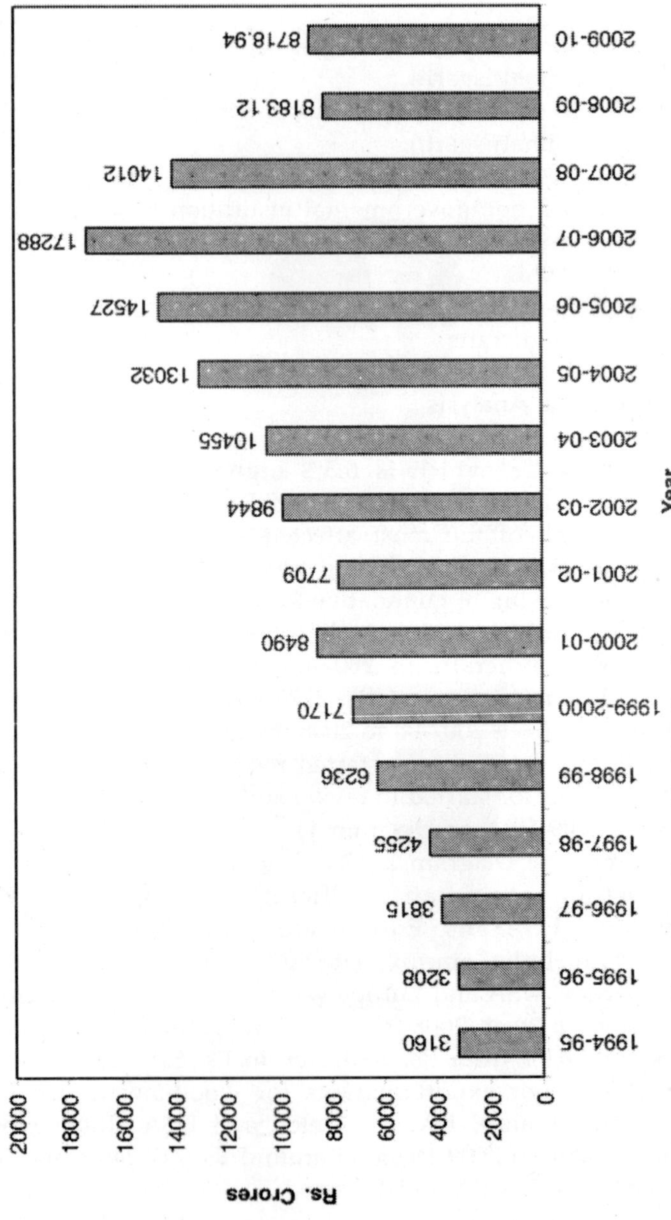

Year	Rs. Crores
1994-95	3160
1995-96	3208
1996-97	3815
1997-98	4255
1998-99	6236
1999-2000	7170
2000-01	8490
2001-02	7709
2002-03	9844
2003-04	10455
2004-05	13032
2005-06	14527
2006-07	17288
2007-08	14012
2008-09	8183.12
2009-10	8718.94

Source : Export Promotion Council for Handicrafts.

DIAGRAM 2

Country-wise Export's Share of Handicrafts during 2009-10

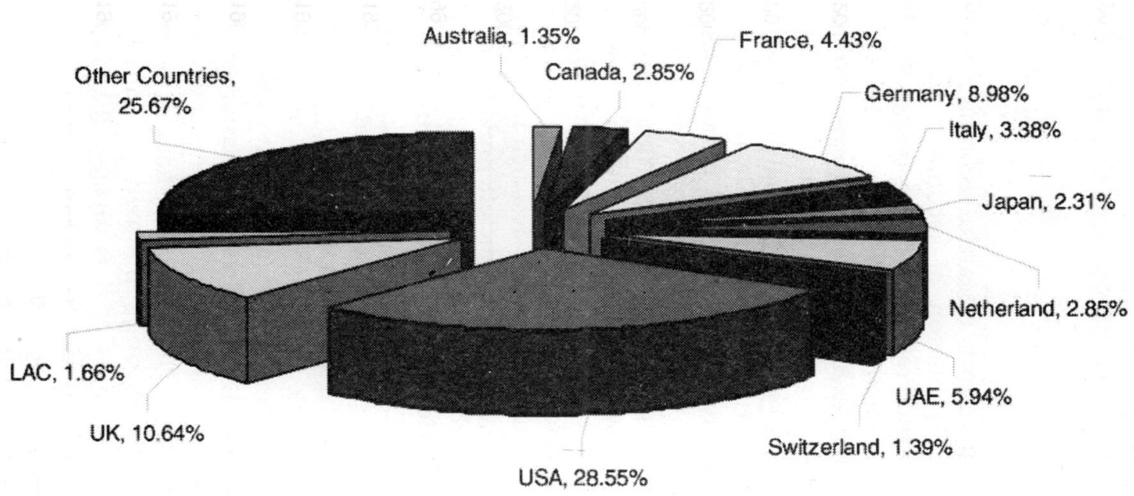

Source : Export Promotion Council for Handicrafts.

Again the major export markets for metal wares are USA and Europe. The total export of artmetalware stood Rs. 1877.64 cr. in the year 2009-10. Export of art metal ware to Europe and USA was approximate Rs. 1129.24 which is about 60% of total metal ware exports.

Analysis

Through exploratory research, secondary research and analysis and interpretation of secondary data it can be concluded that handicrafts of district Bastar (Chhattisgarh) have commercial viability and scope in Delhi. There is both kind of demand domestic and export exist for handicrafts products. In terms of quality, finishing and carving handicrafts of Bastar scores high and quite appreciated among handicraft retailer's exporters and art galleries. Data suggest that brass metal handicrafts are more demanded in comparison of teak wood and other handicrafts of Bastar. Demand for terracotta handicrafts in large emporiums export houses and in art galleries almost non existent. Data suggest that handicraft sector offers decent return on an average 30-40% margin is there in brass and teak wood handicrafts for retailers. According to data and exploratory research handicraft retailers are interested in teak wood handicraft provided that there is adequate supply and good carving and finishing exist in these handicrafts. Handicrafts of Bastar Chhattisgarh face stiff competition from UP and Rajasthan. It is advised by the handicraft retailers that teak wood handicraft should be marketed in tribal figure because it is USP of Bastar. The shortcomings of teak wood handicrafts are finishing and availability there is very less suppliers who offer quality teak handicrafts but wooden handicrafts of Bastar scores high in finishing. As far as availability is concerned there is abundant forest in Chhattisgarh. The need of the hour is to promote teak wood handicrafts through planned marketing strategy. Observation and analysis of data suggest that there is good scope of teak wood handicrafts in Delhi provided that planned marketing strategy is adopted.

There is huge demand of brass metal handicrafts in Delhi. The brass handicrafts of Chhattisgarh are made using lost wax technique which is old-age practice. While metal crafts of

Rajasthan and UP are machinery made that is why metal crafts of Bastar scores high in quality and demanded more.

Data suggests that most of the large handicraft stores of Delhi engaged in brass handicraft business which offers decent average return of 30-40%.

India is major supplier of handicrafts to world market but its share is only 2 percent in world exports lagging far behind China

Major export destination for Indian handicrafts is USA and Europe specifically Germany in Europe. USA and EU account for approximate 60% of Indian exports. Primary data also confirms these facts. USA and Europe were worst affected by recession consequently Indian exports declined by 41% percent in 2008-09 to Rs. 8188 cr from Rs. 14012 cr. As economy started recovering exports of handicrafts also started taking pace and increased by 6.5% to Rs. 8718.94 cr. from Rs. 8188.12 cr.

Tribal people of Bastar, Chhattisgarh, are deft artisans, they create masterpiece. Their handicrafts are depiction of their rhythmic life with jungle. But there are issues which need to be addressed. In absence of institutional support they are being exploited by middleman. These artisans are not getting adequate price for their products. These people need marketing, finance, transportation, training and infrastructure support. Although government extends its support to artisans but still there is a gap to be filled. There is a huge scope in this sector as we know handicrafts are invaluable.

RECOMMENDATIONS

Establishment of common facilities centers for imparting training, marketing, providing market intelligence, establishing linkages with financial institutions, providing raw materials, providing managerial inputs is the need of the hour. Following recommendations are suggested for development of promotion of tribal handicraft in India.

Availability of Raw-material at Reasonable Rates

- It has been observed that in few of the cases, the raw

material for handicraft products is not locally available.

- It has been suggested that raw material depots may be set-up in craft concentration tribal pockets to facilitate uninterrupted supply of standard raw material to craftsperson in appropriate quantity and quality at reasonable rates.
- In such circumstances the state governments may supply raw material like seasoned wood or wool at cheaper rates become incentive for the tribal to continue producing crafts.

Provision for Special Fund

- Development of tribal economy rests with the state governments though funds are released by the government of India unfortunately no special fund is released for development of tribal handicrafts.
- As a result, there is decay in the quality of tribal handicrafts. Special funds need to release from Handloom and Handicrafts Commissioner to the state governments for establishing training and design centers.

Credit Facilities/Financial Support

- The craftsperson mostly working on job/contract basis and they do not have enough capacity to store the requisite raw materials to produce their own products.
- They are dependent upon intermediaries for credit facilities/financial support on higher rate of interest.
- Furthermore, topographically they are residing in remote and far-flung areas having no access to mode of transportation. In view of this some NGOs were of the view that instead of craftsperson approaching to bank individually, the bank should camp at specified place to get completed the documents and distribute the loan to the tribal artisans.

- Further it has been suggested that finance should also be made available through Post Office.

Upgradation of Technology and Production Techniques

- Despite their best quality of products their biggest handicap of these artisans is the absence of assured market for their products partly due to lack of proper communication system which influences the cost of their products. In such a scenario the prices of their products became costly. In addition to the use of mechanized items and use of plastics as substitutes eroded the demand for handicrafts.
- The tools and equipment are also based on very old and traditional techniques leading to higher cost of production thereby causing difficulties in the sale of final products.
- Design development production of new items and improvements, traditional tools and age-old techniques need to be attempted in such items where the originality of the tribal designs is retained. Introduction of new items with improved models may be designed continuously to attract new customers and explore new markets by the present design and training institutes.

Setting-up of Institute of Design in Tribal States

- During interaction with various stakeholders, it has been reported that the artisans are still practicing very old designs. Some of which have become outdated in the market and cost of these designs are also on higher side as such the demand is not appreciable
- It has been suggested that the various State Governments might consider setting up an Institute of Designs under the aegis of Directorate of Small Scale Industries at some suitable place.
- The Design Centre will work with the mission to become professional centre of Design excellence and

innovation, disseminate technical knowledge and develop skill based upon the topographical requirement of rich human resource base of artisans.

- The Design Centre shall also work for value addition of products of artwares leading to higher demand of the products while preserving the traditional skill and excellence and blending with contemporary marketing content and taste for handicrafts. The central or state government should also explore the possibility of involving ITI for skill development and training in handicraft sector especially in tribal districts.

TA/DA to Artisans for Marketing Programs

- The participants expressed deep concern over the withdrawal of TA/DA to artisans/NGOs for attending the market-related programs like Crafts Bazaars, Expos, other Exhibitions etc. Since most of the tribal states are comparatively a poor state.
- It has been recommended that like J&K, North-Eastern States and some districts of Orissa, TA/DA/Transportation charges should be provided to artisans in the old pattern.

Consultancy for Handicrafts Sector

- It has been suggested that consultancy/seminar should be organized from time to time in different crafts for the benefit of artisans/exporters and dissemination of market intelligence, upgradation of technology, standardization of quality and packaging for higher value addition of products.
- Government of India constituted a Tribal Commission and the report was also submitted. In the questionnaire issued to the state governments there was no mention about tribal crafts promotion. In such a scenario an All India survey for promotion of handicrafts need to be carried out and the results of the survey may be utilized for formulating policies while drafting.

Publicity of Tribal Handicrafts

- Since not much documentary literature is available, it has been suggested that facilities for putting handicrafts profile of Tribal state on website and publication of brochures, etc. should be provided to promote the state handicrafts.
- NSTFDC/TRIFED should play more pro-active role for promotion of Tribal Handicraft and market linkage by organizing region-wise periodical promotional activities, opening of display centre's/ sales outlets at important tourists destination/places in the states.

Exemption of Sales Tax/Trade Tax on Tribal Handicrafts Products

Although handicrafts were a manual labour and low capital industry yet Trade/Sales Tax is being levied equivalent to machine-made goods which is unjustified. Therefore, it has been recommended that Sales Tax/Trade Tax on handicrafts should be exempted like many other states.

Implementation of Welfare Schemes for Artisans

It has been reported that welfare schemes like Insurance, Pension, and Workshed-*cum*-Housing, etc. should be monitored and implemented with faster pace.

Creation of Handicrafts Cell in Directorate of Industries

To give the priority to solve problems of crafts and craftspersons of the state, an exclusive handicrafts cell should be created in the Directorate of Industries. For better marketing of their products tribal may form into cooperative societies to supply their products to the state government handicrafts cells to avoid uncertainty.

Setting-up of Urban Haats/Sale-*cum*-Demonstration Centers

- In order to promote handicrafts at the places of tourist spots, Urban Haats may be opened in tribal states to provide regular marketing channel to artisans.

- State governments may establish tribal "hatts" in important cities on permanent basis for continuous interaction.
- A sale-*cum*-demonstration center may be established at important cities with more than fifty lakh population by involving TRIFED. Ministry of Tribal affairs may release fund for the same as part of promotion of tribal handicrafts.
- TRIFED and other handicraft development bodies should set-up tribal emporium at important national and international tourist destinations.

Skill Dissemination by Master Artisans in each District

- It has been suggested that state government should take initiative by identifying master artisans at state-level. Further these master artisans would be entrusted with a responsibility of identifying district-level artisans and impart Training of Trainers (ToT) to them.
- These trained district level master artisans will further visit the clusters and disseminate their newly acquired skills to the local artisans. It will result in skill up-gradation of local artisans which will enable them to get good price for their handicraft items.

ESTABLISHING INTEGRATION WITH OTHER STATE DEPARTMENTS

Central or state government should explore the possibility of establishing a partnership with other state departments like Social Welfare Department, Tourism Department, and Industries Department, etc. for the promotion of tribal handicraft in the state. For example, State Tourism Development Corporations in the state should be actively engaged for the promotion and sales of Tribal Handicraft. Especially in tribal dominated states, the state government must ensure procurement of certain percentage of products developed by Tribal Artisans.

References

Unido Report on Bell Metal, Cluster Devlopment Programme, Year 2002.

Export Promotion Schemes for Handicrafts, *EPCH*.

Status Study of Tribal Handicraft—Planning Commission, Year 2006.

Art of Marketing Village Craft; Challenges in Applying Quantitative Marketing to Resist Recession.

Marketing Reaserch an Appllied Orientation 2007—Naresh K. Malhotra.

www.indianhandicraftexporters.com

www.epch.com

www.indiamirror.com/crafts

CHAPTER

39

Micro-credit Programs Serving the Relatively Poorer Sections of the Society ? Evidence from India

P. MAHENDRA VARMAN

I. INTRODUCTION

Credit is a crucial input for economic development of individuals and households especially in rural areas. Rural households need credit for a variety of reasons. They need credit to meet requirements for working capital, long-term investment in agriculture and other income bearing activities apart from credit needs for food, housing, health, education and other social obligations. If the credit needs of the rural poor are to be met, rural households need access to credit institutions that provide them a range of financial services and provide credit at a reasonable rate of interest. Such institutions in rural India happen to be the micro-finance institutions. One of the major indicators of Social performance of any microfinance institution is poverty outreach (Zeller, 2003). The key objective of microfinance programmes is to provide financial services to poor people who are excluded from such services by the formal banking system. It is in this perspective that governments, development partners and donor agencies continue to provide support to such institutions, to enable them to deepen their poverty outreach system (Kimos *et. al.*, 2009).

A growing body of evidence suggests that very poor households are excluded from accessing microfinance programs (Hulme and Mosley, 1996, Navjas *et al.*, 2002, Datta, 2004). According to these authors many poor people are dropping out of credit programs after having failed to keep up with the repayment installments. Some critics also question the efficacy of micro-credit in reaching the extremely poor people. They argue that, while micro-credit has contributed positively to the well-being of the poor people in general, it has failed to reach the poorest of the poor. Cuong, 2008 in his paper states that, programs that provide credit for the households are not always improving welfare and reducing poverty.

As far as micro-credit in India is concerned, it was conceived to fill the gap existing in the formal financial network and extend the outreach of banking to the poor. The rural credit system in the country has undergone radical changes in respect of focus, structure, and approach over time. The beginning of microfinance in India was made with National Bank for Agriculture and Rural Development's (NABARD) pilot project in Karnataka state (1986-87) of linking SHG's with the formal banks, mediating through NGO's (Tara, 1999). This project known as the SHG-Bank linkage project was expected to be advantageous to the banking sector from both angles of fulfilment of social goals (like reaching out the poor) and achieving operational efficiency (by externalising part of their transaction cost). The SHG-Bank linkage model is the indigenous model of micro-credit evolved in India and has been widely acclaimed as a successful model.

2. THE ISSUE

In spite of the remarkable progress, geographically there has been a skewed development of SHG-Bank linkage programme in India. There is wide regional disparity both in terms of the spread of SHGs linked to banks and cumulative bank loans disbursed under the programme. In March 2008, while the Southern Region accounted for 48.2 per cent of the total SHGs, the share of North-Eastern Region was just 3.4 per cent. In terms of share in the total bank loans to SHGs, the region-wise differential gets further magnified. In this context, it would be pertinent to see whether the programme has adequately reached into the regions where concentration of poverty is higher. (Kumar and Golait, 2009)

The all-India poverty ratio is about 27.6 per cent. While the Northern (15.7 per cent), North-Eastern (19.2 per cent), Southern (19.8 per cent) and Western region (25.8 per cent) had lower than the all-India poverty ratio, Central (35 per cent), and Eastern Region (36.2 per cent) has higher poverty ratios than at the all-India average. The present distribution of the SHG-Bank linkage programme does not appear to have taken into consideration of the extent of poverty. In the Eastern and Central Region, the proportion of total poor in India is significantly higher than the proportion of SHGs linked to banks in these regions. While both the regions taken together accounted for 61.1 per cent of the total poor in India, they accounted for only 31.0 per cent of all SHGs linked to banks in India. Further, their share in total loans to SHGs formed only 17.4 per cent. On the other hand, Southern Region which accounted for 15.3 per cent of the total poor in India had about 48.2 per cent of all SHGs linked to banks and much higher share (71.4 percent) of the total loans to SHGs. A negative correlation of -0.32 between SHG spread (measured in terms of the number of SHGs per lakh population) and poverty ratio implies that the States where the SHG spread is better are also the States where the poverty ratio is lower. States such as Uttar Pradesh, Madhya Pradesh, Bihar, Chhattisgarh, Jharkhand, Meghalaya, Manipur, Tripura, Nagaland, Arunachal Pradesh and Sikkim which have high levels of poverty are also the states with lesser number of SHG spread. On the other hand, one can see that states with comparatively lower poverty ratio such as Andhra Pradesh, Karnataka, Kerala, Tamilnadu and Rajastan are the states with higher SHG spread.

However, this may not mean that micro-financing SHGs are serving to richer section of people because though the overall relationship between the variables—SHGs per lakh population and proportion of BPL Population measured by correlation coefficient is negative (-0.32), the relationship between the same set of variables seems to have a significant positive correlation (0.354) for the southern region which commands almost 50 percent share in the number of SHGs formed in India. Nevertheless the relationship established using secondary data between the variables 'SHGs per lakh population' and 'Proportion of BPL Population' seems not so strong, which makes it ambiguous whether really poor and deserved are served by micro-credit SHGs. Since the key purpose of microfinance programmes is to provide financial services to poor people who are excluded from such services by the formal banking system, it is pertinent from the point of this study to assess whether the microfinance institutions (MFIs) and micro-credit programs are targeting the poor or are they serving the better offs.

Hence, the main objective of this paper is to examine whether relatively poorer section of the local population are being targeted and are beneficiaries of the micro-credit programs.

3. IMPORTANCE OF MEASURING THE DEPTH OF POVERTY OUTREACH

Assessment of the poverty level of microfinance clients is important for both practitioners and donors including the Government. For practitioners, increased understanding of the target market and whether it is being reached can help in the design of financial services better suited to the needs of different groups of clients. For donors seeking to ensure the most effective use of their scarce resources for combating poverty, poverty assessment is used to assist in making decisions on resource allocation. While there is much evidence that microfinance can be a critical input towards the achievement of the Millennium Development Goals and the reduction of poverty (Littlefield *et. al.* 2003), the impacts of microfinance are not automatic. To achieve significant direct impacts on poverty it is essential that Microfinance institutions (MFIs) reach poor and very poor clients, and therefore measurement of poverty outreach becomes an important proxy indicator for the success of microfinance in achieving impacts on poverty.

As the microfinance industry, matures and competition for clients and donor funding increases, the need for a more detailed and sophisticated understanding of the operating environment becomes clear. For MFIs this means improving understanding of the market, and the needs and desires of different groups of potential clients. MFIs must also be clear about their mission and operating priorities. For donors there is increasing emphasis on organisational appraisal, and the need to express donor strategy and priorities through funding decisions. Sophisticated rating and assessment systems have been developed which help donors and managers assess the efficiency and cost-effectiveness of MFIs, and to make necessary changes to improve performance. Poverty outreach, however, is a neglected area. Although it was a key motivation for microfinance, poverty is often an implicit rather than an explicit objective, and is not given detailed consideration as an organisational objective. Poverty outreach and poverty impact are almost never considered amongst the performance criteria for judging a well-functioning organisation, or in donors' decisions to allocate funding. (Simanowitz, 2003). There is therefore a need to be clear who the MFIs wishes to serve and why and at the same time, donors should be making transparent funding decisions based on clear statements as to who that funding is designed to reach, and whether it is effective in doing so. Hence, it is indispensable from this study's point of view to examine if the poor and needy are served by the microfinance SHGs; in particular, it is necessary to examine the depth of poverty levels, i.e., to what extent of poverty levels the microfinance programmes have reached. This study uses a primary survey data to analyse the depth of poverty outreach.

4. SAMPLING TECHNIQUE AND SAMPLING DESIGN

Before going to the analysis of the survey data, we shall

first have a look at the sampling technique and design used for collecting the required data. Since this paper tries to study the depth of micro-credit outreach, it is inevitable for any researcher to have the social and economic profiles of the clients and non-clients. As far as India is concerned, there are no reliable sources of secondary data available on the social and economic profiles of the clients of microfinance programmes at a macro-level. Hence, with the limitations of time and resource constraints, a sample survey of seven villages in Madurai district of Tamilnadu was done. Madurai district was chosen in such a way that it represents the average in number of SHGs formed and functioning among the districts of Tamilnadu. A stratified random sampling was done such that the whole of Madurai District was divided in to seven parts based on the number of Taluks. After dividing Madurai District, seven villages; each representing one taluk, having sufficient number of microfinance-SHG clients was chosen at random. Table 1 gives a glimpse of how the sampling was in fact done. There are altogether seven Taluks in Madurai, viz., South Madurai, North Madurai, Melur, Perayur, Thirumangalam, Usilampatti and Vadipatti. Each of these taluks comprises of several villages under it. Due to time and resource constraints, seven villages viz., Kaithirinagar, Vaagaikulam, Pulipatti, Vittilpatti, Nadukottai, Pullaneri and Vaigasipatti was chosen to represent each of the above mentioned taluks respectively and these villages were chosen in such a way that they had sufficient number of microfinance-SHG client households.

The sample unit in our study is household. From Table 1, we can make out that all the villages in the sample has, on an average, 74 percent of its households were as client households of microfinance SHGs. Since the sample villages had almost 3/4th of the total number of households involved in microfinance programs, (Columns 5 and 6 of Table 1) leaving the remaining 1/4th of households as non-client households, collecting sample client and non-client households proportional to their respective population would not ensure us sufficient number of non-client[1] households to the total samples, restricting us in making statistically valid comparisons among client and non-client households. Therefore, in order to draw

TABLE I

Sampling Design and Sample Distribution of Survey Households

Taluk	*Village*	*# of Total HH*	*Tottal population of Village*	*# and % of SHG-HH*	*# and % of Non-SHG HH*	*10% of Total SHG HH*	*30% of Total Non-SHG HH*	
(1)	*(2)*	*(3)*	*(4)*	*(5)*	*(6)*	*(7)*	*(8)*	
South Madurai	Kaithirinagar	536	1765	403 [75]	133 [25]	41	40	Total Sample Size
North Madurai	Vaagaikulam	353	1487	271 [77]	82 [23]	27	24	
Melur	Pulipatti	344	1533	282 [82]	62 [18]	28	19	
Perayur	Vittilpatti	303	1123	219 [72]	84 [28]	22	25	
Thirumangalam	Nadukottai	447	1794	321 [72]	126 [28]	32	38	
Usilampatti	Pullaneri	410	1597	303 [74]	107 [26]	31	32	
Vaadipatti	Vaigasipatti	365	1488	276 [76]	89 [24]	28	27	
Sample Size						209	205	414

Note : Figures in the brackets indicates the percentage to total households.

Source : Primary Survey Data.

sufficient and equal number of both client and non-client households a smaller percentage (10%) of SHG client households and a comparatively larger percentage (30%) of non-client households were randomly chosen from their respective population (Columns 7 and 8 of Table 1).

Consequently, based on the above mentioned criterion, data was collected from a total of 209 SHG-client households and 205 non-client households putting the total sample size of the study to 414 households. Data from both client and non-client households were collected using a well structured questionnaire enquiring details on family particulars, Socio-economic particulars, household ownership of consumer durables, land ownership, characteristics and facilities of household dwelling, Building materials of house, SHG member details, households banking transactions, and time allocation behaviors of household heads and spouse.

5. PROFILE OF THE SAMPLE VILLAGES

This section would highlight the profile of sample villages in terms of households economic and societal status such as major occupation of households, levels educational attainment, family size and structure, community affiliation and so on in order to have a good understanding of the sample villages. All villages are dominated by Hindus while very few households belong to Christian and Muslim religion. The share of households belonging to Scheduled Caste (SC) is very high in four of the seven sample villages, namely, Vittilpatti, Pulipatti, Nadukottai and Vaigasipatti (50% and more) Households belonging to Most Backward Class (MBC) and Backward Class (BC) have a considerable proportion in all the sample villages while the households belonging to other communities (OC) form a negligible part of the sample survey (Figure 1). Education being one of the important aspects of growth, development and exposure, it is very low in both the villages. Here the number of years of education undergone by the member of the household is considered for level of education and it has been classified into five divisions.[3]

Those who have not got any formal education at all are classified as uneducated, one to seven years of schooling is

FIG. 1

Village-wise Community[2] Affiliation of Head of the Household

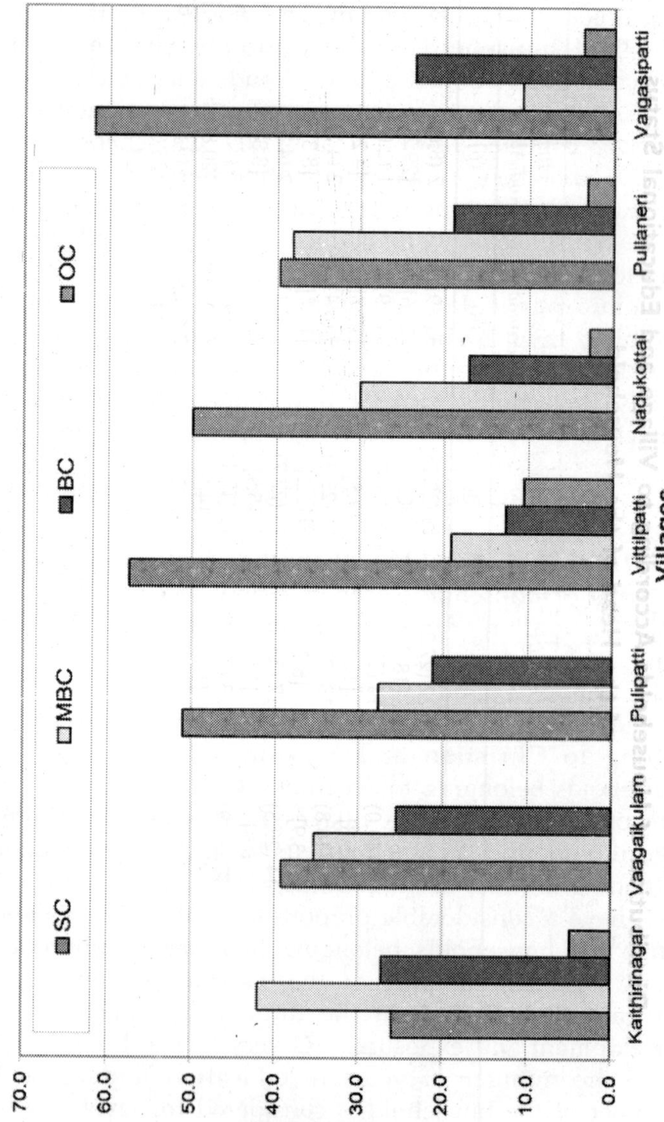

Source : Computed from Primary Survey Data.

TABLE 2

Distribution of Households According to Village and Educational Status of the Head of the Household

Village/Education Level	*Un-Educated*	*Primary*	*Secondary*	*Higher Secondary*	*Graduation and Above*	*Total*
(1)	*(2)*	*(3)*	*(4)*	*(5)*	*(6)*	*(7)*
Kaithirinagar	26 (32.1) [16.5]	22 (27.2) [18.0]	18 (22.2) [22.0]	8 (9.9) [28.6]	7 (8.6) [29.2]	81 (100) [19.6]
Vaagaikulam	25 (49.0) [15.8]	14 (27.5) [11.5]	8 (15.7) [9.8]	3 (5.9) [10.7]	1 (2.0) [4.2]	51 (100) [12.3]
Pulipatti	22 (46.8) [13.9]	10 (21.3) [8.2]	10 (21.3) [12.2]	3 (6.4) [10.7]	2 (4.3) [8.3]	47 (100) [11.4]
Vittilpatti	14 (29.8) [8.9]	17 (36.2) [13.9]	13 (27.7) [15.9]	1 (2.1) [3.6]	2 (4.3) [8.3]	47 (100) [11.4]
Nadukottai	23 (32.9) [14.6]	23 (32.9) [18.9]	12 (17.1) [14.6]	7 (10.0) [25.0]	5 (7.1) [20.8]	70 (100) [16.9]
Pullaneri	24 (38.1) [15.2]	17 (27.0) [13.9]	14 (22.2) [17.1]	2 (3.2) [7.1]	6 (9.5) [25.0]	63 (100) [15.2]
Vaigasipatti	24 (43.6) [15.2]	19 (34.5) [15.6]	7 (12.7) [8.5]	4 (7.3) [14.3]	1 (1.8) [4.2]	55 (100) [13.3]
Total	158 (38.2) [100]	122 (29.5) [100]	82 (19.8) [100]	28 (6.8) [100]	24 (5.8) [100]	414 (100) [100]

Note : Figures in normal brackets indicates percent to total households in each Village.
Figures in square brackets shows the percent to total households in respective Educational Level.

Source : Computed from Primary Survey Data.

classified as primary education, eight to ten years of schooling as secondary, ten to twelve as higher secondary and those who have done collegiate education and above are classified as graduation and above in this context. A major chunk (60%) of household heads have attained only primary education alone or below. Only around 11% of the total household heads have attained an educational level of higher secondary and above (Table 2), however Kaithirinagar alone has sizeable percentage (around 20%) of household heads with higher secondary and above.

With respect to the occupation of head of the households, a major proportion (32%) of households fall into the category of self-employed.[4] In four of the seven sample villages, households engaged in self-employment forms major part of occupation types of the household heads (Table 3). The table also shows that apart from household heads engaged in self-employed activities, there are considerable number of households working as casual labourers and skilled labourers (24% each). The household heads with regular salaried occupation is negligible.

As far as the Households membership in microfinance SHGs is concerned, on an average 75 percent of the total number of households in each of the sample villages are attached to SHGs.

The mean monthly income of the sample households is around Rs. 3300. The normal curve fitted to the frequency distribution of households income in Figure 2 shows that income of the sample households is positively skewed indicating that there are more number of households with lesser income and comparatively lesser number of households with higher income levels. The mode value of the income distribution is 3000 which says that most of the households make their living with just Rs. 3000 a month.

Till now in the current section, Section 5, we have examined some of the socio-economic attributes across the sample villages such as Community affiliation of households, Educational status of head of the households, occupation of head of the households, Income and average family size of the sample households. In a birds eye view, there does not seem to be any significant differences with respect to socio-economic

TABLE 3

Distribution of Households According to Village and Occupational Status of the Head of the Household

Village/Occupation	*Agricultural Labour*	*Casual Labour*	*Skilled Labour*	*Regular Salaried*	*Self Employed*	*Total*
(1)	*(2)*	*(3)*	*(4)*	*(5)*	*(6)*	*(7)*
Kaithirinagar	7 (8.6) [13.2]	18 (22.2) [18.2]	13 (16.0) [12.7]	4 (4.9) [12.9]	39 (48.1) [30.2]	81 (100) [19.6]
Vaagaikulam	7 (13.7) [13.2]	12 (23.5) [12.1]	13 (25.5) [12.7]	4 (7.8) [12.9]	15 (29.4) [11.6]	51 (100) [12.3]
Pulipatti	7 (14.9) [13.2]	12 (25.5) [12.1]	15 (31.9) [14.7]	6 (12.8) [19.4]	7 (14.9) [5.4]	47 (100) [11.4]
Vittilpatti	6 (12.8) [11.3]	13 (27.7) [13.1]	12 (25.5) [11.8]	2 (4.3) [6.5]	14 (29.8) [10.9]	47 (100) [11.4]
Nadukottai	10 (14.3) [18.9]	15 (21.4) [15.2]	23 (32.9) [22.5]	5 (7.1) [16.1]	17 (24.3) [13.2]	70 (100) [16.9]
Pullaneri	12 (19.0) [22.6]	13 (20.6) [13.1]	10 (15.9) [9.8]	6 (9.5) [9.4]	22 (34.9) [17.1]	63 (100) [15.2]
Vaigasipatti	4 (7.3) [7.5]	16 (29.1) [16.2]	16 (29.1) [15.7]	4 (7.3) [12.9]	15 (27.3) [11.6]	55 (100) [13.3]
Total	53 (12.8) [100]	99 (23.9) [100]	102 (24.6) [100]	31 (7.5) [100]	129 (31.2) [100]	414 (100) [100]

Note : Figures in normal brackets indicates percent to total households in each Village.
Figures in square brackets shows the percent to total households in respective Occupation.

Source : Computed from Primary Survey Data.

TABLE 4

Village-wise Share of Sample SHG and Non-SHG Member Households and Average Family Size

Village	*No. and % of HHs in SHGs to total HHs*	*No. and % of HHs not in SHGs to total HHs*	*Average Family Size*
Kaithirinagar	403 (536) [75] {41}	133 [25] {40}	5
Vaagaikulam	271 (353) [77] {27}	82 [23] {24}	4
Pulipatti	282 (344) [82] {28}	62 [18] {19}	5
Vittilpatti	219 (303) [72] {22}	84 [28] {25}	5
Nadukottai	321 (447) [72] {32}	126 [28] {38}	4
Pullaneri	303 (410) [74] {31}	107 [26] {32}	5
Vaigasipatti	276 (365) [76] {28}	89 [24] {27}	4

Note : Figures in normal brackets indicates total number of households in each Village. Figures in square brackets shows the percentage of SHG households to total households in respective Village. Figures in flower brackets are the number of sample households in each village.

Source : Computed from Primary Survey Data.

FIG. 2

Frequency Distribution of Sample Households Monthly Income

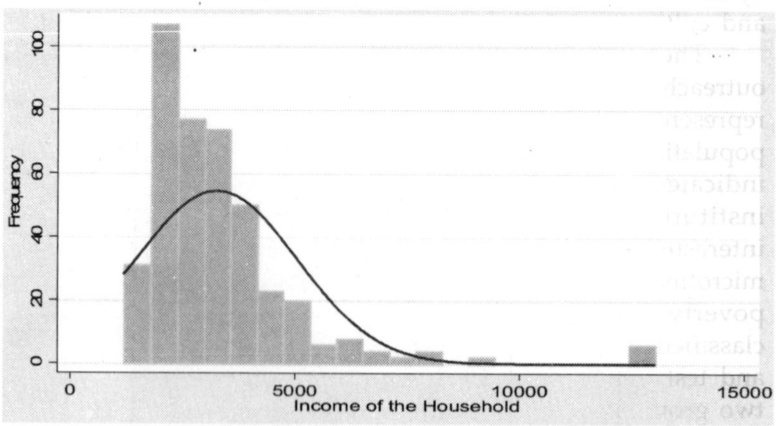

Note : The curve in the figure is the normal curve fitted to the frequency distribution.

Source : Computed from Primary Survey Data.

variables among the villages. All the villages have somewhat similar socio-economic structure with very small variations. However, there might be significant differences in the socio-economic variables especially the assets holdings within the villages, in our point of view between SHG client households and non-client households, that happens to be the main focus of this study and the examination of which will lead us in assessing the depth of poverty outreach of microfinance clients.

6. MULTIDIMENSIONAL NATURE OF POVERTY AND ASSOCIATION OF VARIOUS SOCIO-ECONOMIC VARIABLES WITH SHG CLIENT STATUS

The objective of this study is to examine whether the relatively poorer sections of the population are being benefitted out of the microcredit programs. Poverty according to United Nations is defined as "lack of income and productive resources to ensure sustainable livelihoods; hunger and malnutrition; ill-health; limited or lack of access to education and other basic services; increased morbidity and mortality from illness; homelessness and inadequate housing; unsafe environments and social discrimination and exclusion. It is also characterised by lack of participation in decision-making and in civil, social and cultural life" (United Nations, 1995).

There are two ways in which microfinance poverty outreach could be measured, viz., the extent of outreach is represented by the absolute number of households in the target population reached by the MFIs, whilst the depth of outreach indicates how deep into the pool of the underserved the institution has been able to reach. Here we are mainly interested in the second type i.e., examining the depth of microfinance outreach. To do this, we shall first consider each poverty indicators individually to correlate with the pre-classified set of microfinance clients and non-client households and test whether there is a significant difference between the two groups.

Theoretically, poverty can be reflected by income, consumption and asset holding status of individual households. But the collection of accurate data on income and consumption requires extensive resources for household survey

(Seema *et. al.*, 2006). While economists often use income to measure wealth, welfare, and other indicators of well-being, income data has limitations in both accuracy and measurement, particularly in the context of developing countries. For instance, for people living in informal labour markets incomes are often highly variable. Income can be seasonal, such as when earned from farming or the tourist market, or just variable and lumpy for small-business owners. Taking a snapshot of income at one point in time may therefore produce a less reliable picture of these types of workers than those who receive regular paychecks. Furthermore, they may be engaged in barter and other non-monetary forms of trade. In all of these cases there is a high potential for error in data based on the recollection and value of all sources of income. This means that income itself does not necessarily provide a reliable measure of well-being. Expenditures and consumption are also commonly used to measure well-being (Chen and Ravallion, 2000). Expenditures solve some of the problems of income, such as seasonality. However, a number of the same difficulties of income also apply to expenditure, such as measuring the value of bartered good. However, assets may provide a better picture of long-term living standards than an income snapshot because they have been accumulated over time and last longer.

According to (Amartya Sen, 2001), Poverty is not simply the absence of financial resources. Poverty is the lack of capability to function effectively in society and inadequate education can thus be considered a form of poverty. There is clear evidence that education can reduce poverty. Education can reduce poverty in a number of ways.

Firstly, more educated people are more likely to get jobs, are more productive, and earn more. Secondly, education (particularly of girls) brings social benefits that improve the situation of the poor, such as lower fertility, improved health care of children, and greater participation of women in the labour market. The home background of pupils is the single most important factor influencing educational outcomes. Poverty is strongly correlated with a range of other home background variables, including parental educational attainment, thus it is difficult to separate the effects of limited financial resources from other home background factors.

Educational research has consistently found home background (socio-economic status) to be an important determinant of educational outcomes, and economic research has shown that education strongly affects earnings. Throughout the world it has been found that the probability of finding employment rises with higher levels of education, and that earnings are higher for people with higher levels of education. A better educated household is less likely to be poor (Berg, 2008).

Therefore, checking for differences between clients and non-clients on a number of Home background (socio-economic) characteristics can improve understanding of whether and why differences in poverty levels between the two groups occur and also we can know who the beneficiaries are. Hence, we shall first explore the differences between the two groups, viz., clients and non-clients with respect to various factors that would indicate poverty status. Since differences between groups can be tested using both the t-test of differences between means and the chi-square test for cross-tabulations, determining which test to apply depends on the type of data scale used to measure the variable. Apparently the variables in our survey dataset especially those that are closely related with poverty status of the household are mostly of nominal or categorical in nature and a few are continuous in nature.

Table 5 shows the association of nominal variables of households such as caste, education of the head of the households, access to formal banking, source of drinking water, etc. with SHG membership status, using Chi-Square Test of Independence. Here we try to test the hypothesis that the SHG. Membership status is independent of the selected variables. Accordingly, it is found that except source of drinking water and type of toilet, all other selected variables are associated with SHG membership, i.e., caste, land holdings, family size, education of the head of the households, occupation of the household head, households access to formal banks, food security measured by meals per day, Materials used for roofing and flooring, type of toilet and cooking fuel used are statistically significantly associated with SHG membership status.

Through the Independent sample t-test for differences between two means given in Table 6, it has been established

TABLE 5

χ^2 Test of Independence of Selected Variable with SHG Membership Status

Variables	*N*	*χ^2 (Chi-Square)*	*d.f*	*Sig. (2 tailed)*
Caste	414	14.703 ***	3	0.002
Land Holding (Categorised)	414	178.51 ***	2	0.000
Family Size (Categorised)	414	27.993 ***	2	0.000
Education of the Head of the Household	414	66.015 ***	4	0.000
Occupation of the Head of the Household	414	25.682 ***	4	0.000
Access to Bank measured in terms of a/c Holding	414	15.151 ***	1	0.000
Food Security Measured as Meals per Day	414	5.908 **	2	0.042
Source of Drinking Water	414	0.061	1	0.815
Material Used for Roofing	414	46.33 ***	3	0.000
Material Used for Flooring	414	34.82 ***	3	0.000
Type of Toilet Used	414	1.872	1	0.171
Cooking Fuel Used	414	8.318 ***	2	0.016

Note : (***) Significant at 1% level, (**) Significant at 5% level, (*) Significant at 10 % level.

Source : Computed from Primary Survey Data.

that there exists a significant difference in the mean values of selected variables among the clients and non-clients. All the selected (continuous) variables viz., household income, land holding, family size, age of the head of the household, number of children and number of rooms in the house are associated with SHG membership status and they are statistically significant at 1% level.

Due to the multi-dimensional nature of poverty, reliance on any one poverty dimension, such as monthly/annual household income, food security, access to housing/accommodation or education, is deemed to be inappropriate and inadequate. Poverty is a complex phenomenon whose

TABLE 6

Independent Sample t-Test—Testing the Equality of Means of Clients and Non-Clients

Variables/Client status	*N*	*Mean*	*Standard Deviation*	*t*	*d.f*	*Sig. (2 tailed)*	*Mean Difference*
(1)	*(2)*	*(3)*	*(4)*	*(5)*	*(6)*	*(7)*	*(8)*
Household Income							
Non-Client	205	3612.20	2105.87	4.168***	412	0.000	719.85
Client	209	2892.34	1328.57				
Land Holding							
Non-Client	205	2.3860	1.4960	8.185***	412	0.000	1.3408
Client	209	1.0453	1.8239				
Family Size							
Non-Client	205	4.14	1.25	-5.09***	412	0.000	-0.7101
Client	209	4.85	1.59				
Age of the Head of the Household							
Non-Client	205	48.38	11.07	3.388***	412	0.001	3.61
Client	209	44.77	10.60				
No. of Children							
Non-Client	205	2.3	0.92	4.462***	412	0.005	-1.50
Client	209	3.8	1.44				
No. of Rooms in the House							
Non-Client	205	2.16	1.05	5.963***	412	0.000	0.55
Client	209	1.60	0.83				

Note : (***) Significant at 1% level, (**) Significant at 5% level, (*) Significant at 10% level.

Source : Computed from Primary Survey Data.

dimensions extend far beyond income any other single dimension. Being a multidimensional concept, poverty is not easy to define. It can mean different things to different people. Conceptual and methodological differences in defining poverty can lead to the identification of different individuals and groups as being poor. As a result, identifying the difference in poverty levels between the SHG clients and non-clients vary according to how poverty is defined. A brief discussion on the concept and various approaches in measurement of poverty should help in understanding the discussion on why it is better to give emphases on creating an index using various dimensions poverty.

Four approaches in measuring poverty are highlighted here. These include the monetary approach, the capabilities approach, the social exclusion approach and the participatory approach. The monetary approach is most commonly used to define and measure poverty. A poverty line is defined in terms of the monetary income sufficient to attain a minimal standard of living. While estimating the poverty line, allowance is usually made for both food and non-food expenditure. A person whose income (expenditure) falls below the poverty line is considered poor. Since income provides access to goods and services, it is considered a good proxy for covering many aspects of poverty. However, a limitation of the approach is that it does not capture access to social services, which clearly enhances a person's well-being. The capabilities approach pioneered by Amartya Sen emphasizes that income is only valuable in so far as it increases the capabilities of individuals and thereby permits functioning in a society. The ultimate objective is capabilities such as the ability to lead a long life, to function without chronic morbidity, to be capable of reading, writing and performing numerical tasks and to be able to move from place to place. In accordance with this approach, a person whose capabilities or functioning falls below a minimum acceptable standard is poor. The resources or income required to achieve the same capabilities can vary from person to person. The capabilities approach is much broader and addresses the neglect of social goods in the monetary approach.

The social exclusion approach emphasizes relations among individuals. Social exclusion occurs when individuals or groups are unable to participate fully in the society in which they live. As a result of exclusion, the incomes, capabilities or other characteristics of the poor become unacceptably distant from the norms of their community or reference group. In accordance with the exclusion approach, poverty is a social construct and has little to do with the fulfillment of an individual's minimum needs.

The participatory approach takes into account the views of poor people themselves. The people themselves decide what it means to be poor, and that determines the magnitude of poverty. Here the perspective is that of the poor; it is not imposed from outside. The approach relies on the contextual method of analysis and this helps in understanding the poverty dimensions within the social, cultural, economic and political environment of a locality. Hence, poverty outcomes may not be quantified and standardized. In case all the approaches identify broadly the same people as poor, any single approach can be used to measure poverty. However, the empirical evidence shows that poverty rates in countries differ significantly according to the approach adopted. As a result, policy options can differ depending on the approach selected.

Taking into consideration the above pros and cons of various approaches adopted to capture the level of poverty, its better to develop a measure that would capture the multi-dimensional nature of poverty. This can be done only by consolidating all the dimensions and constructing an index based on which the households can be classified in to different groups of socio-economic status of poverty. To combine various dimensions of poverty, our study, based on World Bank's study on poverty assessment by Henry *et. al.* (2003), uses Principal Components Analysis (PCA) developing a unique socio-economic index to identify the difference in poverty levels between the clients and non-clients.

7. PRINCIPAL COMPONENTS ANALYSIS—AN OVERVIEW OF THE TECHNIQUE

Principal components analysis is a tool or technique to determine the weights for an index of (asset) variables. It is a technique for extracting from a set of variables those few orthogonal linear combinations of the variables that capture the common information most successfully (Filmer and Pritchett, 2001). The weights for each principal component are given by the eigenvectors of the correlation matrix PCA works better not only when variables are correlated but also when the distribution of variables varies across cases or in this instance, Households. It is the variables (assets) that are more unequally distributed between households that are given bigger weights in PCA (Vyas and Kumaranayake, 2006). Variables with low standard deviations would carry a lower weight from the PCA; for example, an asset which all households own or which no households own (i.e., Zero Standard Deviation) would exhibit no variation between households and would be zero weighted, and so of little use in differentiating between households.

Suppose we have a set of N variables, a^*_{1j} to a^*_{Nj}, representing the ownership of N assets by each household j. Principal components starts by specifying each variable normalized by its mean and standard deviation: for example, a1j = (a* 1j–a*1,)/(s*1), where a*1, is the mean of a*1j across households and s*1 is its standard deviation. These selected variables are expressed as linear combinations of a set of underlying components for each household j:

$$a_{lj} = v_{ll} \cdot A_{lj} + v_{l2} \cdot A_{2j} + ... + v_{lN} \cdot A_{Nj}$$

...

$$j = 1,...J$$

$$a_{Nj} = v_{N1} \cdot A_{lj} + v_{N2} \cdot A_{2j} + ... + v_{NN} \cdot A_{Nj}, \qquad ....(1)$$

where the *A*s are the components and the *v*s are the coefficients on each component for each variable (and do not vary across households). Because only the left-hand side of each line is observed, the solution to the problem is indeterminate. Principal components overcomes this indeterminacy by finding the linear combination of the variables with maximum variance—the first principal component A_{1j} – and then finding a second linear combination of the variables, orthogonal to the first, with maximal remaining variance, and so on.

Technically the procedure solves the equations $(R -, \lambda_n I) v_n = 0$ for λ_n and v_n, where R is the matrix of correlations between the scaled variables (the *a*s) and v_n, is the vector of coefficients on the nth component for each variable. Solving the equation yields the characteristic roots of R, $\ddot{e}_n$ (also known as eigen values) and their associated eigenvectors, v_n. The final set of estimates is produced by scaling the v_ns so the sum of their squares sums to the total variance, another restriction imposed to achieve determinacy of the problem. The "scoring factors" from the model are recovered by inverting the system implied by Eq. (1), and yield a set of estimates for each of the N principal components:

$$A_{1j} = f_{11} \cdot a_{1j} + f_{12} \cdot a_{2j} + \ldots\ldots + f_{1N} \cdot a_{Nj}$$

...

$$j = 1,\ldots J$$

$$A_{nj} = f_{N1} \cdot a_{1j} + f_{N2} \cdot a_{2j} + \ldots\ldots + f_{NN} \cdot a_{Nj} \qquad \ldots\ldots\ldots\ldots(2)$$

The first principal component, expressed in terms of the original (un-normalized) variables, is therefore an index for each household based on the expression

$$A_{1j} = f_{11}\ (a^*_{1j} - a^*_1)/(s^*_1)+\ldots+ f_{1N}\ (a^*_{Nj} - a^*_N)/(s^*_N) \qquad \ldots\ldots\ldots(3)$$

Intuitively, the first principal component of a set of variables is the linear index of all the variables that captures the largest amount of information that is common to all the variables and hence forms the index value for each household.

8. CONSTRUCTION OF THE SOCIO-ECONOMIC STATUS (SES)/POVERTY INDEX AND ASSESSING THE DEPTH OF MICRO-CREDIT OUTREACH

The indicators used for the analysis of relative poverty levels of clients and non-clients of microfinance SHGs are classified into four key dimensions of poverty which is listed in Table 7. Under each of these dimensions a couple of indicators that best describe the difference in poverty levels of clients and non-client households are chosen. The choice of the indicators was based on a number of criteria, including the ease and accuracy with which information on the indicators could be elicited in the household survey and the significance of the

TABLE 7

Variables Used to Create Poverty Index

Category	*Variables*	*Measurement of Variables*
Human Resource/ Demographic Variables	(i) Education of Head of the Household	No. of Years of Education
	(ii) Household Size	No. of members in the family
	(iii) Children	No. of Children in the Family
Food Security and vulnerability	(i) Meals taken per Day	Avg. no. of meals taken per day past week
	(ii) Cooking Fuel Used	L.P. Gas used as Cooking Fuel
Dwelling Particulars of Households	(i) High Quality Roofing	Cement Used in Roofing
	(ii) Low Quality Roofing	Thatched/Tiles Roofing Used
	(iii) Separate Kitchen	Having Separate Room for Kitchen
	(iv) No. of Rooms	No. of Rooms in the House
	(v) Drinking Water Source	Having Pipe Connection inside Home
	(vi) Type of Toilet	Existence of Open Toilet System
Wealth/ Ownership of Consumer Durables and Farm Animals	(i) Grinder	Whether Grinder Owned
	(ii) Cooker	Use of Cooker
	(iii) Motor Vehicle	No. of Motor Vehicles Owned
	(iv) Cycle	No. of Cycles Owned
	(v) Cattle	No. of Cattle Owned
Income/ Ownership of Land Assets	(i) Income	Average Monthly Household Income
	(ii) Other Land Assets	Other Land Assets in Acres

Source : Prepared from Primary Survey Data.

association of the individual indicators with average monthly income of the households, which is the poverty benchmark indicator used for this study.

Three indicators relating to the human resource and demographic characteristics of the household were selected. These indicators are : (i) number of children, (ii) household size and (iii) level of education of the head of the household. For assessing the food security and vulnerability of the household two indicators namely, (i) hunger (Average number of times meals are taken per day), and (ii) cooking fuel used were selected for creating a socio-economic index. As far as dwelling and related aspects are concerned, indicators such as : (i) Material used in roofing, (ii) whether the household is having a separate kitchen, (iii) Number of rooms in the residence, (iv) Type of toilet in use, and (v) Source of drinking water are being taken for construction of the Index. Household assets play an important role in the determination of poverty

FIG. 3

Screeplot Showing the Proportion of Variance of the Original Variable Explained by each Principal Component

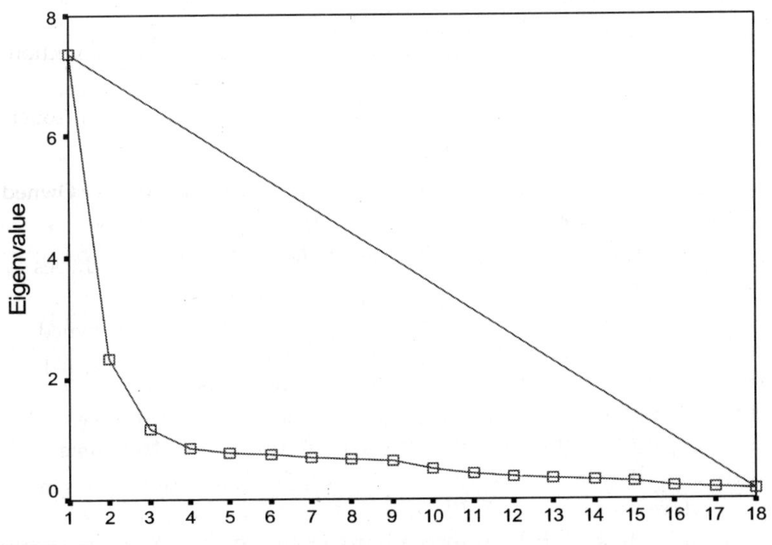

levels of people in the study area. A total of ten assets were selected for constructing the socio-economic index viz., use of Grinder, use of Cooker, number of Motor Vehicles, number of Cycles, number of Cattle and possession of land assets.

The screeplot in Figure 3 shows that only the first three components would sufficiently explain the original variables (Eigen Values > 1). The proportion of variance explained by the first principal component is 40.92%, second component explains 12.88% of the variance and each subsequent component explains a decreasing proportion of variance in the original variables. Component scores are the weight assigned to each variable (Normalized by its mean and standard deviation), in the linear combination of variables that constitute the first principal component. However in the construction of the socio-economic index, only the factor scores (that's the eigenvectors) of the first principal component are used.

Table 8 reports the component scores from the principal components analysis of the 18 indicators. The mean value of the index is 0 and the standard deviation is 1. The asset variables that take only the values 0 or 1 (1 for possession of a particular asset and 0 for not possessing), the weights have an easy interpretation: a move from 0 to 1 changes the index by corresponding factor score. For example, a household that owns a motor cycle has an asset index higher by 0.807 than one that does not; owning a cycle reduces a household's asset index by 0.036 units; using L.P. gas as the main cooking fuel increase the asset index by 0.776 units (Table 8). Like-wise for continuous variables such as land asset holding and household income, an increase in an unit of these two variables would increase the socio-economic index by 0.846 and 0.783 units respectively.

The end result of PCA is the creation of a single index of relative socio-economic status that assigns to each sample household a specific value, called a Factor Score, representing that household's socio-economic status in relation to all other households in the sample. The lower the score, the poorer the household relative to all others with higher scores.

The scores of MFI client households and non-client households are then compared to indicate the extent to which the micro-finance programmes are reaching the poor. Before a

TABLE 8
Summary Statistics and Component Scores

Variable used in Principal—Component Analysis (PCA)	*Mean*	*Standard Deviation*	*Component Score*
Education of Household head	3.81	1.08	0.263
Household Size	4.50	1.47	-0.83
Children	2.86	1.92	0.63
Meals taken per Day	2.06	0.26	0.793
Using Gas as Cooking Fuel	0.776	0.417	0.621
Cement used in Roofing	0.23	0.425	0.73
Thatched/Tiled Roof	0.44	0.36	-0.47
Separate Kitchen	0.183	0.53	0.51
No. of Rooms	1.88	0.98	0.704
Water Pipe Connection	0.52	0.43	0.663
Open Toilet System	0.72	0.30	-0.788
Grinder	0.18	0.38	0.719
Cooker	0.821	0.383	0.644
Motor cycle	0.14	0.36	0.807
Cycle	0.39	0.51	-0.036
Cattle	0.44	0.65	-0.089
Household Income	3248.79	1791.45	0.846
Land Assets Holding	1.791	1.071	0.783
Socio-Economic Status (SES)	0.000	1.000	

Source : Computed from Primary Survey Data.

comparison is made, the share of the local population that is likely to fit the assessment's definition of poor must be decided on. In this assessment methodology, a cutoff of 33 percent (terciles) is used to define the poor, middle and richer groups within the local population. This cutoff is proposed by most of the studies that assess poverty, including Henry *et. al.*, (2000) in their work on 'Assessing the Relative Poverty of Microfinance Clients.

To use the socio-economic index (SES)[5] for making comparisons, the non-client sample is first sorted in an

Fig. 4
Determining Cutoff Scores for Poverty Ranking

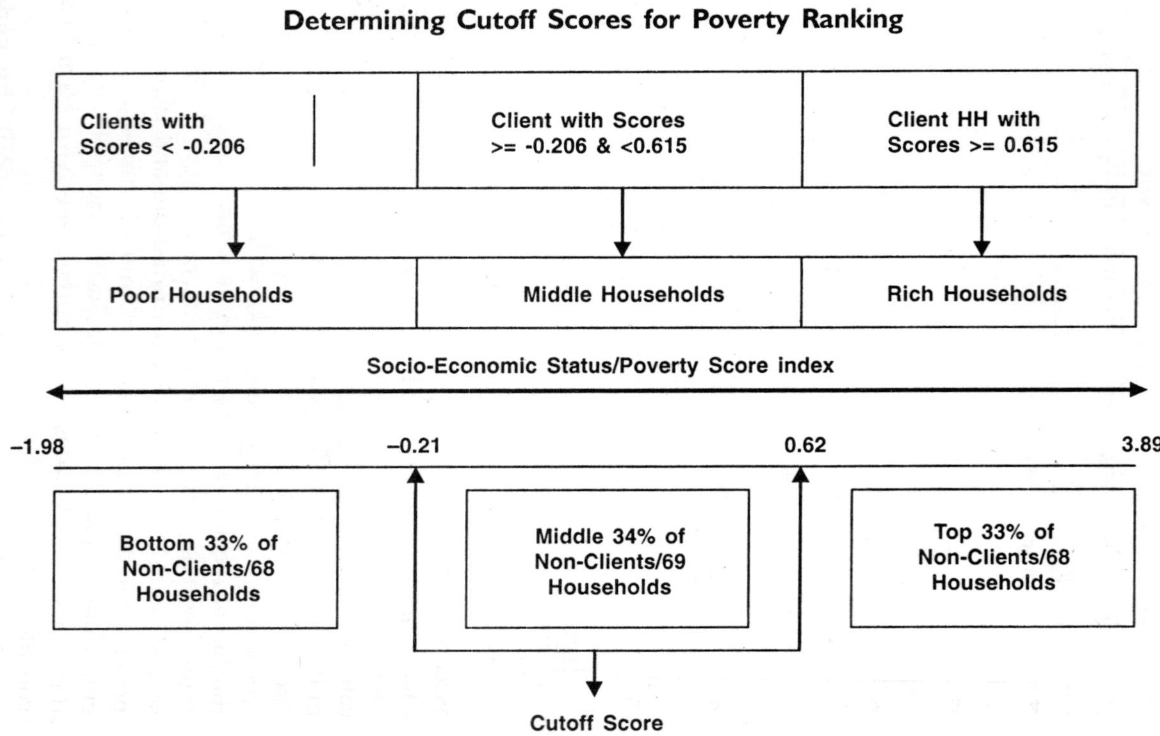

FIG. 5

Box-Plot Showing the Distribution of Poverty Levels of Households with SHG-Member, without SHG Members and Overall Sample Households

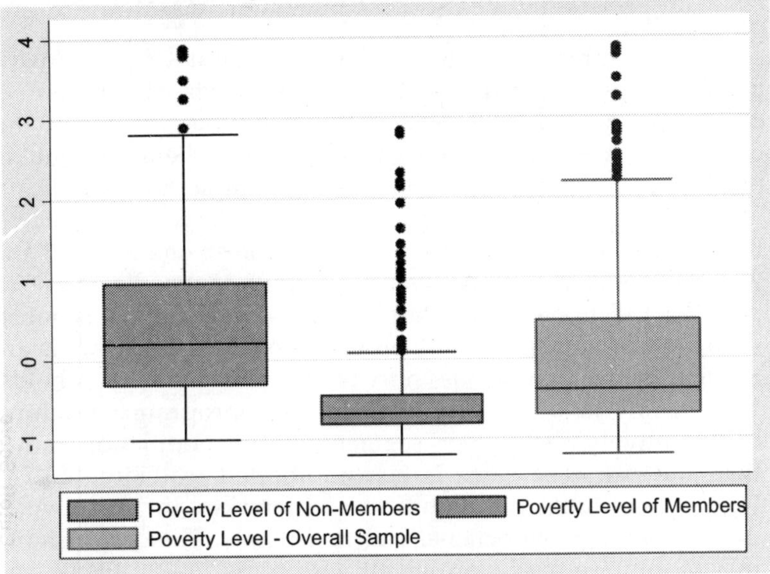

ascending order according to its index score. Once sorted, non-client households are divided in terciles based on their SES index score: the top one-third of the non-client households are categorized as "richer households", followed by the households in the "middle" and finally the bottom one-third categorized as "poor households". Since there are 205 non-clients, each group contains 68 households. The cutoff scores for each tercile define the limits of each poverty group. Client households are then categorized into the three groups based on their household scores. Figure 4 illustrates the use of cutoff scores to create poverty terciles from non-client households. Accordingly the cutoff scores of -0.21 and +0.62 were calculated from the survey data. The households with SES scores below -0.21 are considered as poor, between -0.21 and +0.62 are considered as moderate or middle class and those households that has a SES score above +0.62 are considered as rich.

In Figure 5, the horizontal line in the middle of each box indicates the median SES index score of each group of households revealing that on an average the households with SHG members/clients are having a SES index score much lower compared to that of households with non-SHG members or overall.

In all three cases the medians are situated at a lower position (not in Middle) indicating that their distribution is positively skewed. The tails of the box plots show that the dispersion in poverty levels is less in case of client households but higher for non-clients and over all sample households.

The χ^2 test given in Table 9, shows the percent of client households who are as poor as the poorest one-third of the non-client population. Around 68 percent of the clients are as poor as the poorest one-third of the non-client households. Also we can notice that only 8 percent of client households are as rich as the richest one-third of the non-client households. This result is a sign of microfinance programs reaching comparatively the poorer segment of the population rather than the rich. The same is being indicated by Figure 6. The Figure highlights the significant differences in the poverty distribution between clients and non-clients. The graph shows that clients are over-represented within the lowest tercile and

TABLE 9

Distribution of Households According to Client Status and Poverty Group

SHG Client Status	*Poverty Group on the Basis of SES Index*			*Total*
	Poor	*Middle*	*Rich*	
Non-Clients	68 (33.2) [32.22]	69 (33.6) [53.49]	68 (33.2) [91.89]	205 (100) [49.52]
Clients	143 (68.42) [67.78]	60 (28.71) [46.51]	6 (2.87) [8.11]	209 (100) [50.48]
Total	211 (50.97) [100]	129 (30.91) [100]	74 (18.12) [100]	414 (100) [100]

Value of χ^2 = 80.04; Sig. (2 Tailed) = 0.000 d.f = 2 N = 414

Source : Computed from Primary Survey Data.

FIG. 6

Percent Breakdown by Poverty Score Tercile of MFI Clients and Non-Clients

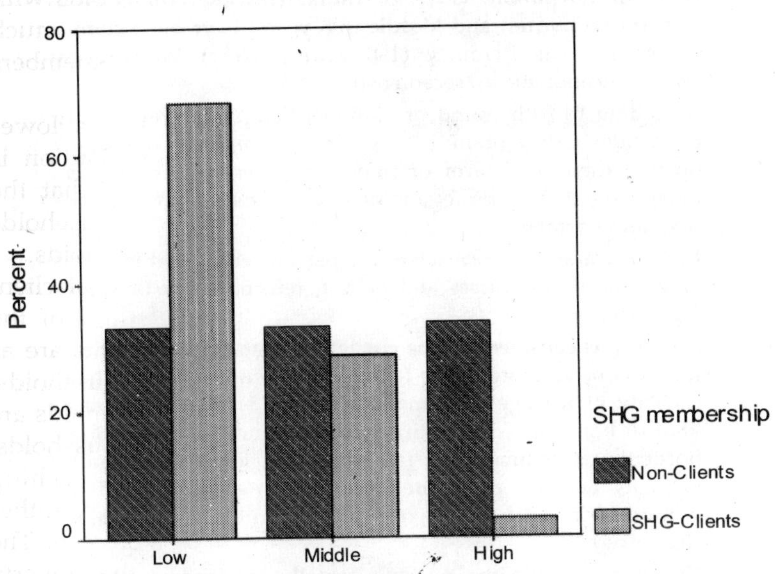

underrepresented in the highest tercile. Hence, we can conclude that the microfinance programs in India have social considerations by way of reaching a larger share of poorest households than what is found in the population in general.

NOTES AND REFERENCES

1. A larger sampling size for non-clients captures the presumably larger variance among non-clients with respect to any poverty indicator than the variance found among clients. Because of MFI targeting and self-selection of clients, the client group is likely less heterogeneous (has less variance) than the Non-client group (Zeller *et. al.*, 2000).
2. In India, people are identified and classified into four main social classes viz., Scheduled Caste/Tribe (SC/ST), Most Backward Class (MBC), Backward Class (BC) and Other Caste/Forward Caste (OC),

SC/ST being considered most deprived and under-privileged community.

3. The classification is on the basis of National Educational Policy, Dept. of Education, Govt. of India. However for convenience, Primary (1-5 years) and Middle (6-8 years) have been clubbed and categorised as Primary (1-8 years). Refer Website http://www.education.nic.in/secedu.asp.
4. According to 64th round of National Sample Survey (NSS) Socio-Economic Survey of India, 2007-08, *Self-employed* are person who operate their own farm or non-farm enterprises or are engaged independently in a profession or trade on own account or with one or a few partners.

   Regular salaried/Wage employees are persons employed in others farm or non-farm enterprises and getting return, salary or wages on a regular basis.

   A person is considered to be engaged as *agricultural labour* if he/she follows one or more of the following agricultural occupation in the capacity of a wage paid manual labour, whether paid in cash or kind or both: (i) fFrming, (ii) Dairy Farming (iii) production of any horticultural commodities, (iv) Raising of livestock or poultry and (v) Any activity performed on a farm as incidental to or in conjunction with farm operations (including forestry and timbering).

   Casual wage labour are persons casually engaged in other farm or especially non-farm enterprises and getting in return wage according the terms of daily or periodic work contract. Usually in rural areas a type of casual labourers can be seen who normally engage themselves in 'public works' or those activities sponsored by government or local bodies for construction of roads, buildings, bunds, digging of canals and ponds, etc.
5. Socio-Economic Status Index (SES) constructed using Principal Components Analysis is used here as Relative Poverty Index and the terms are used synonymously.

References

Achia *et. al.* (2010), A Logistic Regression Model to Identify Key Determinants of Poverty Using Demographic and Health Survey Data, *European Journal of Social Sciences*, Vol. 13, No. 1.

Adams, D.W. and Von Pischke, J.D. (1992), "Micro-enterprise Credit Programs : Déjà vu", *World Development*, Vol. 20, No. 10.

Caroline Moser and Andrew Felton (2007), The Construction of an Asset Index—Measuring Asset Accumulation in Ecuador, Chronic Poverty Research Centre, Working Paper 87.

Conning, Jonathan (1999), Outreach, Sustainability and Leverage in Monitored and peer-monitored lending. *Journal of Development Economics*, 60:51-77.

Cull, R., A. Demirguz-Kunt and J. Morduch (2007). Financial Performance and Outreach: A Global Analysis of Lending Microbanks. *Economic Journal*.

Hashemi, Syed M., Sidney Ruth Schuler, and Ann P. Riley (1996). "Rural Credit Programs and Women's Empowerment in Bangladesh"; *World Development*, Vol. 24, No. 4, pp. 635-53.

Hema Bansal (2003), SHG-Bank Linkage Program in India, *Journal of Microfinance*, Volume 5, Number 1.

Hulme, D., and P. Mosley (1996), Finance Against Poverty, Volumes 1, Routledge: London.

Joseph Kimos Adjei and Thankom Arun (2009), Microfinance Programmes and the Poor: Whom Are They Reaching? Evidence from Ghana, Brooks World Poverty Institute, Working Paper 72.

Kaladhar, K. (1997), "Micro-finance in India; Design, Structure and Governance", *Economic and Political Weekly*, Oct. 18, pp. 2687-06.

Kumaran, K.P. (2001), "Self help groups of the rural poor in India : An analysis", *National Bank News Review*, April to June, pp. 31-37.

Lapenu, C. and Zeller, M. (2002), Distribution, growth, and performance of the microfinance institutions in Africa, Asia and Latin America: A Recent Inventory. *Savings and Development*, No. 1-XXVI: 87-111.

Leland, H.E. and Pyle, D.H. (1977), "Informational asymmetries, financial structure and financial intermediation", *Journal of Finance*, Vol. 32, No. 2, pp. 371-87.

Madheswaran, S. and Amita Dharmadhikary (2001), "Empowering Rural Women through SHGs : Lessons from Maharashtra Rural Credit Project", *Indian Journal of Agricultural Economics*, Vol. 56, No. 3, July-Sept. pp. 427-43.

Majumdar, N.A. (1997), "Overhauling the somnolent rural credit system", *Economic and Political Weekly*, Oct. 18, pp. 270-10.

Mersland, Roy (2007), "The agenda and relevance of recent research in Microfinance", MPRA Paper No. 2433 (http://mpra.ub.uni-uenchen.de/2433/).

Morduch, J. (2000), "The micro-finance schism", *World Development*, Vol. 28, No. 4, April.

Muhammad Abdul Latif (2001), "Micro-credit and Savings of Rural Households in Bangladesh", *The Bangladesh Development Studies*, Vol. 27, March.

NABARD (2002), "Ten Years of SHG-Bank Linkage; 1992-2002", *NABARD and Micro Finance*.

Nanda, Y.S., (2001), "Significance of Establishing Linkages with Self-Help Groups and Banks", NABARD, India, refer Web *http://www.gdrc.org/icm/nanda-link.html*

Narayana, D. (2003), "Why is the credit-deposit ratio low in Kerala ?" WP No. 342, Centre for Development Studies.

Narayana, D. (2000), "Banking sector reforms and the emerging patterns in commercial credit deployment in India", *Review of Development and Change*, Vol. 5, No. 2 July-Dec.

Pankaj Kumar and Ramesh Golait (2009), Bank Penetration and SHG-Bank Linkage Programme: A Critique, *Reserve Bank of India Occasional Papers*, Vol. 29, No. 3.

Rajaram Dasguptha (2001), "Working and impact of rural self help groups and other forms of micro-financing : An informal Journey through self-help groups", *Indian Journal of Agricultural Economics*, Vol. 56, No. 3, July-Sep.

Rajasekhar, D. (2000), "Micro Finance programmes and women's empowerment : A study of two NGOs, from Kerala", *Journal of Social and Economic Development*, Vol. 3, No. 1, Jan.-June, pp. 76-94

Ramachandran, V.K. and Madhura Swaminathan (2001), "Does Informal Credit Provide Security?" Rural banking policy in India, International Labour Office, Geneva, Refer Website.

Reddy, Y.V. (2005), Microfinance—Reserve Bank's Approach, a Lecture delivered as Governor, Reserve Bank of India at the Micro-finance Conference, organised by the Indian School of Business, Hyderabad, India, on 6th August http://www.bis.org/review/r050826b.pdf

Sainath, P. (2008), *The Hindu*, Article in the Front-Page titled "4,750 Rural Bank Branches closed down in 15 years" in India's National News Paper, Friday, 28th March, 2008.

Shahidur, R. Khandker (2000), "Savings, Informal Borrowings and Micro-finance", *Bangladesh Development Studies*, Vol. 26, No. 2 and 3, June-Sept.

Sham, K. Bhat, V. Nirmala and N. Ranjini (2000), "Impact of Financial Reforms on the banking sector in India", *Journal of Institute of Public Enterprises*, Vol. 23, No. 1.

Tara, S. Nair (2000), "Rural Financial Intermediation and Commercial Banks : Review of Recent Trends", *Economic and Political Weekly*, Jan. 29.

Villi, C. (2005), A Study on "The Role of NGOs in Social Mobilizations in the Context of SGSY", Project funded by National Institute of Rural Development, India.

Vinod Vyasulu and D. Rajasekhar (1991), "Rural Credit System in the 1990's", Working Paper, ISEC, Bangalore.

Zeller *et. al.* (2000), Assessing the Relative Poverty of Microfinance Clients—A CGAP Operational Tool, IFPRI, Washington, D.C., U.S.A.

CHAPTER

40

Rural Development, its Importance and Current Scenario with Limitations of the Farm Sector

N. SARAVANAKUMAR

INTRODUCTION

The poorest as well as the more prosperous 'Green Revolution' states of Punjab, Haryana and Uttar Pradesh have recently witnessed a slow-down in agricultural growth. Some of the factors hampering the revival of growth are poor composition of public expenditures, public spending on agricultural subsidies is crowding out productivity-enhancing investments such as agricultural research and extension, as well as investments in rural infrastructure, and the health and education of the rural people. In 1999/2000, agricultural subsidies amounted to 3 percent of GDP and were over 7 times the public investments in the sector. Over-regulation of domestic agricultural trade while economic and trade reforms in the 1990s helped to improve the incentive framework, over-regulation of domestic trade has increased costs, price risks and uncertainty, undermining the sector's competitiveness. Government interventions in labor, land, and credit markets. More rapid growth of the rural non-farm sector is constrained by government interventions in factor markets labor, land, and credit and in output markets, such as the small-scale reservation of enterprises.

ISSUES AND CHALLENGES

- *Inadequate infrastructure and services in rural areas* : Weak Framework for Sustainable Water Management and Irrigation.
- *Inequitable allocation of water*: Many states lack the incentives, policy, regulatory, and institutional framework for the efficient, sustainable, and equitable allocation of water.
- *Deteriorating irrigation infrastructure*: Public spending in irrigation is spread over many uncompleted projects. In addition, existing infrastructure has rapidly deteriorated as operations and maintenance is given lower priority.

Inadequate Access to Land and Finance

- *Stringent land regulations discourage rural investments*: While land distribution has become less skewed, land policy and regulations to increase security of tenure (including restrictions or bans on renting land or converting it to other uses) have had the unintended effect of reducing access by the landless and discouraging rural investments.
- *Computerization of land records has brought to light institutional weaknesses*: State government initiatives to computerize land records have reduced transaction costs and increased transparency, but also brought to light institutional weaknesses.

- *Rural poor have little access to credit*: While India has a wide network of rural finance institutions, many of the rural poor remain excluded, due to inefficiencies in the formal finance institutions, the weak regulatory framework, high transaction costs, and risks associated with lending to agriculture.

Weak Natural Resources Management

One quarter of India's population depends on forests for at least part of their livelihoods.

- *A purely conservation approach to forests is ineffective*: Experience in India shows that a purely conservation approach to natural resources management does not work effectively and does little to reduce poverty.
- *Weak resource rights for forest communities*: The forest sector is also faced with weak resource rights and economic incentives for communities, an inefficient legal framework and participatory management, and poor access to markets.

Weak Delivery of Basic Services in Rural Areas

- *Low bureaucratic accountability and inefficient use of public funds*: Despite large expenditures in rural development, a highly centralized bureaucracy with low accountability and inefficient use of public funds limit their impact on poverty. In 1992, India amended its Constitution to create three tiers of democratically elected rural local governments bringing governance down to the villages. However, the transfer of authority, funds, and functionaries to these local bodies is progressing slowly, in part due to political vested interests. The poor are not empowered to contribute to shaping public programs or to hold local governments accountable.

PRIORITY AREAS FOR THE WORLD BANK SUPPORT

Enhancing Agricultural Productivity, Competitiveness, and Rural Growth

Enhancing productivity: Creating a more productive, internationally competitive and diversified agricultural sector would require a shift in public expenditures away from subsidies towards productivity enhancing investments. Second, it will require removing the restrictions on domestic private trade to improve the investment climate and meet expanding market opportunities. Third, the agricultural research and extension systems need to be strengthened to improve access to productivity enhancing technologies. The diverse conditions across India suggests the importance of regionally differentiated strategies, with a strong focus on the lagging states.

Improving Water Resource and Irrigation/Drainage Management: Increase in multi-sectoral competition for water highlights the need to formulate water policies and unbundle water resources management from irrigation service delivery. Other key priorities include: (i) modernizing Irrigation and Drainage Departments to integrate the participation of farmers and other agencies in irrigation management; (ii) improving cost recovery; (iii) rationalizing public expenditures, with priority to completing schemes with the highest returns; and (iv) allocating sufficient resources for operations and maintenance for the sustainability of investments.

Strengthening rural non-farm sector growth: Rising incomes are fueling demand for higher-value fresh and processed agricultural products in domestic markets and globally, which open new opportunities for agricultural diversification to higher value products (e.g. horticulture, livestock), agro-processing and related services. The government needs to shift its role from direct intervention and overregulation to creating the enabling environment for private sector participation and competition for agribusiness and more broadly, the rural non-farm sector growth. Improving the rural investment climate includes removing trade controls, rationalizing labor regulations and the tax regime (i.e. adoption of the value-added tax system), and improving access to credit and key infrastructure (e.g. roads, electricity, ports, markets).

Improving Access to Assets and Sustainable Natural Resource Use

Balancing poverty reduction and conservation priorities: Finding win-win combinations for conservation and poverty reduction will be critical to sustainable natural resource management. This will involve addressing legal, policy and institutional constraints to devolving resource rights, and transferring responsibilities to local communities.

Improving access to land: States can build on the growing consensus to reform land policy, particularly land tenancy policy and land administration system. States that do not have tenancy restrictions can provide useful lessons in this regard. Over the longer term, a more holistic approach to land administration policies, regulations and institutions is necessary to ensure tenure security, reduce costs, and ensure fairness and sustainability of the system.

Improving access to rural finance: It would require improving the performance of regional rural banks and rural credit cooperatives by enhancing regulatory oversight, removing government control and ownership, and strengthening the legal framework for loan recovery and the use of land as collateral. It would also involve creating an enabling environment for the development of micro-finance institutions in rural areas.

STRENGTHENING INSTITUTIONS FOR THE POOR AND PROMOTING RURAL LIVELIHOOD

Promoting Community-Based Rural Development: State Government efforts in scaling up livelihood and community-driven development approaches will be critical to build social capital in the poorest areas as well as to expand savings mobilization, promote productive investments, income generating opportunities and sustainable natural resource management. Direct support to self-help groups, village committees, user's associations, savings and loans groups and others can provide the initial 'push' to move organizations to higher level and access to new economic opportunities. Moreover, social mobilization and particularly the empowerment of women's groups, through increased capacity for collective action will provide communities with greater "voice" and bargaining power in dealing with the private sector, markets and financial services.

Strengthening Accountability for Service Delivery: As decentralization efforts are pursued and local governments are given more prominence in the basic service delivery, the establishment of accountability mechanisms becomes critical. Local governments' capacity to identify local priorities through participatory budgeting and planning needs to be strengthened. This, in turn, would improve the rural investment climate, facilitating the involvement of the private sector, creating employment opportunities and linkages between farm and non-farm sectors.

CONCLUSION

The volatility of the prices of most food raw materials has increased, as a result of the presence influence the global market in the medium and even the long-run. Heightened uncertainty poses the risk of discouraging production and investment in the agricultural sector, thereby weakening security challenge of the poorest families. Agricultural production requires long-term investments, but the market has not so far offered income protection mechanisms that go beyond one or two crop cycles. It therefore appears that government participation is needed to ensure income for sensitive groups, including both producers and consumers, according to its own policy objectives. The heterogeneity of the productive sector is such that a long-term approach is required for its analysis, one that takes into account all of the obstacles and risks it faces, so that policies and programs are adopted in a consistent and openminded way, with consideration given to the impact on consumers, with a focus on sensitive groups, to avoid overburdening government finance, and with a view to reducing possible distortions in the allocation of productive resources. The short-term objective and the long-term effects should be adequately weighed by government when it adopts policies, so that the tools applied represent effective progress in reducing the vulnerability of affected population segments, and measures that could ultimately prove to be counter-productive by promoting greater protectionism or increased distortions are avoided. Using incentives that encourage participation of the private sector (individual producers, organizations, and businesses) in initiatives designed to control the risks of their activity is one way to work together for the common good Establishment of the Special Safeguard Mechanism in WTO, so that developing countries have a way to protect themselves during times of price crises, one reason for strengthening the openness of countries to trade, and countries should fight for this in the Doha Round.

References

Livingstone, Ian, 1991. "A Reassessment of Kenya's Rural and Urban Informal Sector," *World Development, Elsevier*, Vol. 19(6), pp. 651-70, June.

McPherson, Michael A., 1996. "Growth of micro and small enterprises in Southern Africa," *Journal of Development Economics*, Elsevier, Vol. 48(2), pp. 253-77, March. Ahmed, Raisuddin and Hossain, Mahabub, 1990. "Developmental impact of rural infrastructure in Bangladesh:," Research Reports 83, *International Food Policy Research Institute* (IFPRI).

Pernia, Ernesto M. and Pernia, Joseph, M., 1986. "An Economic and Social Impact Analysis of small industry Promotion: A Philippine experience," *World Development, Elsevier*, Vol. 14(5), pp. 637-51, May.

Park, Albert and Johnston, Bruce, 1995. "Rural Development and Dynamic Externalities in Taiwan's Structural Transformation," *Economic Development and Cultural Change*, University of Chicago Press, Vol. 44(1), pp. 180-208, October.

CHAPTER

41

Rural Non-Farm Employment : Indian Perspective

HARINDRA KISHOR MISHRA

Growing income inequality inevitably leads to divergence within society, create income in security and aggravates poverty. A man with income insecurity can not be saved from destitution. One who has no other means of livelihood but labour power cannot be protected from distress if he loses the opportunity to make use of the same. The world Investment Report, 1999 admits the lopsided character of the economic system. It says" because of their technology intensity and competitive behaviour they are likely to generate smaller number of jobs than other firms of equal output and size." Very few workers in the world benefit from unemployment protection. Workers in the rural areas and also in the urban informal sector have virtually no protection against unemployment. It is ell understood today that globalisation has significant implication for social protection in the countries with open economy who are exposed to the cyclical movement of the global market, Also, it is found that there are instances of reduction of social security expenditure in the countries where the share of trade in GDP has increased (Dasgupta, 2000, p. 7.)

Unemployment has always it roots much deeper in the economic system than is commonly supposed. There is no problem so fundamental to economic development as the problem of unemployment. So long as the sari faction of human needs in the prime objective of all economic activity, the prevalence of unemployment an underemployment will stand as an index to economic distress and primary poverty. The larger the opportunities of employment, the greater the scope for the people to increase their-prosperity and augment production of goods and services, thereby national welfare. Industrialization was the answer to problems of economic backwardness, unemployment and poverty in the developed countries. However, since the less developed countries depend much more heavily on agriculture and are faced with much faster growth of population and labour force than the industrialized countries faced in their pre-industrial phase. It is widely maintained that higher priority must be given to agriculture and agro-based industries in order to provide not only adequate food and fibre but also sufficient employment and income to the poor. But the recent studies revealed that though the new technology in agriculture did need more labour, the increase was less than proportionate to the increase in yield. As such, the growth of agricultural employment, both in absolute terms and relative to increase in agricultural output was far less then expected. It is also evident that in peasant economies, typically characterised by continuing population pressure, an ever declining land-man ratio, small and fragmented agricultural holdings, highly inequitous land distribution structures, increasing application of labour-saving farm production technologies, etc., agriculture, alone cannot provide the ultimate answer for rural unemployment and under-employment. That underscores for employment generation in the rural areas. This brings the development of non-farm sector into focus (Prakash and Srivastava, 2000, pp. 378-79).

Persistent rural poverty is one of the major challenges India has to cope with in the new millennium. Although India achieved great success in bringing down rural poverty . At the national level, the rural poverty ratio declined very modestly from 39.1% in 1987-88 to 37.3% in 1993-94, When person-days of employment in central schemes shot up from 675 million to 1,073 million. But the rural poverty ratio then dipped sharply to 27.1% in 1999-00 despite the fact that person days of work plummeted to 546 million. There is little connection between poverty and wage schemes on a long-term basis. Sen (1996), on the other hand, argues that the declining trend in poverty ratio witnessed since the early 1970s has been arrested in the post-reform period. Increase in relative prices of food and decline in public expenditure are some of the reasons mentioned for the adverse impact of reforms on the poor. According to this study, the growth in non-farm employment had taken place in the 1980s as a result mainly of larger public expenditure and its multiplier effects on the rural economy, and was responsible in considerable measure for the observed decline in rural poverty in the second half of the 1980s despite drought years like 1987. In the post-reform period, however, non-farm employment fell owing to the expenditure cuts following from the stabilization and structural adjustment policies and despite a run of good harvest years rural poverty started rising. The study also shows that real wages have not increased in the post-reform period because of increase in, relative food prices (Dev, 2000, pp. 830-31).

An important objective of development planning in India has been to provide for increasing employment opportunities not only to meet the backlog of the unemployed but also the new additions to the labour force. The increasing diversification of the economy together with acceleration in economic growth has resulted in structural changes in the nature of the job market. Economic reforms in the areas of abolishing quantitative restrictions (QRs), reducing tariffs, reforming labour laws and abolishing SSI reservations have aimed at fostering labour-intensive production in India.

The Table 1 reveals that the average annual growth rate of overall employment (in both the organised and unorganised sectors) was 2.73% per annum in the period 1972-73 to 1977-78

TABLE I

Rate of Growth of Population, Labour Force and Employment

Period	*Rate of Growth of population (percent per annum)*	*Rate of Growth of Labour force (UPSS) (percent per annum)*	*Rate of Growth of employment (UPSS) (percent per annum)*
(1)	(2)	(3)	(4)
1972-73 to 1977-78	2.27	2.94	2.73
1977-78 to 1983	2.19	2.04	2.17
1983 to 1987-88	2.14	1.74	1.54
1987-88 to 1993-94	2.10	2.29	2.43
1993-94 to 1999-2000	1.93	1.03	0.98

Source : *Economic Survey*, 2001-02, p. 240.

but declined to 1.54% per annum in 1983 to 1987-88. However, the growth rate of employment increased to 2.43% per annum over 1987-88 to 1993-94 and declined to 0.98% per annum in the period 1993-94 to 1999-2000. The labour force participation rate declined in 1999-2000 as compared with 1993-94, which is reflected in a sharp deceleration in the growth of the labour force from 2.29% per annum in 1987-88 to 1993-94 to 1.03% per annum in the period 1993-94 to 1999-2000. The deceleration in employment is therefore also associated with a sharp decline in the growth of the labour force.

Rural employment schemes have been institutionalised in India to tackle poverty. In the post-monsoon season, employment schemes are devised at the minimum wage to help the needy.

TABLE 2

Fall of Rural Employment Schemes Person-days of Work

(Millions)

1990-91	875
1993-94	1073
1994-95	1226
1995-96	1232
1996-97	803
1997-98	868
1998-99	792
1999-2000	547
2000-01	486

Source : The Times of India, New Delhi, August 11,2002, p. 8.

The above table shows that person-days of work generated rose steadily to a peak of 1,232 million mandays in 1995-96, but then decline sharply. This was due partly to budgetary stringency, partly to disillusionment with the use of rural employment as an anti-poverty tool.

The structure of the workforce in an economy is reflected of the allocative efficiency of workers in different sectors. If a disproportionately large number of workers are engaged in the primary sectors, particularly when these sectors are not very productive, it is an indicator of other sectors not being able to absorb labour fat enough and/or workers finding themselves skill-wise redundant outside agriculture. The population growth compounds the problem, particularly in subsistence agriculture. This is a clear condition of loss of welfare to the society due to structural under development. The Indian position is somewhat similar, where the labour productivity in agriculture is only about one third of that outside it and the wages/earnings are lower in it when compared to those prevalent in other sectors (India Rural Development Report, 1999, p. 55).

As per the National Commission for Enterprises in the Unorganised sector (NCEUS), based on data from NSS 55 Round Survey on Employment and Unemployment, the estimated total employment in the country during 1999-2000 was 396.8 million and among them the estimated informal sector workers were 342.6 million. The estimates of total employment and employment of informal sector workers as per 61 Round Survey during 2004-05 were 457.5 million and 394.9 million respectively. As per both surveys more than 86 per cent of total employment was in the informal sector.

NCEUS, 2007 gives several recommendations for improvement in the growth of non- agricultural enterprises in India. These are: (a) improvement in the credit flow to the non-agricultural sector; (b) encouraging SHGs and micro finance institutions for livelihood promotion; (c) creation of a Nation Fund; (d) upscaling cluster development through growth poles; (e) expanding employment through strengthening self employment programmes; (f) Universalising and strengthening National Rural Employment Guarantee Programme (NREGP); and (g) increasing employability through skill development.

NON-FARM SECTOR IN DEVELOPING COUNTRIES

According to World Bank estimates in early 1980, most of the 15 developing countries where the statistics are available the percentage of rural labour force primarily engaged in non-farm work fails between 20 to 30 percent. As economic development advances there is a continuous shift in the spatial distribution of the growth of non-farm employment in developing countries. In the earlier stages of development rural areas and town appear to provide over three quarters of all non-farm employment opportunities as compared with approximately two-thirds in Southern and Eastern Asia and one third in Latin America. A large number of empirical studies have shown that a movement of labour from agriculture to industry and services is characteristic of process of economic growth in all countries. This lends support to the view that industrialization is a key to economic growth in developing countries so that rapid creation of wage employment is an effective answer to the employment problem. (Rao, 2000, p. 17).

In South Asia, a large proportion of the population lives in rural areas, and this population continues to grow at a substantial rate. Given the limits to the cultivable land in' the region, the fast-proliferating rural labour force will not be productively absorbed in the agricultural sector. There is necessity to shift the workforce from agricultural to non-agricultural occupations in order to increase labour income. The rural non-farm sector can be an escape way for agricultural workers in South Asia. There has been a significant increase in the share of the rural non-farm sector since the early 1970s. However, it is not clear whether it is due to pull or push factors. In recent times various studies have come up on non-farm sector in developing countries.

Okafor (1983) in the basis of his study of 36 Nigerian villages, argues that employment diversification has received scanty attention. Moreover, involvement in non-farm activities by adults and children alike is a common affair and the intensity of this involvement is positively correlated with seriousness of population pressure.

A study on Bangladesh by Alauddin and Tisdell (1995) reveals that the employment generating effect of the green revolution is slowing down—in other words that agriculture at the margin has been becoming progressively less labour intensive in recent years. Changes in the occupational structure may reflect a relocation of surplus labour from the farm to the non-farm sector (where, too, it is largely surplus) rather than herald a real turning point. "This is true of Nepal's sitution.

Haggblade, Hazell and Brown (1987) look specifically at different types of issues with regard to rural non-farm sector in Sub-Saharan Africa. Using cross-section and time series data, they found that services commerce, and food processing activities are the main and rapidly growing non-farm activities in rural towns of the region.

Oshima (1986) in his study of post-war Asian growth (Japan, Taiwan, South Korea and South East Asia) of off-farm employment and income finds that the agricultural development of regional infrastructure and industrialization are important for the growth of off- farm employment. Further, institutional forces were also found to be responsible for off-farm employment.

Some studies on India have shown that : (a) the rural non-agriculture sector relates mainly to modem products and services; (b) the role of markets and rural entrepreneures has clearly over shadowed that of government policies and programmes; and (c) the shift of workers toward non-agricultural activities was an outcome of growth and not merely 'distress' phenomenon reflecting non-absorption of surplus workers in agriculture. On the other hand; Sen (1998) argues that the growth in non-agricultural employment during the 1980s was mainly due to public expenditure and that the decline in public expenditure during the reform period was responsible for the decline in the growth of such employment. If public expenditure (including non-capital account) is the main factor that raises questions about the sustainability of the growth of such employment over time.

Apart from above studies, several scholars have studies the problems and prospects of non-farm sector in their respective countries. The results of these studies broadly disclose certain common factors which explain the growth of non-farm sector, for example, shortage of capital, backward technology, small size operations, seasonal involvement of population, etc. Some of the studies found that labourers are receiving low returns and inverse relation between non- farm employment and farm size.

REAL WAGE RATES IN AGRICULTURAL AND NON-AGRICULTURAL ACTIVITIES

The real wage rates in agriculture and non-agriculture are live indicators of the buoyancy of the economic system. An increasing real wage rate in agriculture would employ an excess of demand over supply of labour in the sector. This would mean that there is no compulsion for agricultural labour to spill over to the non-agricultural sector creating a 'distress' employment situation in the latter (Unni, 1997). However, the existence of the phenomenon at the aggregative national level has been disputed on the ground that real wage rate in agriculture have been rising (Vaidyanathan, 1994; Basant, 1994). The idea that non- gricultural employment is 'residual' is now some what discredited because not only are average wages seen to be higher in such employment than in agriculture, but, more importantly, because agricultural wages have increased as non-agricultural employment has grown suggesting that what is involved is a pull factor which tightens the agricultural labour market (Sen, 1997, p. 87). Nonetheless, NSS data show that the actual picture is more complicated and suggests that 'distress' movement into non-agriculture has continued to be important for a significant section of rural workers, as well as that the main dynamic source of rural employment generation over this period has been the external agency of the state rather than forces internal to the rural economy.

Many studies have shown that from the mid-seventies to the mid-eighties, the real wage rates in agriculture tended to rise slowly and steadily (Unni, 1988; Jose, 1988; Bhalla, 1993; Vaidyanathan, 1994). The prime mover of this rise in agricultural wage rates has been found to be the diversification of the workforce, into the non-agricultural sector, rather than growing labour productivity (Bhalla, 1993; Mukherjee, 1996). During the 1990s, however, it has been observed that the trend of rising agrarian wages is not being sustained (Mukherjee, 1996; Parthasarathy, 1996). The real wage rates of adult males in agriculture rose in most states up to 1990 and fell in 1991 and 1992. The wage rates rose in 1993 and then more or less stagnated thereafter in some states. The non-agricultural real wage rates in fact tended to fall in the 1990s. The ratio of non-agricultural to agricultural real wage earnings of adult males also showed a declining tendency in the 1990s. It appears that excess supply or lack of demand for labour has been putting pressure on the real age rates and ratios in the non-agricultural sector (Unni, 1997).

The arguments against distress induced activities in the non-agricultural sector have been framed in terms of rise in real wages. This is true to the extent that the poorest of the poor, without adequate human and physical capital will offer himself for work in the wage labour market. A large proportion of such workers without adequate demand for labour would depress the real wages. However, there is a large section of self-employed workers in the non-agricultural sector. In 1983, while about 11% of rural male workers were self-employed, only 4.7% were casual employees and 6.5% were regular employees in the non- agricultural sector. The option to undertake self-employment in non- agricultural activities will reduce the pressure on the real labour market, The argument of declining wages due to distress induced increase in supply of wage labour to the labour market will not apply to this segment of the self-employed workforce (Unni, 1998, p. A.-41). Rural non .. farm sector not only provides direct employment but also indirectly raises agricultural wages', Real rural wages are higher and poverty is lower in non-farm sector than in agriculture. This reveals that this sector offers more productive jobs and is not the dumping ground of agricultural workers under distress.

EMPLOYMENT IN NON-FARM SECTOR : ALL INDIA ESTIMATES

Independent studies show that the elasticities of labour absorption in several sectors have fallen in the recent decades. Absorption outside agriculture is becoming increasingly difficult In the 1990s, construction activity, the most labour using sector too has ceased to grow—The limited possibility to absorb labour outside agriculture also stems from three inter-related factors. Non-agricultural activities grow more than proportionately to growth in agriculture, but agriculture itself has not shown a spectacular growth recently particularly in the

1990 and generally in areas outside some green revolution belts, except perhaps WB. It is also seen that infrastructure growth has been very tardy Particularly in the rural areas of semi-arid and eastern states. As a result most non-farm activities are non- viable there. The Rakesh Mohan Committee on infrastructure has been most explicit on this aspect. Governments have had fewer funds to spare for such activities following the fiscal squeeze in 1990s. Finally the quality of human resources is poor since basic schooling is yet to reach the hinterland, not to speak of its quality and relevance. It is now established that private rates of return to education are higher at the primary levels in rural areas than in any other category. Yet, investments in this sector are low and the quality poor. This urban bias and a low priority accorded to spending on social sectors (expenditure on education is less than three percent of the GDP in India compared to a figure double this East and south East Asia) is a significant retarding factor in occupational diversification (India Development Report, 1999, p. 57).

The rural labour processes of the 1980s have revealed considerable contribution of non-farm work to farm household incomes to the extent of 50% in Malaysia, 46% 10 Sri Lanka, 38% in Japan and 32% in Philippines, as compared to 19% in India (Islam, 1987; Mooiji, 2000). Some pertinent observations of the Planning Commission of India contained in Eighth Five Year Plan, Vol.1 may be noted in this context. Rural non - farm activation which engage over 20% of rural workers in India have demonstrated a steady growth of employment over the recent years. Among these manufacturing has recorded about 3% annual growth, construction 11%, transport 7% and trade 4% per annum during 1977-78 to 1987-88. Manufacturing and services respectively accounted for 32% and 24% of rural non-farm employment, trade accounted for 18% and construction 15% during 1978-88.

The Table 3 reveals that the rate of growth of employment in the non-farm sector was fairly high during 1972-73 to 1983-87. These rates of growth raised expectations of a structural breakthrough and when the rate decelerated steeply, there was an pervasive scepticism of the role of the non-farm sector.

TABLE 3

Growth of Employment in Agriculture and Non-Agriculture Sectors in Rural India

Period	*Agriculture (UPS+SS)*	*Non-Agriculture (UPS+SS)*	*All Rural (UPS+SS)*
1972-73 to 1977-78	1.76	5.33	2.99
1977-78 to 1983	1.43	3.83	1.97
1983 to 1987-88	0.12	5.79	1.14
1987-88 to 1993-94	2.32	1.41	2.10
1993-94 to 1999-2000	0.20	2.34	0.66

Source : Bhalla (2001).

The problem of unemployment is worst in females than males. Women constitute one-half of our population but they still continue to languish far behind. It may be surprising to note that women contribute two-third of total human labour hours but they get one-length of the world's property and income.

The Table 4 indicates that there has been no trend rise in the share of the rural female workforce, that their predominant source of work still remains agriculture and that their share in the non-farm sector is on the decline. It may be because of higher wages or skill requirements. Both sectors seem to lead male preference. But there are evidences of Women entering into agricultural activities left by migrating men or even into temporary bondage in certain circumstances.

TABLE 4

Share of Women in Rural Workforce

Nature of work	*1977-78*	*1983*	*1987-88*	*1993-94*	*1999-2000*
Agriculture	38.4	39.9	39.1	39.4	38.7
Non-agriculture	25.9	25.2	25.3	22.9	21.3
All Workers	36.3	37.1	36.1	35.8	34.6

Source : Visaria (1996); NSSO (2000).

Industrial distribution of employment indicates that there has been significant increase in the share of RNF employment for males. It increased from 19.4 per cent in 1977-78 to 33.5 per cent in 2004-05 around 14 percentage points increase in 27 years (0.52 percentage increase per annum). On the other hand, diversification of employment has been very slow for females. Share of RNF employment for females rose from 11.9 per cent in 1977-78 to 16.7 per cent only in 2004-05, around 4.8 percentage points increase in 27 years (0.18 percentage point increase per annum). For males, significant increase in the shares occurred in construction and trade, hotels and restaurants. There is a mismatch between growth and employment. Growth is occurring in services and industry but they are not able to absorb employment. Data for the year 2004-05 show that among the rural workers, 66.5 per cent of male workers are still primarily in agricultural activity. For female workers, the share of agriculture was 83 per cent.

EMPLOYMENT IN NON-FARM SECTOR : STATE-WISE ESTIMATES

It may be beneficial to look at the pace of rural non-farm expansion in individual states as well the progress of non-farm expansion has been quite uneven among individual states. The states where non-farm employment for male workers expanded fast, especially during the 1970s and 1980s are: Punjab, Haryana, Gujarat, Himachal Pradesh, Rajasthan and Tamil Nadu. The states witnessing a modest expansion are Karnataka, Maharashtra, Orissa and West Bengal. It has been an experience of a slow expansion in other states, most notably in Bihar, Kerala and Madhya Pradesh. In a fairly general way, the above pattern seems to suggest an intimate relationship between fast agricultural growth and high non-farm expansion. The non-farm expansion owes itself, inter alia: to local demand and supply interlinkages with agriculture which tend to grow stronger as agricultural growth picks up, and rural-urban space friction gets mitigated (Chadha, 1997, pp. 196-97). In the phase of halting non-farm expansion, 1987-88 to 1993-94, the proportion of rural male workers engaged in non-farm activities witnessed a sizeable growth only in Haryana and

West Bengal. In most other states it remained constant or increased marginally.

Besides the share of the non-agricultural sector in the rural workforce, Sheila Bhalla (1993 and 1997) has computed a diversification index at the state level for 1961, 1971 and 1981. This is a more sensitive measure, which takes into account the changing composition of the non-agricultural segment as well as agriculture's and non-agricultural's shares m the total workforce.

According to her, rural India has experienced over the two decades an increase in the complexity and balance within the non- cultivating segment rather than a significant widening of rural workforce diversification. Of the three sectors of the rural economy in India, the secondary sector the second fastest, while the primary sector, taken as a whole; has scarcely diversified at all. Within the primary sector, however, the other than cultivation segment has diversified the most, even more than he tertiary sector, in eight out of 15 states.

Secondary sector workforce, in eight out of 15 states became increasingly concentrated in the decade 1971-81, including Gujarat, Maharashtra and Karnataka. Again rural manufacturing sub-sector has also increased concentration during 1971 to 1981. Rural workforce diversification is most remarkable in the tertiary sector, with only three states showing decline in degree of diversification.

Workforce data at the district level generate spatial patterns both far more complex and far more interesting analytically, than those produced by the corresponding state level information. The state level figures not only average out pronounced intra-state variations, but also tend to conceal the common workforce development patterns of neighbouring districts which happen to lie on opposite sides of state boundaries. Analysis of more highly disaggregated district-wise data reveals the existence of three special kinds of regions whose distinctive workforce development profits set them apart from all other areas of the country. The first kind of region is concentrated geographically in Bihar. There, nine districts display the distressing symptoms of agricultural involution. The workers of one district, in Madhya Pradesh also labour under the same depressing conditions. A second

TABLE 5

Percentage of Workforce in Non-agricultural Activities in India

Sl. No.	State	Rural			All		
		1993-94	1999-2000	2004-05	1993-94	1999-00	2004-05
(1)	(2)	(3)	(4)	(5)	(6)	(7)	(8)
1.	Andhra Pradesh	20.70	21.2	28.2	32.9	34.6	41.5
2.	Assam	20.80	32.3	25.7	29.0	39.8	34.0
3.	Bihar	15.70	19.4	24.3	24.9	26.9	31.2
4.	Gujarat	21.30	20.2	22.7	37.7	40.5	45.1
5.	Haryana	28.10	31.5	35.9	43.4	47.2	49.7
6.	Himachal Pradesh	19.70	26.4	30.4	23.4	30.4	35.9
7.	Karnataka	18.80	17.9	19.00	34.3	37.5	39.3
8.	Kerala	43.60	51.7	58.0	51.5	61.5	64.5
9.	Madhya Pradesh	10.20	12.9	16.4	22.3	26.1	30.8
10.	Maharashtra	17.40	17.4	20.0	40.8	43.7	46.8
11.	Orissa	19.10	21.8	31.0	26.9	29.3	37.6
12.	Punjab	25.30	27.4	33.1	42.6	46.7	52.4
13.	Rajasthan	20.10	22.3	27.1	31.3	34.0	38.3
14.	Tamil Nadu	29.50	32.1	34.6	44.8	53.6	58.7
15.	Uttar Pradesh	20.00	23.8	26.9	32.2	36.5	39.1
16.	West Bengal	36.70	36.4	37.3	52.8	53.3	54.3
	All India	21.60	23.7	27.3	36.0	39.6	43.3

Note : Principal and subsidiary workers.

Source : NSSO, various Rounds on Employment and Unemployment.

path of workforce development has been followed by some of the high farm productivity districts of Punjab and coastal Andhra , where increasing workforce concentration in agriculture has been associated with distinctive patterns of rural and urban workforce restructuring. Last but not least, there are now large blocks of diversified districts, clustered around industrial cities, or forming long geographical corridors, one thousand Kilometres long or more, linking large urban conglomerations (Bhaila, 1997, pp. 161-62).

The Table 5 reveals percentage of workforce in non-agricultural activities across states. At the all India level, it increased from 21.6 percent in 1993-94 to 27.3 per cent in 2004-05 around 5.7 percentage points increase in 11 years (0.52 percentage points increase per annum). The share in rural areas is more than 30 per cent in seven states viz., Haryana, Himachal Pradesh, Kerala, Orissa, Punjab, Tamil Nadu and West Bengal. Share of non-farm in rural employment increased in poorer states also. There was 10 percentage points increase between 1993-1994 and 2004-05 in Bihar, Himachal Pradesh and Orissa.

CONCLUSION

According to J.M. Keyn (1936), "the idea of economists and political philosophers, both when they are right and they are wrong, are more powerful than is commonly understood. Indeed, the world is ruled by little else, practical men who believe themselves to be quite exempt from any intellectual influences, are usually the slaves of some defunct economist."

In the mid 1970s John Mellor (1976) Postulated that as a result of the emerging green revolution technologies increasing agriculture productivity and incomes of farmers, a demand-led growth of both the agricultural and non-agricultural sectors would take place. This could occur through multiple linkages of agriculture with the rural non-farm sector, including both consumption and production linkages.

The consumption linkages operated through an increase in income of the rural farmers. This would result in an increase in demand for goods and services produced in nearby villages and towns. Agricultural wages were also expected to rise with increase in agricultural productivity. So that the labourers would also have an increased demand for food and other non-food items. Production linkages, both backward and forward, would also emanate from the agricultural sector. Backward linkages from agricultural to non-agricultural sector was the demand from farmers for inputs produced in the non-farm sector, such as ploughs, engines and other tools. Forward linkage reflected the need to process agricultural goods so that agro-processing industries, e.g. rice-milling, tobacco Processing, would develop. The growth of the non-farm sector would in turn stimulate the growth of agricultural productivity, via investment of surplus back into agriculture. Thus a 'virtual spiral' of demand-led growth would be set in motion both in the agricultural and non-agricultural sectors (Unni, 1998, P.A.-39). Vaidyanathan (1986) suggested that when the absorptive capacity of agriculture and urban areas is limited, rural non- farm activities ted to act as a sponge for the surplus labour. This has been termed the residual sector hypothesis and suggests a distress-induced growth of the non-farm sector.

There will be more employment opportunities in the rural non-farm sector if higher agricultural growth and improvements in rural infrastructure and credit facilities were forthcoming. On the other hand, if rural non-agricultural activities are dependent on subsidies, reservations and public expenditure there may not be much growth in the post-liberalisation period. It is understood that these activities are dependent on subsidies, reservations and public expenditure, there may not be much growth in the post-liberalisation period. It is understood that these activities have to withstand competition because of globalisation The small scale sector can play an important role in generating employment. In India, where it accounts for nearly 40% in industrial output, 35% of exports and 80% of total industrial employment, the role of the small-scale sector is unlikely to diminish in the near future. Nevertheless, the sector is undergoing significant structural changes, with greater ancillerisation and provision of services. In the industrial sector some activities are reserved for the small-scale sector. These reservations would be removed in the future and small industries would have to face competition from multinationals. Subsidies also reduced or eliminated over time. In this way, there is need for restructuring for industry, labour law reform, social security and retraining for workers to encourage labour-intensive production.

It is recognized that the rural non-farm sector is important both in generating productive employment and alleviating poverty in rural areas as agriculture and urban areas cannot absorb the increasing workforce. Within agriculture and allied activities, there seems to be some diversification towards non-cereal crops. However, risk and uncertainty is associated with diversification. Technology, infrastructure and market have to be improved in order to shift the farmers to non-foodgrain crops. By any standards the unutilized potential of food-processing in India is enormous. An expansion of this sector is an ideal way of bringing industry to rural areas, expanding the value chain of agricultural production, providing assured markets for farmers, enabling them to diversify into higher value horticultural crops and expanding employment by creating high quality non-agricultural work opportunities in rural areas. There cannot be one policy package for the entire rural non-farm sector. Sub-sectoral policies in different regions are needed. In general, development of the manufacturing sector is important for absorbing labour force productively. Right now many workers are absorbed in low productive services sector. Encouragement to women and training and improvement in skills would enhance employment opportunities (Dev, 2006, p. 1513).

The recent findings by the Labour Bureau's Employment and Unemployment Survey (2009-10), disaggregated data make it evident that rural India should remain the focus of policy-makers. It is home to 72.26 per cent of the country's 238 million households, and 10.1 per cent of its labour force is unemployed, compared with 7.3 per cent in urban areas. The double-digit rural unemployment rate points to both an excess of labour that needs to be moved from agriculture and allied activities to manufacturing and services, and the declining role of the primary sector as an employment provider. In rural areas, agriculture continues to be the dominant employer (57 per cent), followed by construction (7.2 per cent), manufacturing (6.7 per cent) , and community services (6.3 per cent). Although there is a case for strengthening well-conceived schemes guaranteeing minimum employment, they could at best be bridge solutions. Upgrading the skills of the rural workforce to enable it to match the requirements of the manufacturing and services sectors would provide more lasting results. Equally important is the survey's observation that the spread of social security cover is woefully inadequate only a paltry 163 persons per 1,000 benefit from the various social security measures. This low coverage should remind policy-makers that a lot more needs to be done for a workable social security system to be in place. Despite the recent initiatives, the challenges on the employment generation and the social security fronts remain largely unaddressed.

Although the global economic downturn offers itself as a ready explanation, the malaise runs deeper: the inability to come up with an effective rural employment strategy. Despite industrialization, rural India remains the main employment provider. That the situation has not changed drastically in the two decades of reforms is evident from the survey results. First, it brings out the magnitude of unemployment; secondly, it turns the spotlight on the low availability of social security benefits for the Indian workforce.

The ongoing process of globalization is not only non-reversible but also its adverse impacts on the poor can not be completely arrested. The ongoing attempt to generate employment in non-farm activities through globalisation of industrial and trading activities, information technology and advancing knowledge based economy, is not going to provide employment to more than a friction of the rural workforce in the near future. Their better survival can be ensured through diversifying agriculture and increasing productivity in agriculture through latest technical and technological know-how; generate gainful wage employment on a large scale in the rural infrastructure. sector; and providing better education, drinking water, shelter, health services- and social security measures.

Poverty will be chronic and persistent if employment potential has no priority. Infact, intensified global competition and freer flow of goods may wipe out rural industries, especially if import / export duties are to be adjusted in favour of global giants. Gandhiji was in favour of production by the masses and not mass production. Global operators reverse this emphasis. Let us encourage labourisation to increase productivity and product quality to go up in the ladder of economy and find a commanding place in the competitive global market. Infact, India has always stood for the ' transition of mankind from an ordinary collection of nations' to 'mankind as a global community.' The global development depends upon the global well-being of the individuals all the world over. Vedanta also says that, "Mayall the people be happy, Mayall the people be healthy, My all see only the good things and no body should be unhappy".

References

Alauddin, M. and C. Tisdell (1995), Labour Absorption and Agricultural Development : Bangladesh's Experience and Predicament", *World Development*, Vol. 23, No. 2.

Basant, Rakesh (1994), Economic Diversification In Rural Areas : A Review of Process With Special Reference to Gujarat", *Economic and Political Weekly*, Vol. 29, No. 39.

Bhalla, Sheila (1993), The Dynamics of Wage Determination and Employment Generation in Indian Agriculture", *Indian Journal of Agricultural Economics*, Vol. 48, No. 3, July-September

———, (1997), The Rise and Fall of Workforce Diversification Process in Rural India", in G.K. Chadha and Alakh N. Sharma (eds.), Growth, Employment and Poverty Change and Continuity in Rural India, Vikas Publishing House Pvt. Ltd., New Delhi.

———, (2001), "Workforce Diversification and the Performance of Unorganised Sector Enterprises", Workshop on Rural Transformation in India, Institute for Human Development, New Delhi

Chadha, G.K. (1997), "Access of Rural Households to Non-Farm Employment : Trends, Constraints And Possibilities" in G.K. Chadha and Alakh N. Sharma (eds.) *op. cit.*

Dev, S. Mahendra (2000), "Economic Reforms, Poverty, Income Distribution and Employment", *Economic and Political Weekly*, March 4.

———, (2006), "Policies and Programmes for Employment", *Economic & Political Weekly*, Vol. XLI, No. 16, April 22.

Dasgupta, Gurudas (2000), "Globalisation, Economic Reform and Indian Workers", *Mainstream*, Vol. XXXVIII, No. 50. December, 2.

Islam, R. (1987), "Rural Industrialisation and Employment in Asia: Issues and Evidence", in Islam, R. (ed.), Rural Industrialisation and Employment in Asia, ILO-ARTEP, New Delhi.

India Rural Development Report (1999), "Regional Disparities in Development and Poverty", *National Institute of Rural Development*, Hyderabad.

Jose, A.V. (1988), "Agricultural wages in India", *Economic and Political Weekly*, Vol. 23, No. 26, June 25.

Mellor, John (1976), "The New Economics of growth: A Strategy for India and the Developing World", Carnell University Press.

Mukherjee, Anindita, "Liberalization and Rural wages and Employment in India", in K. Raghvan and Leena Sekhar (eds.), Poverty and Employment", New Age, International Publishers, New Delhi.

NSSO (2000), Employment and Unemployment in India 1999-2000 : Key Result (NSS 55th Round, July 1999-June 2000), Report No. 455, December

Parthasarthy, G. (1996), "Recent Trends in Wages and Employment of Agricultural Labourers", *Indian Journal of Agricultural Economics*, Vol. 51, Nos. 1 & 2, January-June.

Prakash, Brahm and Sushila Srivastva (2000), " Women's Participation in Rural Non-Farm Employment in M. Koteswar Rao (ed.), Rural Employment The Non-Farm Sector, Deep & Deep Publications Pvt. Ltd., New Delhi.

Rao, M. Koteswar (2000), "Rural Non-Farm Employment in India : A Note", in M. Koteswar Rao (ed.), *op. cit.*

Sen, Abhijit (1997), Structural Adjustment and Rural Poverty: Variables That Really Matter", in G.K. Chadha and Alakh N. Sharma (eds.), *op. cit.*

———, (1998), "Rural Labour and Poverty", in R. Radha Krishna and A.N. Sharma (eds.) Empowering Rural Labour in India; Market, State and Mobilisation, Institute for Human Development, Delhi.

Unni, Jeemol (1998), "Agricultural Labourers in Rural Labour Households, 1956-57 to 1977-78 : Change in Employment, Wages and Income", *Economic and Political Weekly*, Vol. 23, 0.26, June 25.

———, (1997), "Employment And Wage Among Rural Labourers: Some Recent Trends". *Indian Journal of Agricultural Economics*, Vol. 52, No. 1.

———, (1998), "Non-Agricultural Employment and Poverty in Rural India: A Review of Evidence", *Economic and Political Weekly*, Vol. 13, March 28.

Vaidyanathan, A. (1994), "Employment Situation: Some Emerging Perspectives", *Economic and Political Weekly*, Vol. 29, No. 50, December 10.

Visaria, P. (1996), "Structure of Indian Workforce, 1961-94", *The Indian Journal of Labour Economics*, Vol. 39, No. 4.

Index